Alain Krief · László Hevesi

Organoselenium Chemistry I

Functional Group Transformations

With 262 Schemes

Springer-Verlag
Berlin Heidelberg NewYork
London Paris Tokyo

Professor Dr. Alain Krief
Professor Dr. László Hevesi

Facultés Universitaires Notre-Dame de la Paix
Département de Chimie
B – 5000 Namur (Belgium)

ISBN-13: 978-3-642-73243-0   e-ISBN-13: 978-3-642-73241-6
DOI: 10.1007/978-3-642-73241-6

Library of Congress Cataloging-in-Publication Data
Krief, A. (Alain), 1942 – Organoselenium chemistry I. Bibliography: p. 1. Organoselenium com-
pounds. I. Hevesi, L. (László), 1941 –. II. Title.
QD412.S5K75 1988   547'.05724   87-36926
ISBN-13: 978-3-642-73243-0

Typesetting: Friedrich Pustet, Regensburg. Printing: Kutschbach, Berlin.
Bookbinding: Lüderitz & Bauer, Berlin.
2152/3020-543210

Foreword

During the last fifteen years organoselenium chemistry underwent a spectacular mutation: from an exotic area of science practised by a few specialists it became a relatively well mastered and widely used methodology by synthetic organic chemists. The key to this success is that a fair number of selenium based reagents and reactions have been discovered, which are able to perform specific transformations selectively and often under very mild conditions. The most popular of these are:

 i) oxidation of various types of substrates by selenium dioxide,
 ii) oxidations using benzeneseleninic anhydride,
 iii) selenoxide syn elimination leading to olefins,
 iv) [2,3] sigmatropic rearrangement of allylic selenoxides and selenimides giving rise to allyl alcohols and allylic amines respectively,
 v) electrophilic selenium-induced ring closures leading to lactones, to carbo- and heterocycles followed by reductive or oxidative deselenylation,
 vi) carbon-carbon bond forming reactions using selenium-stabilized organo-metallics or carbocationic species,
 vii) radical cyclisation triggered by homolytic C-Se bond cleavage.

Although organoselenium compounds have been known for more than a century, it is only since the discovery of selenium dioxide (SeO_2) by Riley in 1931 that this first selenium reagent was introduced into organic synthesis. Since that time SeO_2 has been used for the oxidation of olefins and of carbonyl compounds in the alpha position. The reaction was initially used for structure elucidation and later as a key step in the synthesis of natural products. For the next forty years selenium dioxide as well as elemental selenium and potassium selenocyanate were the only selenium containing reagents used. Since 1973 and the pioneering work of Barton, Clive, Reich, Sharpless, Sonoda, as well as of our laboratory several inorganic and organic reagents containing an active selenium atom have been proposed. These have proved to be particularly powerful and are being extensively used in organic chemistry especially for the synthesis of natural products.

The aim of this Volume is to present the most frequently used selenium containing reagents now available to chemists, to mention the scope as well as the limitations of their reactions, and to compare those which possess similar reactivities. We will present the reagents in the order of increasing oxidation level around the selenium atom. Except for particularly relevant cases no effort has been made to compare these reagents to those devoid of selenium but able to perform the same transformations. References to related methods have however been inserted in the text and are preceded with a sign ($@$) whereas references marked with an asterisk designate review articles.

In this Volume I we have restricted our review to those reactions which do not involve isolatable selenium containing intermediates and therefore which are operationally "one step" reactions.

On the other hand, transformations carried out in two or more separate steps and which usually involve the isolation (and eventually the purification) of stable organoselenium intermediates will be described in the forthcoming Volume II. In this latter Volume II will also be included a detailed review of the preparation of various selenium-based reagents.

We thank Drs J. L. Desiron and R. Menzies (Societe Generale des Minerais, Bruxelles) who gave us invaluable informations on industrial aspects of selenium.

We are particularly grateful to Professor S. V. Ley (Imperial College, London) who reviewed and corrected the entire manuscript.

We owe special thanks to Mrs Anne Krief for her patient contribution in processing the references; to Mr Alain Burlet for his skill in drawing the Schemes as well as to Mrs Evelyne Boca-Bastaits and Miss Veronique De Beys for their courage and competence in typing the manuscript.

Namur, September 1987   László Hevesi
                                                              Alain Krief

Contents

Contents

Contents

Chapter 1

Introduction

1.1 Historical Review

The initial recorded discovery of selenium is attributed [1] to two Swedish chemists J. J. Berzelius and J. G. Gahn who observed a curious residual slime during the oxidation of sulfur dioxide from copper pyrites. It is also possible that selenium was discovered earlier: in the "Rosarius Philosophorum" of Arnold of Villanova, written in the fourteenth century, there is reference to a red sulfur deposit (sulfur rubeum) which formed on the walls of the oven after the condensation of crude sulfur.

1.2 Abundance and Distribution of Selenium

Among the elements, selenium is the 68th in crustal earth abundance [2]. Its weight percentage [3] ($7 \times 10^{-5}\%$) is closely related to that of cadmium and antimony. It is widely but unevenly distributed in rocks and soils and often found associated with sulfur, copper, iron and silver. Normal soil content is estimated at 0.2 ppm. Its abundance in lunar and terrestrial basalts is virtually constant probably reflecting saturation with iron in the source regions. The amount of selenium can be in some cases lower (< 0.1 ppm in New Zealand) [4, 5] or even much higher (1200 ppm in Ireland) [6] than normal. High concentrations of selenium are found in clay sediments, in volcanic materials and in deposits containing high proportions of organic residues such as coal [7] (10 to 20 times higher than normal). This concentration is believed to be due to the assimilation of selenium by the organisms from which the coal was formed [8].

1.3 Selenium in Plants [*9]

In the United States, selenium has received much notoriety as a poisonous constituent in vegetation. It is suspected to have poisoned several cavalry horses in Nebraska in 1860 and sheep ($> 15\,000$) in the summers of 1907 and 1908 in Wyoming. Many other similar cases of acute or chronic poisoning occurred in 15 Western States [*10]. It was later discovered that when feed grains or grass were grown on certain soils, domestic plants could become toxic to livestock. Some plants are able to absorb selenium. However the ability of a plant to absorb selenium from the soil depends on the chemical form and solubility of the

1

selenium containing material, and the moisture content of the soil [6, 7, *11]. Some plants [*12] require selenium to grow and possess a high content of seleniferous material [*Astragalus* [13, 14] (1000 ppm Se), *Haplopappus*, *Zylorhiza* (120 ppm Se)]. Others, such as *Aster* (72 ppm Se) do not appear to require selenium for their growth but will accumulate the element when grown on soils of high available selenium content. A third group of plants which includes the grasses and grains does not normally accumulate selenium in excess of about 50 ppm.

1.4 Selenium Toxicity in Animals [*15]

Animals grazing on seleniferous vegetation or fed with seleniferous grain develop a variety of conditions due to acute or chronic selenium poisoning. Acute poisoning occurs when plants with high selenium content (10.000 ppm Se) are consumed by the animals, usually under poor grazing conditions. It was described [*12] that high mortality occurred amongst 557 calves which erroneously received 100 mg (0,5 mg/kg live weight) of sodium selenite subcutaneously (number of dead calves/time after the injection: 18/24h — 38/48h — 75/72h — 376/5 weeks). Chronic diseases occur when feedstuff containing more than 5 ppm, but usually less than 40 ppm of selenium, are continuously ingested. However the structures of the selenium containing molecules responsible for such toxicity have not been fully described [16].

On the other hand, at very low concentration the signs of selenium deficiency are manisfest [*12, *17]. It is now clear that selenium is a necessary trace element [16, *17, 18] for the growth and fertility of many avian and mammalian species. Selenium deficiency causes liver necrosis in rats, exudative diathesis in chicks, white muscle disease in young livestock and hepatosis dietica in pigs. Infertility and birth defects in cattle and sheeps also have been associated with insufficient dietary selenium [*19]. It has been estimated that in selenium deficient areas, selenium or selenium-vitamin E combination added to animal feed can prevent annual losses in beef, dairy cattle and sheep valued at 545 million dollars and poultry and swine losses valued at 82 million dollars [*12].

Selenium reduces the toxic effects of arsenic, cadmium and mercury, and vice versa [20–22]. Its cancer protective effects have often been claimed [23].

1.5 Enzymic Role of Selenium [*24, *25, 26]

The nutritional requirement for selenium is recognized to be closely related to vitamin E [17, 27, 28]. The latter acts as antioxidant [28] and probably protects the unsaturated lipid components of cell membranes from oxidative damage which can lead to disruption of cell integrity. The difficulty in explaining the distinct nature of some selenium and vitamin E deficiencies was largely resolved with the discovery by Rotruck et al. [29] that selenium was an essential component of the enzyme glutathione peroxidase. This enzyme catalyses [30], in the presence of reduced glutathione (GSH), the reduction of hydrogen peroxide as well as that of lipid hydroperoxides which are transformed to the corresponding alcohol derivatives. Vitamine E may prevent the formation of lipid peroxides whereas glutathione peroxidase removes lipid hydroperoxides once formed [31].

This enzyme has been studied in several mammalian and avian species and has been obtained in highly purified form from bovine [32, 33], ovine [34] and human [35] erythrocytes and rat liver [36]. Values reported for the molecular weight vary between 76.000 and 92.000. The enzyme is constituted [*24, 36] of four identical subunits each containing 180 to 183 amino acids, the selenium being located in one amino acid per subunit (selenocysteine). The amino acid composition was determined by sequential Edman degradation on rat liver glutathione peroxidase, and elucidation of the three-dimensional structure of bovine erythrocyte glutathione peroxidase was achieved [32] after treatment of the reduced form of the enzyme with H_2O_2.

Other selenium dependent enzymes include [*24, *25, 26, *37]: formate dehydrogenase, clostridial glycine reductase, nicotinic acid hydroxylase and xanthine dehydrogenase.

1.6 Ecological Aspect of Selenium in Human Health
[*11, *38, *39]

No unequivocal proof has yet appeared that selenium is an essential trace element for humans, although it is well established as being an essential element for several other mammalian species [*11]. On the other hand, epidemiological studies in seleniferous areas of the U.S. present descriptions of the selenium toxicity in humans in the form of tooth disorders, vertigo, fatigue and chronic gastrointestinal diseases.

Apart from hydrogen selenide, other commonly used compounds do not seem to be particularly harmful. Elemental selenium is apparently harmless when ingested, and although workers exposed to selenium fumes developed conjunctivitis and rhinitis, followed by varying degrees of bronchitis, these symptoms cleared within 3 days without residual effects. Contact of skin with SeO_2, an industrial product has been mentioned to produce painful irritation which can be avoided by rapid washing with a reducing agent, such as 10% aqueous solution of sodium thiosulfate [40]. In the same group of workers, even before the institution of environmental control, the mortality rate from common diseases was not increased in comparison with that for a control population in spite of the chronic selenium exposure.

Hydrogen selenide is one of the most toxic and irritating selenium compounds. This gas is formed by the action of acids, or in some cases water, on inorganic selenides such as aluminium selenide. 5 µg per liter of hydrogen selenide cause considerable eye and nasal irritation. Although hydrogen selenide is considered 15 times more dangerous than hydrogen sulfide especially due to its acidity which is close to that of formic acid, it has never caused death or illness lasting more than ten days in human beings [41]. The reason is that hydrogen selenide is very easily oxidized back to red selenium on the surface of the mucous membranes of the nose, and probably also in the alveoles of the lung.

1. Introduction

Selenium, inorganic selenium derivatives and organoselenium compounds have been classified in the past as toxic derivatives, and this has, for many years, discouraged intensive research in that area. The highly unpleasant odor of some selenium compounds has been often erroneously associated with their toxicity. However, it has been described [44] that selenides of the type shown in Scheme 1 are useful as perfumes. The little information given above on some of those compounds which have been pointed out as being eminently toxic, speaks much of the real situation.

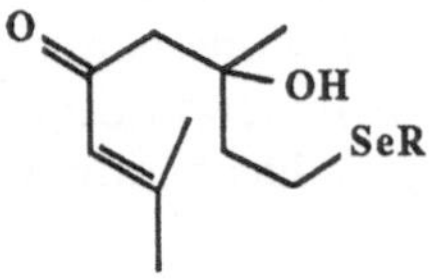

R=Ph or 2-NO$_2$Ph

Scheme 1 [44]

Numerous organoselenium derivatives have been prepared over the last century. Their number has dramatically increased during the last decade, since several of them have been used as intermediates in organic synthesis. Nowadays, organoselenium compounds are prepared in several laboratories, without the necessity of taking particular precautions. One may even mention that some of them are potential chemotherapeutic agents [*42, 43].

To the authors' knowledge there is no report of acute or chronic toxicity of this family of compounds in the normal environment of chemical laboratories, and although care must be taken to work in a well ventilated hood and to take the usual precautions that any chemist must take with all organic compounds, organoselenium chemistry does not seem to be exceptionally dangerous. For ecological reasons, it is more reasonable to throw away wastes in tanks and not in the sink.

As already mentioned, some of the most volatile organoselenium derivatives have a highly unpleasant odor. Reaction of the wastes, dissolved in THF, with hydrogen peroxide or sodium hypochlorite often discharges the odor by oxidizing the selenium atom to ˙a higher oxidation level, usually transforming organoselenium derivatives thereby into colorless, water soluble compounds.

1.7 Extraction of Selenium

Selenium containing minerals [such as Berzelianite (Cu$_2$Se), Blockite (NiSe$_2$), Eucarite (AgCuSe), Weibulite (PbBi$_2$(SSe)$_4$)] are not found in appreciable quantities [45], and are not usually good commerical sources of selenium. However, it has been shown that the recovery of selenium by flotation from certain sandstone formations of New Mexico is both pratical and economical. Most of the selenium is now recovered [45] from copper refinery anode slimes through, *inter alias,* cuppelation, roasting or sulfatizing processes. The recycling of seleniferous dechet is now being increasingly used [45]. The overall capacity of production of selenium in 1985 was estimated to lie around 2300 tons (t) which can be divided as follows: North America: 1225 t; South America: 80 t; Asia: 550 t; Europe: 430 t; Africa: 25 t.

In fact, around 1000 tons were probably produced in 1984, mainly by three companies:

- Noranda Mine Ltd., Commerce Court, West Toronto, Ontario, Canada (600 t production).
- Metallurgie Hoboken Overpelt, Greiner Straat 14, Hoboken, Belgium (240 t production).
- Boliden Metall A.B., Sturegaten 22, Box 5508, Stockholm, Sweden.

A pound of selenium metal (99.5%) was quoted[1] at $ 6.6–7.0 in January 1985.

1.8 Industrial Uses of Selenium

Selenium has been mainly used for both its interesting optical and electrical properties. Selenium can be categorized as a semiconductor with low carrier mobility. Exitation with electromagnetic radiation causes marked increases in conductivity (photoconductivity). This particularity has been used in photocell devices [*46] as well as in xerography [47].

Selenium is also used as a blender in glass production [48, 49]. Its association with cadmium in cadmium selenide (CdSe) confers antifungal properties and is being extensively used in antipellicular shampoos [43].

Finally, selenium is now used as a food additive for animals, but also for humans (150–200 µg/day of selenium as "selenium yeast" preparation is presently considered an adequate amount for ingestion [23]).

1.9 Selenium: Structure, Physical and Chemical Properties

Selenium exhibits several allotropic, crystalline and amorphous forms [*50]. Black powered selenium is usually the most readily available. Red amorphous selenium often deposits during reactions involving organoselenium compounds and by air-oxidation of reduced forms of selenium.

Selenium has an atomic weight of 78.96. Six stable isotopes have been isolated [mass (ratio%) 74 (0.87), 76 (9.02), 77 (7.58), 78 (23.52), 80 (49.82), 82 (9.19)]. This makes the mass spectra of organoselenium compounds particularly rich in peaks, characteristic of this element (Scheme 2). Several unstable isotopes, including those of atomic number 70, 72, 73, 75, 79, 81, 83, 85, 86 and 87 have been identified. Among these [75] Se is a commonly available radionuclide, emitting gamma rays, and possessing a half-life of 122 days. It is extensively used as a tracer in biochemical studies and as a radiopharmaceutical agent for diagnostic purposes [*24].

---

[1] Metal Bulletin (London). Section: free market selenium 99.

1. Introduction

a) Mass spectrum of methyl phenyl selenide (recorded on a quadrupole HP 5995 GC-MS spectrometer, 100°C, 70ev.).

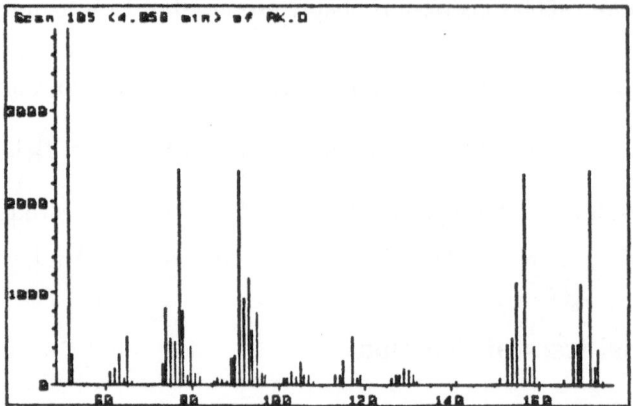

b) Enlargement of groups of peaks corresponding to M⁺ and M⁺-CH₃ in methyl phenyl selenide (see above) and reflecting the natural abundances of stable selenium isotopes.

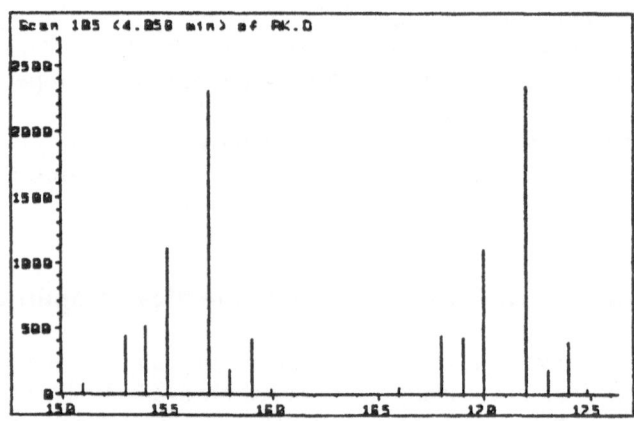

m/z	abund.	m/z	abund.
150.95	29	166.00	18
153.00	188	168.00	183
154.00	218	169.00	180
155.00	476	170.00	465
157.00	1000	172.00	1000
158.00	76	173.00	74
159.00	174	174.00	160
159.90	12	174.90	13

Scheme 2

6

The [77] Se isotope possesses a nuclear spin of ½, and the magnetic moment (μ) associated with this spin has been determined [*51] to have a value of + 0.53326. This isotope has low sensitivity (6.93×10^{-3}) and receptivity (5.26×10^{-4}) relative to that of protons. However the receptivity of [77] Se is 2.98 times that of ^{13}C, [52] which makes it suitable for NMR experiments [*53]. This technique is very useful, and often allows to monitor reactions involving organoselenium derivatives. The exceptionally wide range of chemical shifts covered ($> 2000 \, ppm$) presents an added advantage. A few specific examples are disclosed in Scheme 3.

Broad band proton decoupled ^{77}Se NMR spectra of some organoselenium compounds, taken on a Jeol FX 90Q instrument, 17.04 MHz for ^{77}Se. (ppm relative to Me$_2$Se used as sandard, solvent)

MeSeH (-116.5, CDCl$_3$)

PhSeH (148.8, CDCl$_3$)
PhSeMe (209, CDCl$_3$)(207, THF -H$_2$O)
PhSe(=O)Me (827,CH$_2$Cl$_2$)(842, THF-H$_2$O)
PhSe(=O)$_2$Me (984,CDCl$_3$)(988, CH$_2$Cl$_2$)

PhSeO$_2$Na (1159, H$_2$O-NaOH)
MeSe-SeMe (263.5, CDCl$_3$)   PhSeO$_3$K(1030, H$_2$O-KOH)

Scheme 3

Selenium lies in row 4 of the periodic table, between arsenic and bromine. It belongs to group VI and lies between sulfur and tellurium. One of its most obvious characteristics is its schizophrenic personality [54], behaving as a metallic non-metal or a non-metallic metal.

Its inner shells ($1s^2$, $2s^2$, $2p^6$, $3s^2$, $3p^6$, $3d^{10}$) are completely filled, and its outer shell ($4s^2$, $4p^4$), makes selenium divalent in hydrogen selenide **1**, selenols **2**, selenides **3** and selenenyl halides **4** (Scheme 4). Compounds in which one lone pair (such as selenonium salts **5, 6,** selenoxides **7** and seleninic acids **8**) or two lone pairs (such as selenones **9,** selenonic acid **10**) are engaged in bonds are known.

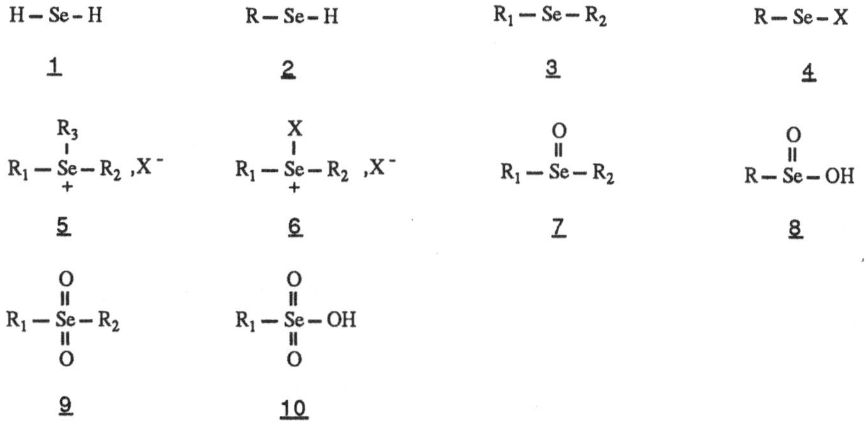

Scheme 4

1. Introduction

Depending upon the compounds in which it is engaged, the selenium atom can have oxidation levels from −2 to +6.

The 4d unfilled orbitals are quite close in energy to the 4p orbitals, and can be quite easily populated like in the case of the postulated tetraphenyl selenane[1] **11** [55] or of the stable bis (4,4'-dimethyl-2,2'-biphenylylene) selenane[1] **12** [56] shown in Scheme 5.

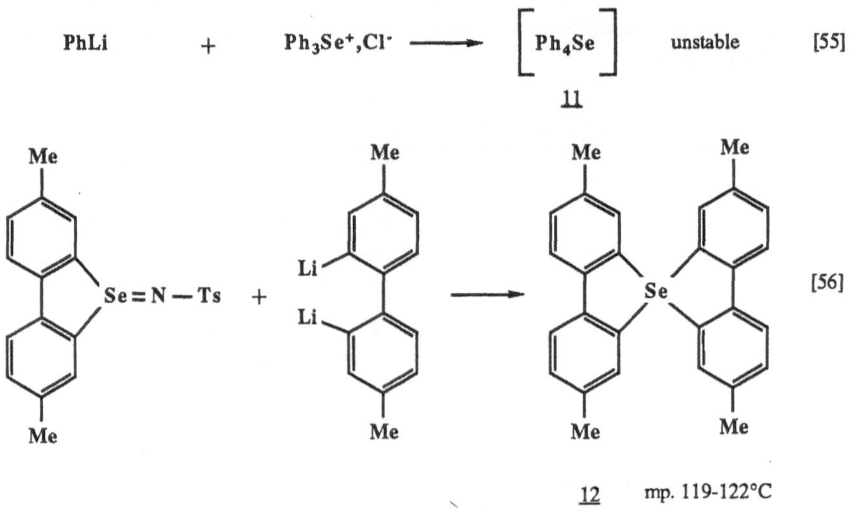

PhLi + Ph$_3$Se$^+$,Cl$^-$ ⟶ [Ph$_4$Se] unstable [55]

11

[56]

12 mp. 119-122°C

Scheme 5

The carbon-selenium bond (234 KJ mole^{-1}) is weaker than the carbon-sulfur (272 KJ mole^{-1}) and the carbon-bromine bonds (285 KJ mole^{-1}). On the other hand, the electronegativity of selenium is close to that of carbon [57, 58, 59]. Some of the proposed values are gathered in Scheme 6, and compared to those of sulfur and bromine.

Electronegativities of Se, C, S and Br.

Se	C	S	Br	Method	
2.55	2.55	2.58	2.96	Pauling method	[57]
2.48	2.50	2.44	2.74	Allred-Rochow values	[58]
2.23	2.63	2.41	2.76	Mulliken type values	[59]

Scheme 6

Organic selenides are ambiphilic derivatives which can act as nucleophilic or electrophilic species, depending upon the reagent used. For example, they act as nucleophilic reagents toward halogens or alkyl halides and produce [*54, *60] the corresponding selenonium salts (Scheme 7a). On the other hand, selenides react

---

[1] Recommended nomenclature, see Ref. 113, p. 518.

with alkyllithiums [*61] to produce a novel organolithium compound and a new selenide (Scheme 7b). The first type of reaction involves the lone pair present on the selenium atom, whereas the second transformation is thought to occur by the addition of the organolithium on the selenium atom of the selenide leading to the intermediate formation of an "ate" complex which then decomposes to give the products. The "ate" complex formation can only be explained [*61] if one assumes that the lone pair present on the carbanionic center can be accomodated in one of the unfilled 4d orbitals of the selenium atom of the selenide.

a) $R_1\text{-Se-}R_2 + R_3X \longrightarrow R_1\text{-}\overset{R_3}{\overset{|}{Se}}\text{-}R_2 , X^-$ [*54, *60]

b) $R_1\text{-Se-}R_2 + R_3Li \longrightarrow \left[R_1 - \overset{-}{\underset{R_3}{Se}} - R_2, \overset{+}{Li} \right] \longrightarrow R_1\text{-Se-}R_3 + R_2Li$ [*61, *66]

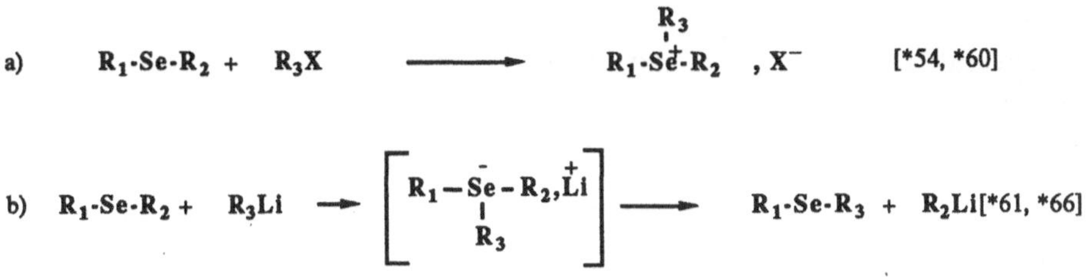

Scheme 7

Selenides are sensitive to oxidation, and are readily transformed to selen-oxides [*54, *61, *62, *63, *64] but further oxidation to selenones is more difficult [65], and is dependent, *inter alias,* on steric hindrance.

The seleno moiety is able to stabilize carbanions [*61, *66, *67], carbenium [*67, *68] ions, as well as carbon radicals [*68], when it is directly attached to the charged or to the radical center.

Some selenium containing reagents such as selenium dioxide (SeO_2), perse-leninic acids ($ArSeO_3H$), seleninic anhydrides or selenoxides are powerful oxidants (Chapt. 4 to 8). Others like selenols or hydrogen selenide are not only acidic derivatives [69–75] (Scheme 8) (pKa of H_2Se is close to that of formic acid) but are also powerful reducing agents (Chapt. 2) and particularly good nucleophi-les [76, 77].

Literature pKa values for Selenols and Thiols

	X=Se	X=S	
H_2X	3.74	7.0	[70, 71]
8-quinoline-XH	4.94	7.68	[73]
$H_3\overset{+}{N}CH_2CH_2XH$	5.0	8.3	[72]
$HO_2CCH(NH_2)CH_2XH$	5.24	8.25	[74]
PhXH	5.9	6.25	[75]

Scheme 8

This brief survey suggests the versatility of selenium containing molecules or reagents. The discovery, in the early 1970's, that selenoxides eliminate selenenic acids to form olefins under very mild conditions quickly resulted in the extensive development of organoselenium chemistry. Over one thousand publications, utilizing new organoselenium reagents have appeared since that time. The novel reactions are often milder than those (using closely related methods) used before to perform similar transformations, while other reactions are completely original. Many of the reagents were found to be exceedingly useful in the synthesis of complex natural products. A book written by K. C. Nicolaou and N. A. Petasis entiteled "Selenium in Natural Products Synthesis" recently appeared [*78], and shows how rapidly these novel reagents have been adopted by the scientific community.

Several review articles and manuals have covered various fields of organoselenium chemistry during the last forty years:

Mayor Y., 1940 [*79], Les Applications de l'Oxyde de Selenium a l'Oxydation des Composes Organiques.

Clark C. W. et al., 1945 [*80], Selenium Dioxide. Preparation, Properties and Use as Oxidizing Agent.

Rabjohn N., 1949 [*81], Selenium Dioxide Oxidation.

Campbell T. W. et al., 1952 [*82], Some Aspects of the Organic Chemistry of Selenium.

Rheinboldt H., 1955 [*60], Methoden zur Herstellung und Umwandlung Organischer Selen- und Tellur-Verbindungen.

Gosselck J., 1963 [*83], Aus der Chemie der Organoselenverbindungen.

Trachtenberg E. N., 1969 [*84], Oxidation, Techniques, and Applications in Organic Synthesis.

Jerussi R. A., 1970 [*85], Selective Organic Transformations.

Okamoto Y., Gunther W. H. H., 1972 [*86], Organic Selenium and Tellerium Chemistry.

Klayman D. L., Gunther W. H. H., 1973 [*54], Organic Selenium Compounds: Their Chemistry and Biology.

Zingaro R. W., Cooper W., 1974 [*87], Selenium.

Sharpless K. B. et al., 1975 [*88], The Utility of Selenium Reagents in Organic Synthesis.

Rabjohn N., 1976 [*89], Selenium Dioxide Oxidation.

Schmid G. H., Garratt D. G., 1977 [*90], Electrophilic Additons to Carbon-Carbon Double Bonds.

Bulka E., 1977 [*91], Selenocyanates and Related Compounds.

Barton D. H. R., Ley S. V. [*92], Design of a Specific Oxidant for Phenols.

Clive D. L. J., 1978 [*62], Modern Organoselenium Chemistry.

Clive D. L. J., 1978 [*93], Selenium Reagents for Organic Synthesis.

Reich H. J., 1978 [*63], Organoselenium Oxidations.

Reich H. J., 1979 [*64], Functional Groups Manipulation Using Organoselenium reagents.

Magnus P. D., 1979 [*94], Organic Selenium and Tellurium Compounds.

Comasseto J. V. et al., 1979 [*95], Reagents De Seleno Em Sintese Organica.

Sharpless K. B., Verhoeven T. R., 1979 [*96], Metal Catalyzed Higly Selective Oxygenations of Olefins and Acelylenes with *tert*-Butyl Hydroperoxide. Practical Considerations and Mechanisms.

Ley S. V., 1980 [*97], Organosulphur, Organoselenium and Organotellurium Chemistry.

Krief A., 1980 [*61], Synthetic Methods Using α-Heterosubstituted Organometallics.

Nicolaou K. C., 1981 [*98], Organoselenium — Induced Cyclizations in Organic Synthesis.

Pennanen S. I., 1981 [*99, *100], Organoseleeniyhdisteet Kemistin Tyovalineina Orgaanisessa Synteesissa.

Witczak Z. J., Whistler R. L., 1982 [101], Carbohydrates Containing Selenium.

Comasseto J. V., 1983 [*102], Vinylic Selenides.

Witczak Z. J., 1983 [*103], Nucleosides Containing Selenium.

Nicolaou K. C., Petasis N. A., 1984 [*78], Selenium in Natural Products Synthesis.

Liotta D., 1984 [*104], New Organoselenium Methodology.

Krief A., Hevesi L., 1984 [*67], Selenoacetals and Seleno-orthoesters, Valuable Reagents in Organic Synthesis.

Krief A., 1986 [*105], Synthesis and Synthetic Applications of 1-Metallo-1-seleno Cyclopropanes and Cyclobutanes and related 1-Metallo-1-silyl cyclopropanes.

Patai S. and Rappoport Z., 1986 [*113], The Chemistry of Organic Selenium and Tellurium Compounds, Vol. 1.

Patai S. and Rappoport Z., 1987 [*114], The Chemistry of Organic Selenium and Tellurium Compounds, Vol. 2.[1]

A specialist Periodical Report on Organic Compounds of Sulfur, Selenium and Tellurium has reviewed the literature published since April 1969 [*106–*111].

More specific reviews discussing the synthesis and the reactivity of selenium containing heterocycles have also been published but will not be presented here. Leading references in this field can be found in "Comprehensive Heterocyclic Chemistry" by Katritzky A. R. and Rees C. W. (1984) [*112], in "The Chemistry of Organic Selenium and Tellurium Compounds" edited by Patai S. and Rappoport Z. (1986) [*113], and in "Synthesis of Tetraheterofulvalenes and of Vinylene Triheterocarbonates – Strategy and Practice" by Krief A. (1986) [*115].

Leading references on inorganic selenium derivatives can be found in "Comprehensive Inorganic Chemistry: Selenium, Tellurium and Polonium" by Bagnall K. W. [*116].

Selenium and Tellurium Abstracts are available from the Selenium-Tellurium Development Association (P.O. Box 3096 Darien, Conn 06820 U.S.A.). These abstracts exist since 1955. Vol. 1 to 7 cover 1955 to 1966 and have been prepared by the Battelle Memorial Institute. Since 1967 Selenium and Tellurium Abstracts is a publication of the Chemical Abstracts Services, published monthly by the American Chemical Society.

---

[1] Two new monographs have appeared after the completion of this compilation: i) Paulmier, C., Selenium Reagents and Intermediates in Organic Synthesis, Organic Chemistry Series, Ed. by Baldwin, J. E., Pergamon Press, Oxford, 1986.; ii) Liotta, D., Ed., Organoselenium Chemistry, Wiley Interscience., New-York, 1987.

Chapter 2

Reactions Involving Hydrogen Selenide, Selenols and Related Compounds

Hydrogen selenide, selenols and their salts are not only valuable for the introduction of the selenium atom into organic molecules, they are also particularly efficient reagents for functional group transformations which do not produce a selenium containing end-product.

The high nucleophilicity of the selenium atom in these reagents as well as their reducing properties have been *inter alias* successfully used:

(i)   for dealkylation of various compounds, for example, conversion of esters to acids, of quaternary alkylammonium salts to amines and of more substitued to less substitued amines;

(ii) for the reduction of epoxides to olefins, of sulfoxides to sulfides and of carbonyl compounds to alcohols;

(iii) for the synthesis of various mixed alkylselenocuprates in the case of cuprous selenide.

2.1 Reactions Involving the Nucleophilicity of Hydrogen Selenides, Selenols and Related Compounds

2.1.1 N-dealkylation of Quaternary Ammonium Salts

Metal selenolates, neat or in ethanol, DMF, HMPA or THF-HMPA, proved to be particularly potent nucleophiles and have been used *inter alias* for the N-demethylation [117] of quaternary ammonium salts of alkaloids whose nitrogen atom is present in one ring only [such as 3-ethyl-morphine methochloride and (±)-laudanosine methochloride (Scheme 9a, b)]. Those in which the nitrogen atom is part of two rings, such as seneciphylline are demethylated either at a much slower rate [117] or they are not demethylated at all [117].

Demethylation does not take place in papaverine methochloride [117] (Scheme 9c), where the nitrogen atom is part of the conjugated isoquinoline system [117]. Although the yields are in some cases better, the method does not seem to have very big advantages over the closely related technique using the thiophenoxide anion [@117].

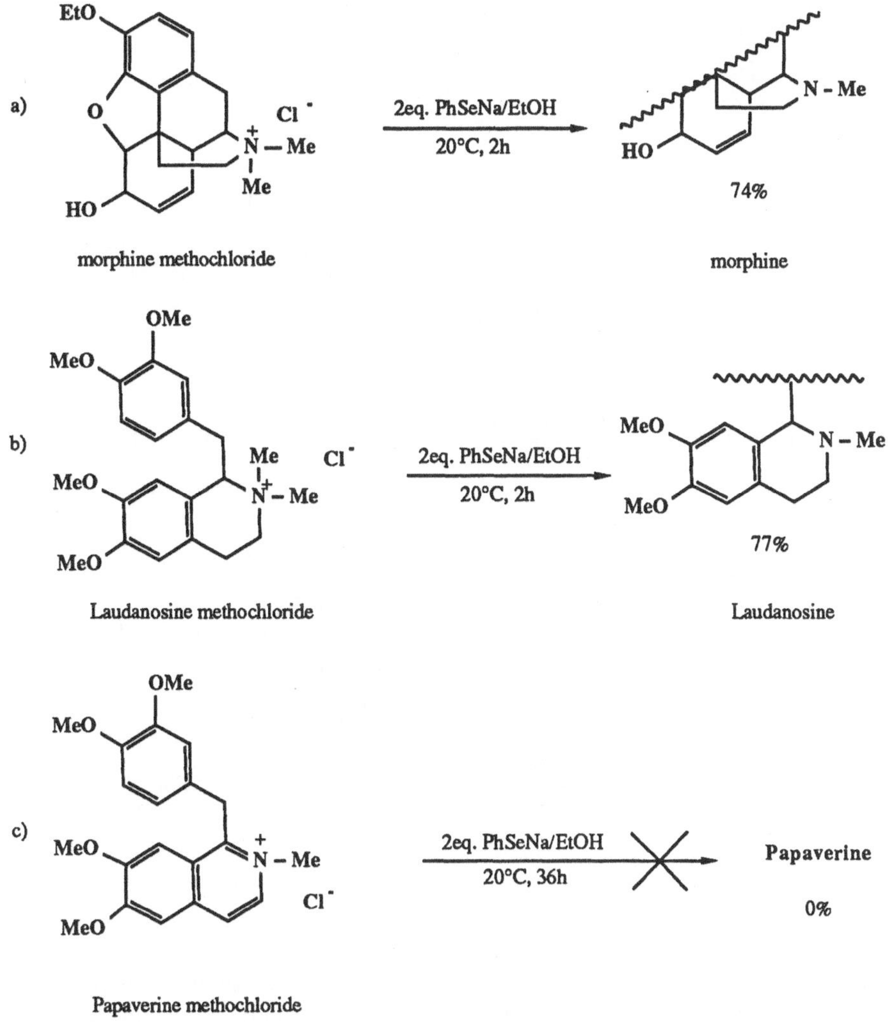

a)

morphine methochloride

2eq. PhSeNa/EtOH
20°C, 2h

morphine

74%

b)

Laudanosine methochloride

2eq. PhSeNa/EtOH
20°C, 2h

Laudanosine

77%

c)

Papaverine methochloride

2eq. PhSeNa/EtOH
20°C, 36h

Papaverine

0%

Scheme 9 [117]

2.1.2 N-dealkylation of Amines

Benzeneselenol at 150 °C can be used to bring about the nucleophilic dealkylation of alkylamines [69], a process which differs fundamentally from other
available procedures [@ 69] in that it does not involve the formation of any new
bonds to nitrogen other than the N-H bonds. The procedure is applicable to the
whole range of alkyl amines (tertiary to primary) and works well for some
compounds which previously could not be dealkylated in reasonable yield [@ 69].
The reaction takes advantage of two properties of selenols: their relatively high
acidity and the great nucleophilicity of their conjugate bases (selenolate anions).
This permits the easy formation of ammonium selenolates on simple mixing of the

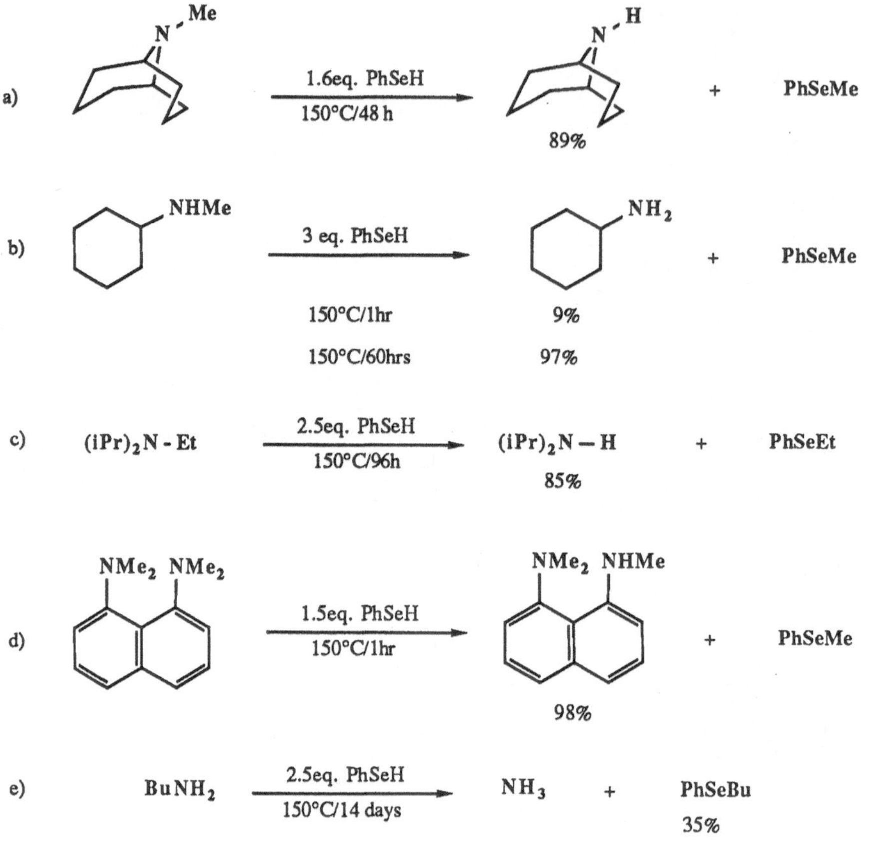

2. Reactions Involving Hydrogen Selenide, Selenols and Related Compounds

a) [bicyclic amine with N–Me] → $\xrightarrow[150°C/48\,h]{1.6eq.\ PhSeH}$ [bicyclic amine with N–H] 89% + PhSeMe

b) [cyclohexyl-NHMe] → $\xrightarrow{3\ eq.\ PhSeH}$ [cyclohexyl-NH$_2$] + PhSeMe

150°C/1hr   9%

150°C/60hrs   97%

c) $(iPr)_2N$ - Et $\xrightarrow[150°C/96h]{2.5eq.\ PhSeH}$ $(iPr)_2N-H$ + PhSeEt

85%

d) [naphthalene with NMe$_2$ NMe$_2$] $\xrightarrow[150°C/1hr]{1.5eq.\ PhSeH}$ [naphthalene with NMe$_2$ NHMe] + PhSeMe

98%

e) $BuNH_2$ $\xrightarrow[150°C/14\ days]{2.5eq.\ PhSeH}$ NH_3 + PhSeBu

35%

Scheme 10 [69]

amine and selenol. Pyrolysis of these salts results in nucleophilic dealkylation of the amine if an alkyl group susceptible to S_N2 displacement is present (Schemes 10, 11). The resulting dealkylated amine and alkyl phenyl selenide are often readily separable due to their different solubility in aqueous acid solution. Primary amines undergo dealkylation very slowly (Scheme 10e) and sterically encumbered amines are dealkylated markedly faster than unhindered ones (Scheme 11). The reaction suffers from severe limitations if the amine is not sufficiently basic to be significantly protonated by benzeneselenol (N,N-dimethyl aniline and N,N-dimethyl acetamide are not demethylated at a useful rate at 150 °C [69]) or when the alkyl group is not susceptible to S_N2 displacement (piperidine is essentially inert but N-methyl pyrrolidine undergoes a second dealkylation [69]). Sequential dealkylation is possible and the difference of reactivity between methyl and ethyl groups is sufficient to allow selective demethylation [69]. However, the amine starting material and product must survive the drastic experimental conditions [69]. This may explain why benzyl amines, nicotine and laudanosine [118] are not cleanly dealkylated [69]. In the

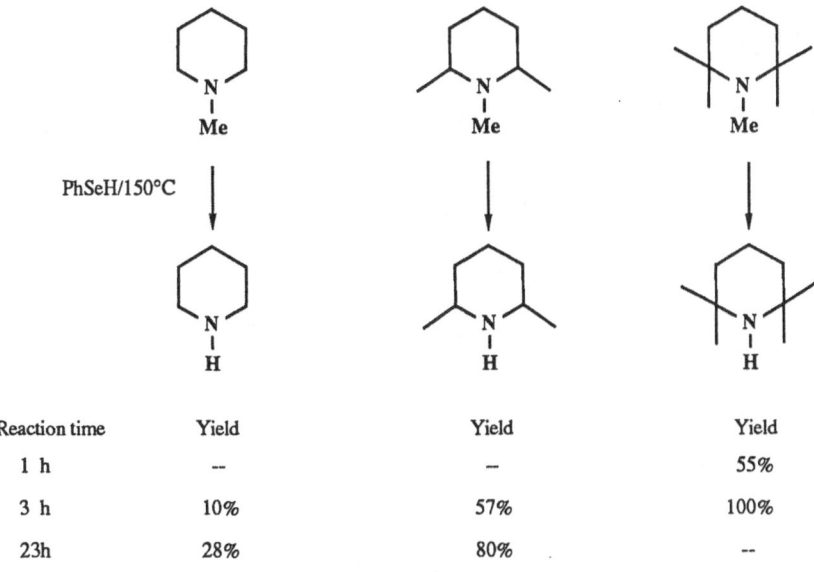

Reaction time	Yield	Yield	Yield
1 h	--	--	55%
3 h	10%	57%	100%
23h	28%	80%	--

Scheme 11 [69]

case of laudanosine competitive demethylation of the aryl methyl ether moiety also occurs [69] (see also Sect. 2.1.3 and Scheme 13) [118].

2.1.3 Dealkylation of Alkyl Aryl Ethers, -Sulfides and -Selenides

Metallo-alkaneselenolates (RSeM: R = Me [119, 120] or PhCH$_2$ [118]) in hot DMF [118, 119, 121] or in HMPA [119, 120] perform the dealkylation of aryl methyl ethers [118, 120] (Schemes 12, 13), aryl methyl sulfides [119, 120] (Scheme

	X		
a)	H	DMF/reflux/8h	68%
b)	H	HMPA/reflux/8h	73%
c)	3-MeO	idem	65%
d)	4-MeS	idem	60%
e)	2-C(O)Me	idem	71%

Scheme 12 [120]

15

2. Reactions Involving Hydrogen Selenide, Selenols and Related Compounds

a)

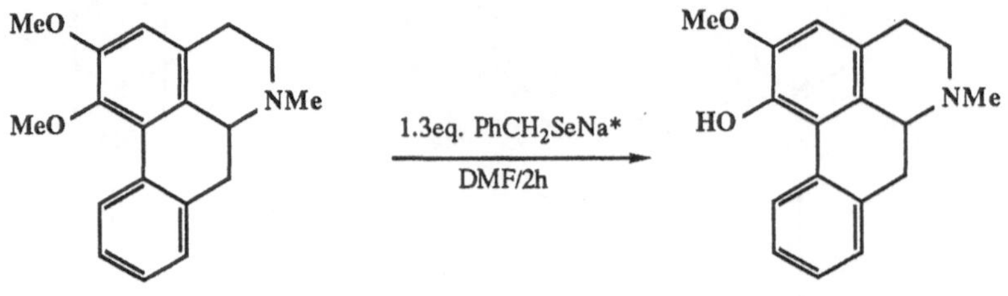

nuciferine

1.3eq. PhCH$_2$SeNa*

DMF/2h

75%

b)

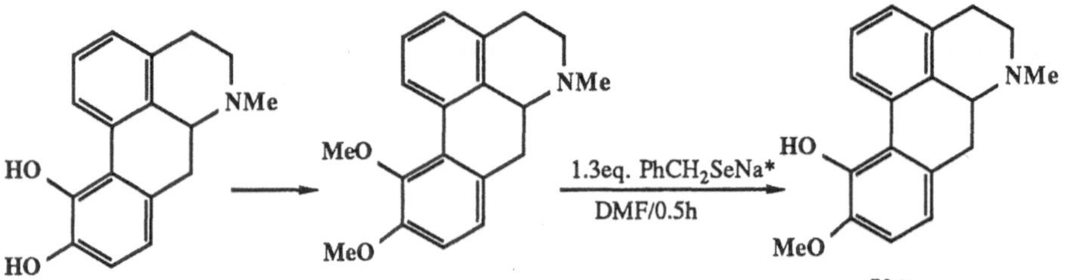

apomorphine

1.3eq. PhCH$_2$SeNa*

DMF/0.5h

79%

apocodeine

* from (PhCH$_2$Se)$_2$ and NaBH$_4$

Scheme 13 [118]

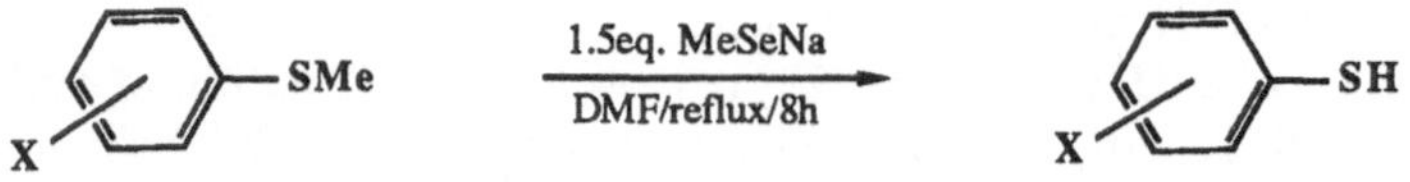

1.5eq. MeSeNa

DMF/reflux/8h

	X	
a)	H	55%
b)	2-CH$_2$OH	42%
c)	2-C(O)Me	60%

Scheme 14 [120]

14) and aryl alkyl selenides [119, 120] (Scheme 15) under much less forcing conditions than those reported for amines (see Sect. 2.1.2). Sodium phenyl-selenolate also reacts with compound **13** (Scheme 16) and leads [122], after (dilute) acidic work-up to the furan **14**. This would result [122] from an initial demethylation of **13** followed by an intramolecular opening of the oxirane ring present, as shown in the Scheme 16. It is however surprising that the epoxide ring is not attacked by the selenolate ion during the process (Vol. II).

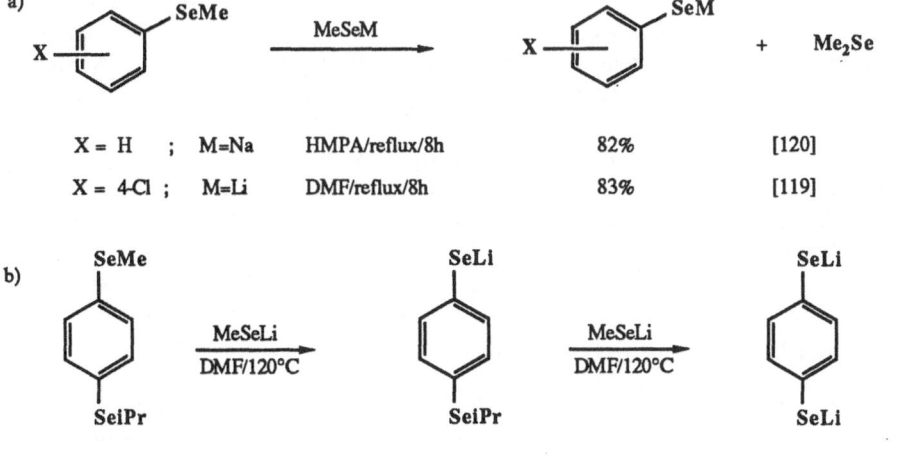

a)
SeMe
X
MeSeM
SeM
X
+
Me_2Se
X = H ; M=Na HMPA/reflux/8h 82% [120]
X = 4-Cl ; M=Li DMF/reflux/8h 83% [119]
b)
SeMe
SeiPr
MeSeLi
DMF/120°C
SeLi
SeiPr
MeSeLi
DMF/120°C
SeLi
SeLi
77%
Yield not reported [119]

Scheme 15

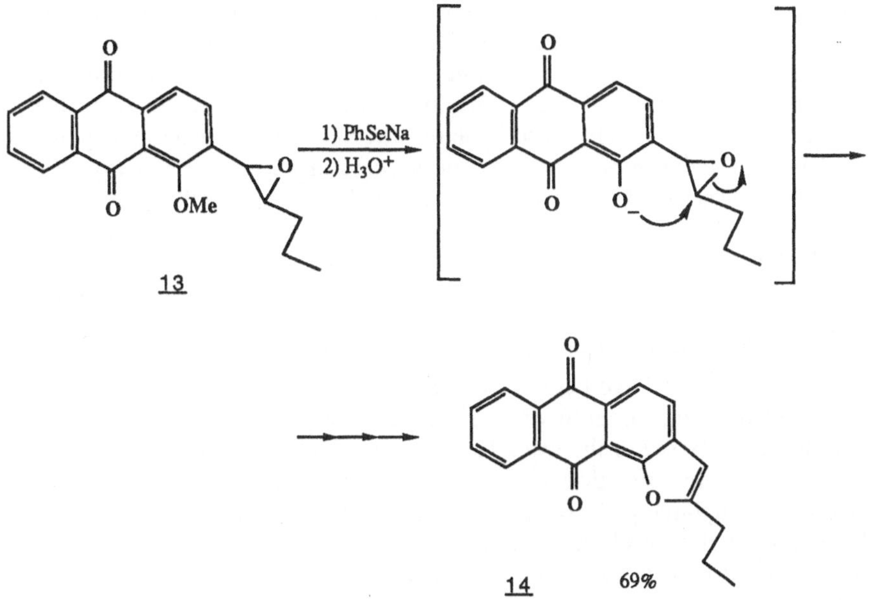

O
O
OMe
O
13
1) PhSeNa
2) H_3O^+
O
O
O
O
O
O
O
14
69%

Scheme 16 [122]

This dealkylation reaction even occurs in the presence of other functional groups such as N-methyl tertiary amines [118] which require much more drastic conditions [69] to be demethylated whereas the methylene dioxy group [118], the acetyl group [120] and an aryl halide [119, 120], are unreactive. With lithium

methyl selenolate in DMF, the ease of dealkylation follows the order [119] ArSeMe > ArOMe > ArSMe. This is different from the order observed when the reaction is performed instead with sodium in HMPT [@119] (ArSeR > ArSR > ArOR), conditions under which it is thought to occur by electron transfer. A difference of reactivity between these two reagents has been noted on several previous occasions. Thus p-(methylseleno)phenyl isopropyl selenide is selectively demethylated [119] with lithium methylselenolate in DMF, whereas cleavage of the methyl and the isopropyl group occurs equally if the reaction is performed with sodium in HMPT [119].

2.1.4 Synthesis of Carboxylic Acids from Esters

Hydrolysis of esters can be accomplished by heating the compounds either in aqueous acids or bases. An alternative approach employs the cleavage of esters via nucleophilic attack at the carbinol carbon. Since the last transformation involves an S_N2 process, it is quite sensitive to steric hindrance at the site of attack [@123], and has typically been successful only with methyl esters [@123, @124]. Selenolate anions, whose nucleophilicity can be adjusted properly by varying the counterion and the degree of solvation of the anion [123, 125], were found to be particularly efficient reagents for carrying out the S_N2 ester cleavage reaction [120, 123, 125]. Closely related results have been observed with the phenyl-selenotrimethylsilane/zinc iodide reagent [126]. The reaction also applies to lactones [123, 125, 127–132] which are transformed to ω-seleno acids (see Volume II). In the case of benzeneselenolates, the manner in which the reagent is prepared has a crucial importance for the success of its further reaction. For instance, the following gradient of reactivity of benzeneselenolate anion was observed [123, 125]: NaSePh/18-crown-6/THF > NaSePh/HMPT-THF > LiSePh/HMPT-THF > LiSePh/THF — LiSePh/ether — PhSeSePh/NaBH4/ THF-ethanol. Only the first three reagents were found to be nucleophilic enough to permit the cleavage of esters [123, 125] but most of the results have been obtained with NaSePh / HMPT / THF at reflux of the solvent [123, 125].

High yields of acids have been observed even in cases where the acyl group shows a relatively high degree of steric hindrance (Scheme 17a, b). This is particularly significant with the ester **15** (Scheme 17a) which is quite unreactive under the usual saponification conditions [@123, @133] but which can be selectively demethylated [123] with sodium benzeneselenolate. It must be recalled that sodium benzeneselenolate also cleaves aryl methyl ethers (Sect. 2.1.3) but such a functional group, if present in the molecule, remains untouched [@123] under these conditions. As expected for an S_N2 reaction the rate of the alkyl — oxygen cleavage process decreases as the substitution pattern at or near the carbinol carbon increases [123] (Scheme 17c, d, e). This should permit the selective cleavage of methyl esters in the presence of longer chain O — alkyl esters [123]. On the other hand amides are completely inert [123] towards PhSeNa (Scheme 18a) whereas urethanes undergo alkyl — oxygen cleavage and subsequent decarboxylation to give the corresponding amine [123] (Scheme 18b). Although most of the reactions are believed to proceed by attack of the ben-

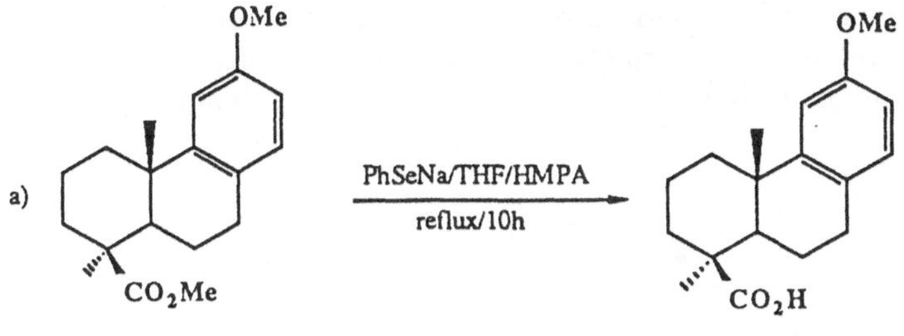

2.1 The Nucleophilicity of Hydrogen Selenides, Selenols and Related Compounds

a)

$\xrightarrow[\text{reflux/10h}]{\text{PhSeNa/THF/HMPA}}$

15   100%

b) $Me_3C\text{-}CO_2Me$ $\xrightarrow[\text{6h}]{\text{idem}}$ $Me_3C\text{-}CO_2H$

96%

$Ph\text{-}CO_2R$ $\xrightarrow{\text{idem}}$ $Ph\text{-}CO_2H$

c)   R=Me   08h   98%

d)   Et   18h   99%

e)   iPr   96h   92%

f)   CH$_2$Ph   08h   96%

g) $Me_2CH\text{-}CH_2CO_2(CH_2)_2CH(Me)_2$ $\xrightarrow[\text{12h}]{\text{idem}}$ $Me_2CH\text{-}CH_2CO_2H$

92%

h) $\begin{array}{c} Me-CH\text{-}CO_2Me \\ | \\ HNC\text{-}Ph \\ \| \\ O \end{array}$ $\xrightarrow[\text{10h}]{\text{idem}}$ $\begin{array}{c} Me-CH\text{-}CO_2H \\ | \\ HNC\text{-}Ph \\ \| \\ O \end{array}$

94%

Scheme 17 [123]

a)   PhCONHMe $\xrightarrow[\text{reflux/24h}]{\text{PhSeNa/THF/HMPA}}$ ✕   PhCONH$_2$

b)   Ph-N(Me)CO$_2$Me $\xrightarrow[\text{10h}]{\text{idem}}$ Ph-NHMe

93%

Scheme 18 [123]

19

zeneselenolate anion on the carbinol carbon [123], there is, when this carbon is hindered, strong evidence for competitive acyl — oxygen cleavage reaction followed by subsequent hydrolysis of the resulting selenolester [123] (Scheme 19).

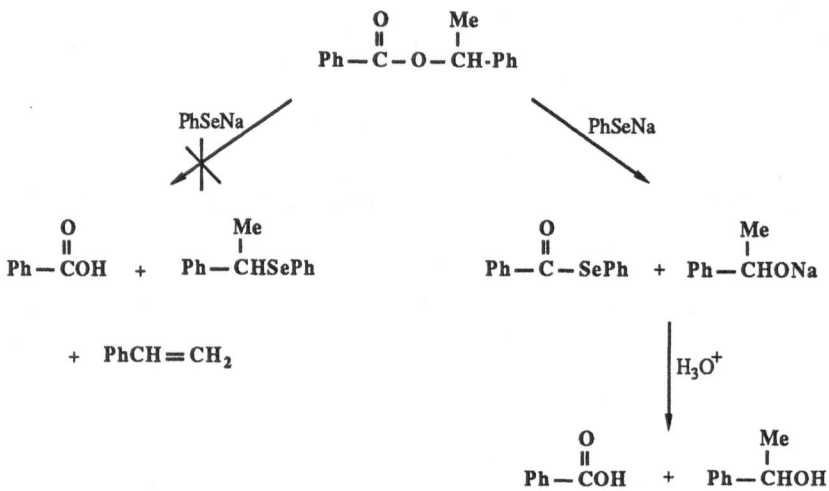

Scheme 19 [12]

2.2 Reduction Reactions Involving Hydrogen Selenide, Selenols, Selenocyanates, Triphenylphosphine Selenide and Related Compounds

Although hydrogen selenide, selenols, sodium selenide, alkane-, areneselenolates and trialkylsilylselenides, metalloselenocyanates and triphenylphosphine selenide are particularly good nucleophiles, they act in several instances as powerful reducing agents.

Some of the reagents presented above have been used for the reduction under mild conditions of:

A benzyl halides to aryl alkanes [134];
B α-halogenoketones [135] and α-selenoketones [136, 137] to ketones;
C β-diheterosubstituted alkanes (such as β-dihalogenoalkanes [138–141], β-hydroxy bromoalkanes [142] and their esters [143]) to olefins;
D epoxides [144–147] to olefins;
E sulfoxides [132, 148], selenoxides [148] and telluroxides [148] to sulfides, selenides and tellurides respectively;
F benzyl sodium thiosulfate [149] to the corresponding disulfide or thiol;
G disulfides to thiols [132, 149];
H imines or carbonyl compound/amine mixtures to amines [150, 151];
I oximes of aromatic aldehydes to N-benzyl hydroxylamines [152];

J   carbonyl compounds to alcohols [153, 154];
K   nitro- [155, 156], nitroso- [155], azo- [155], azoxy- [155], hydrazo- [155], and hydroxylamino- [155], aromatic compounds to aromatic amines [155, 156];
L   arenediazonium fluoroborates to arylhydrazines [157].

2.2.1 Reduction of Benzyl Halides to Aryl Alkanes and of Iodo- and Selenoketones to Ketones

Sodium selenide, sodium alkane- and areneselenolates and potassium selenocyanate react with alkyl halides to produce selenides [*54, *60, *62] (see Volume II) and alkyl selenocyanates [*91, 158] respectively. Ammonium methane-selenolate also produces benzyl selenide on reaction with benzyl iodide but in this case toluene (20%) is formed concomitantly [134]. This side reaction becomes the major one with polynitrobenzyl halides, especially with iodides, which are reduced to the corresponding polynitrotoluenes [134] (Scheme 20). Thiolates also exhibit the same behaviour but proved to be less efficient than selenolates [134].

R_1	R_2	X	Yield(%)	17	:	18
H	H	I	87	18		82
NO_2	H	Br	89	43		57
NO_2	H	I	97	80		20
NO_2	NO_2	I	75	100		0

Scheme 20 [134]

Reactivity similar to that just reported was later observed between potassium benzeneselenolate and α-halomethyl ketones [135] (Scheme 21). α-Chloroacetophenone, 1-adamantyl bromomethyl ketone and 14-bromo-N-(trifluoroacetyl) daunorubicin (19, Scheme 21, X = Br) lead to good yield of the corresponding α-phenylselenoketones, whereas reduction to the methyl ketones was exclusively observed [135] with analogous iodomethyl ketones. Best results are observed when the reductions are carried out at room temperature with a 2:1 selenol:iodomethyl ketone ratio, the methyl ketone being usually readily purified from the diphenyl diselenide concomitantly formed [135]. Thiolates exhibit similar behaviour [135]. The mechanism of these reductions remains somewhat speculative: either electron transfer from the nucleophile to the

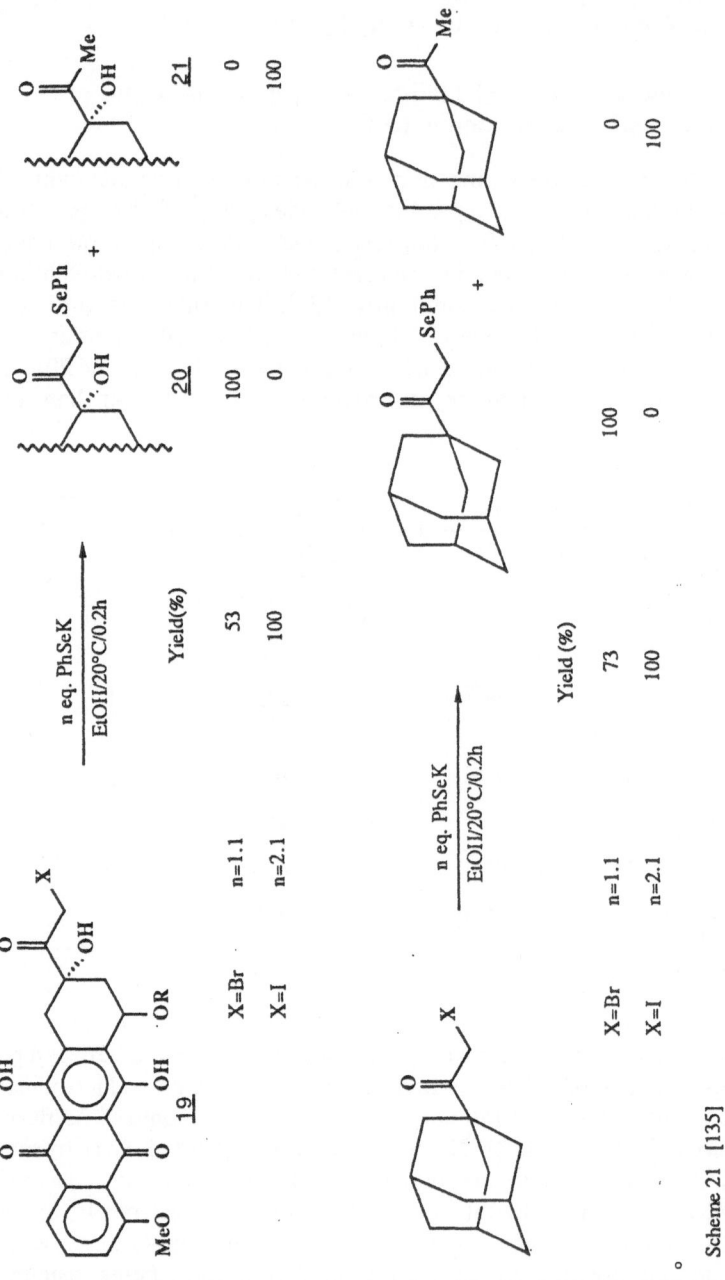

Scheme 21 [135]

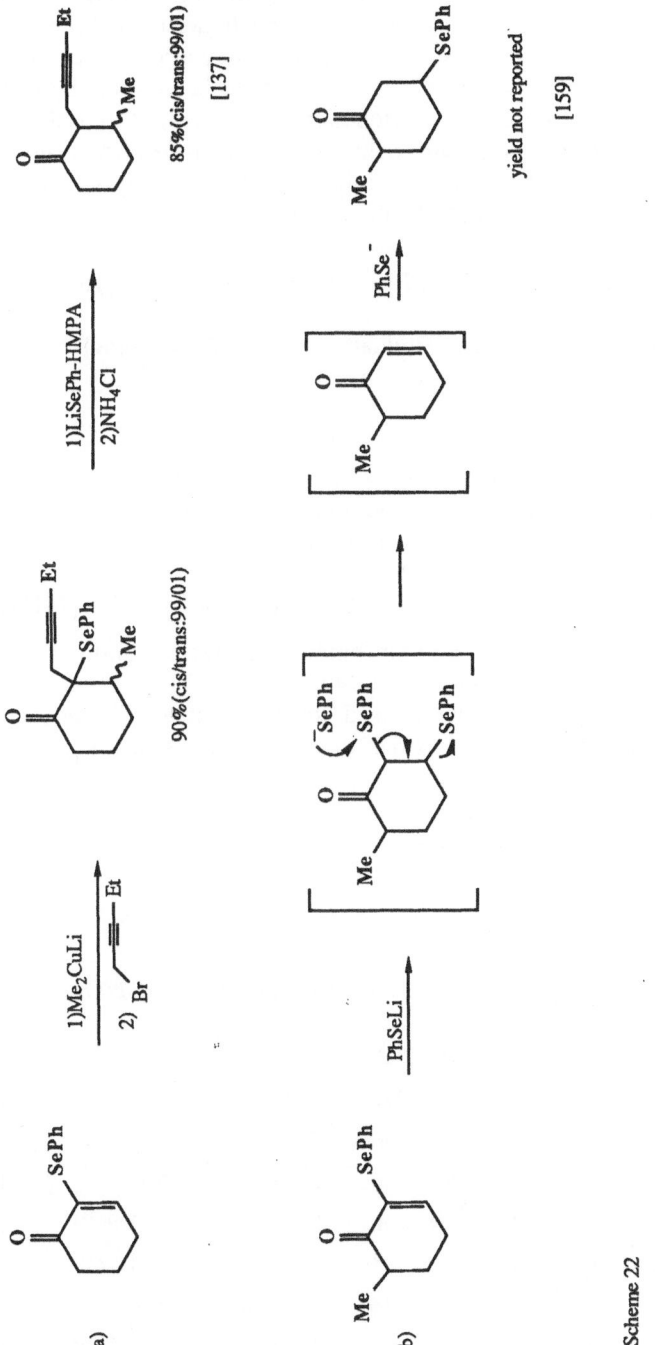

Scheme 22

23

benzylic moiety or attack of the soft selenolate ion on the soft halogen atom have been alternatively suggested by Hevesi [134] for the reduction of benzyliodide. The latter mechanism was also proposed by Israel [135] for the reduction of iodomethyl ketones. This last transformation may also involve the formation of an α-selenoketone intermediate and in fact it has been found by Reich [159] and Liotta [136, 137] that α-phenylselenoketones are smoothly reduced to ketones with lithium benzeneselenolate (Scheme 22a); the transformation of 6-methyl-2-phenylselenocyclohex-2-en-1-one into 6-methyl-3-phenylselenocyclohexanone was proposed [159] to proceed similarly (Scheme 22b).

2.2.2 Reduction of vic-Diheterosubstituted Alkanes to Alkenes

2.2.2.1 Selenolate Mediated Reduction of vic-Dihalogenoalkanes and Bromohydrins to Olefins

Vic-dihalogenoalkanes are smoothly reduced to olefins:

A by sodium selenide in DMF or in DMSO [138, 139] at 70 °C (Scheme 23),
B by sodium benzene- or methaneselenolate in THF-HMPA (3 : 1) [140] at room temperature (except dichloro derivatives which require 80 °C) (Schemes 24, 25),
C by potassium selenocyanate in DMF or in ethanol at reflux [141] (Scheme 26).

As a general trend, diiodo- and dibromoalkanes were found to be more reactive than their dichloro analogues [138, 140].
    These reactions have been used for the synthesis of terminal and α,β-disubstituted olefins from a series of dihalogenoalkanes [138–140] (Schemes

	X	R_1	R_2	R_3	R_4	Solvent/T(°C)	Yield (E/Z ratio)	
a)	I	H	H	H	H	DMF/75	83	[138]
b)	Br	H	H	H	H	DMF/75	77	[138]
c)	Cl	H	H	H	H	DMF/75	75	[138]
d)	Br	H	H	Me	H	DMF/75	53	[138]
e)	Br	H	Me	Me	H	DMF/75	44-60 (99/1)	[139]
f)	Br	H	Me	H	Me	DMF/75	44-60 (14/86)	[139]
g)	Br	H	Me	H	Me	DMF/130	44-60 (40/60)	[139]
h)	Br	H	Me	H	Me	DMSO/75	44-60 (11/89)	[139]

Scheme 23

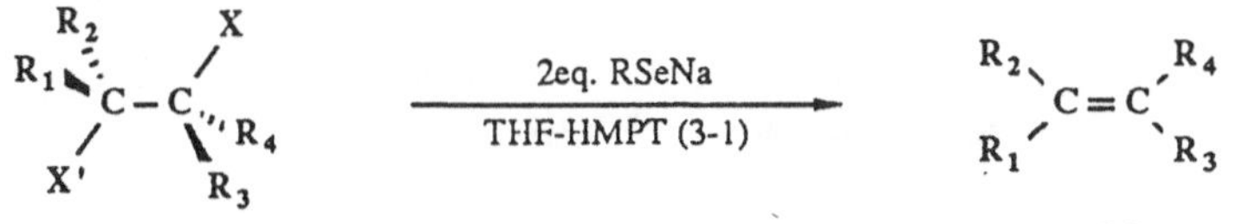

$$\underset{\substack{X'\\ R_3}}{\overset{\substack{R_2 \quad X\\ R_1}}{C-C}}\quad\xrightarrow[\text{THF-HMPT (3-1)}]{\text{2eq. RSeNa}}\quad \underset{R_1}{\overset{R_2}{C}}=\underset{R_3}{\overset{R_4}{C}}$$

22

	R	R$_1$	R$_2$	R$_3$	R$_4$	X	X'	Conditions	Yield (%)	E/Z ratio
a)	Ph	Dec	H	H	H	Br	Br	25°C,0.5h	89	
b)	Me	Oct	H	Oct	H	Br	Br	25°C,2h	87	2/98
c)	Me	Oct	H	H	Oct	Br	Br	25°C,2h	95	100/0
d)	Ph	Pent	H	Me	H	Br	SePh	25°C,2h	95	0/100
e)	Ph	Pent	H	Me	H	Cl	SePh	25°C,2h	95	0/100
f)	Ph	Pent	H	Me	H	Cl	I	25°C,2h	92	3/97
g)	Me	Pent	H	H	Me	Cl	Br	25°C,3h	91	94/6
h)	Ph	Pent	H	H	Me	Cl	Br	25°C,3h	97	53/47

Scheme 24 [140]

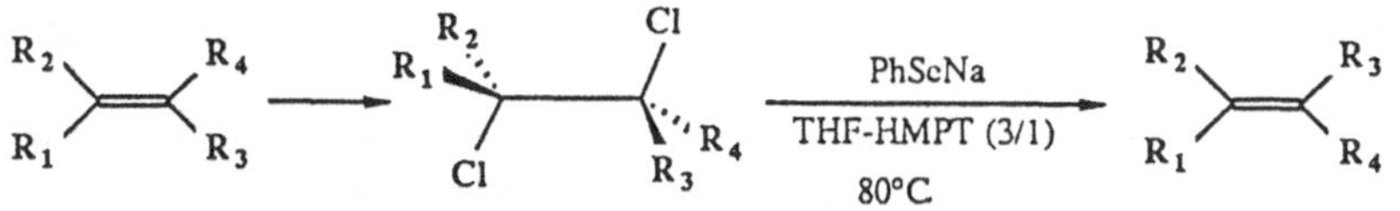

22   23   24

					Yield in 24 (%)	E/Z ratio
	R$_1$	R$_2$	R$_3$	R$_4$		
	Oct	H	Oct	H	92	95/5
	Pent	H	H	Me	77	0/100

(the middle column values: 4h and 2h under reaction between 23 and 24)

Scheme 25 [140]

23–25), for the preparation of styrenes, stilbenes and α,β-unsaturated carboxylic acids and esters from the corresponding saturated dibromo derivatives [141] (Scheme 26) and for the synthesis of unsaturated sugars from the corresponding vic-ditosylates [145] (Scheme 27).

The stereochemistry of all these reactions has been carefully investigated. The elimination of the two halogens occurs in an anti fashion on reaction of sodium selenide with 2,3-dibromobutanes [139] (ss >70%)* (Scheme 23) and with

---

* ss = stereoselection = $\dfrac{\%\text{ isomer via anti elimination} - \%\text{ isomer via syn elimination}}{\%\text{ isomer via anti elimination} + \%\text{ isomer via syn elimination}}$

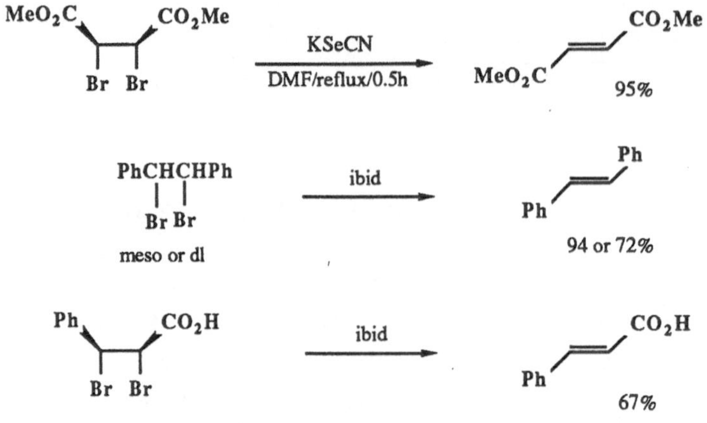

2. Reactions Involving Hydrogen Selenide, Selenols and Related Compounds

Scheme 26 [141]

much higher selectivity when dibromoalkanes (ss > 92%) or bromochloroalkanes (ss > 94%) are reacted with sodium selenolates [140] (Scheme 24). Under the latter conditions the requirement for two equivalents of the sodium selenolate suggests that β-halogeno alkylselenides are intermediates, but these were never isolated. It was, however, found [140] that β-bromo alkyl selenides and their chloro analogues are stereoselectively transformed to olefins under similar conditions (Scheme 24d, e).

The stereochemical outcome of selenolate mediated elimination of *vic*-dihalogeno alkanes dramatically depends upon the nature of the starting material [140]. For example *anti* elimination still occurs but to a lesser extent with β-bromochloroalkanes and methaneselenolate (ss > 88%) (Scheme 24. Entry g) than is the case for *vic*-dibromoalkanes, and to an even lesser extent (ss: 6%) when benzeneselenolate is used (Scheme 24, Entry h) in place of methaneseleno-late [140]. The case of *vic*-dichloroalkanes deserves further comment, since only *syn* elimination is observed [140] (ss > 90%) (Scheme 25). Trans addition of chorine to olefins and treatment of the resulting *vic*-dichlorides with selenolates

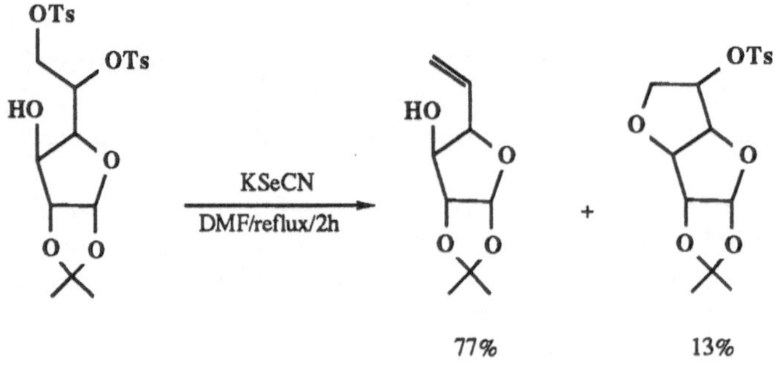

Scheme 27 [145]

26

permits the stereoselective isomerisation of olefins in high yield, as outlined in Scheme 25. Finally whereas β-bromoalkylselenides lead to alkenes on reaction with sodium selenolates [140], no elimination was found when β-hydroxy- or β-phenylthio alkylbromides are reacted with te same reagent and β-hydroxyalkylselenides or β-phenylthioselenides respectively are produced instead [140]. Finally, E stilbenes and E α,β-unsaturated esters and acids are exclusively formed regardless of whether *erythro* or *threo* dibromo derivatives are reacted with potassium selenocyanate (Scheme 26).

Sodium hydrogen selenide in ethanol readily cleaves 2-halogenoethyl esters and produces ethylene and the carboxylic acid salts [143] (Scheme 28).

$$R-\overset{\overset{\displaystyle O}{\|}}{C}-OCH_2CH_2X \xrightarrow[\substack{\text{from Se}^\circ + \text{PhNaBH}_4 \\ 20^\circ C/1h/\text{reflux}/0.1h}]{\text{1.2eq. NaHSe/EtOH-H}_2O} RCO_2Na + CH_2{=}CH_2 + NaBr + Se^\circ$$

X=Br	R=Ph or Undec	99 or 93%
X=Cl	R=Ph	95%

Scheme 28 [143]

Bromohydrins are for their part reduced [142] to olefins on reaction with potassium selenocyanate in DMF or in ethanol at $80\,^\circ C$ (Scheme 29). α,β-Disubstituted olefins are stereoselectively (66–98% ss) obtained from the corresponding bromohydrines by formal *syn* elimination of the two heteroatomic moieties. An episelenide [160–162], a species which is known to be particularly unstable and expected to decompose rapidly and stereoselectively to an olefin, has been postulated [142] as an intermediate in this reaction. In fact, a closely related reaction occurs with potassium thiocyanate, which instead leads to a stable episulfide [142]. Reaction of bromine in DMSO on olefins followed by reaction of the resulting bromohydrins with KSeCN allows the stereoselective (66–98 ss) (Z→E) or (e→Z) isomerisation of disubituted olefins [142] (Scheme 29).

2.2.2.2 Reduction of Epoxides and Thiiranes to Olefins

Episelenides have also been postulated as intermediates in the deoxygenation of epoxides to olefins which occurs on reaction with potassium selenocyanate [144, 145, 160, 163] (Scheme 30), with triphenylphosphine selenide / trifluoroacetic acid [138] or with related compounds such as 1-phenyl-3,4-dimethylphospholeselenide or phosphole-3-selenide [164], with sodium diethyl phosphite / selenium [165, 166], with 3-methyl-2-selenoxobenzothiazole [@167, 168] or with selenocarboxamides [@169] (readily available [170] from nitriles, carbon monoxide and selenium) the reactions being again performed in the presence of trifluoroacetic acid (Scheme 31). Further support in favor of the formation of an episelenide comes from the observation of signals attributed, to the episelenide when the reaction between 1,2-oxidooctene, tributylphosphine selenide and

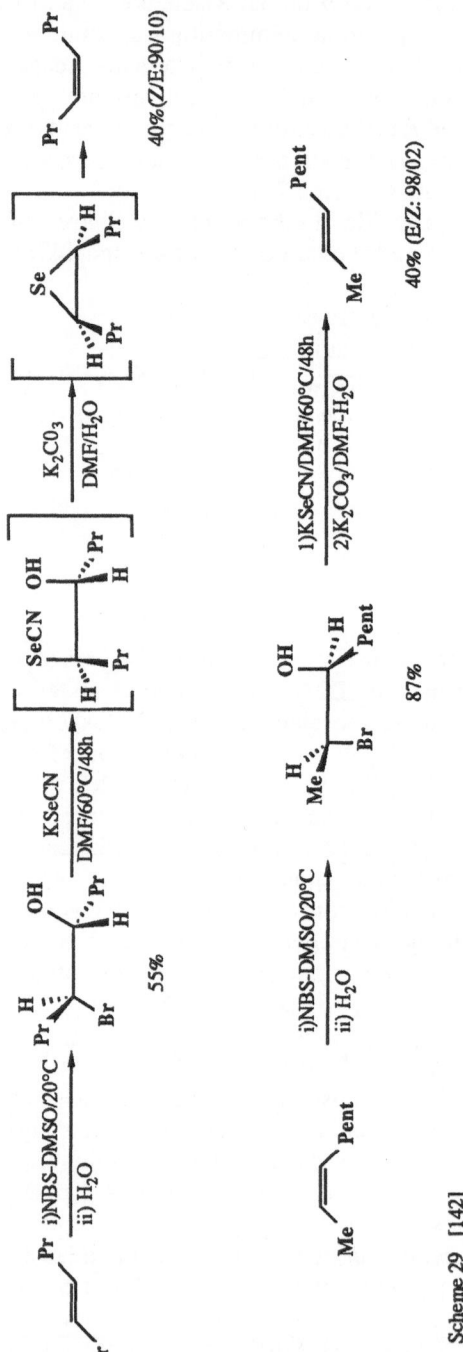

Scheme 29 [142]

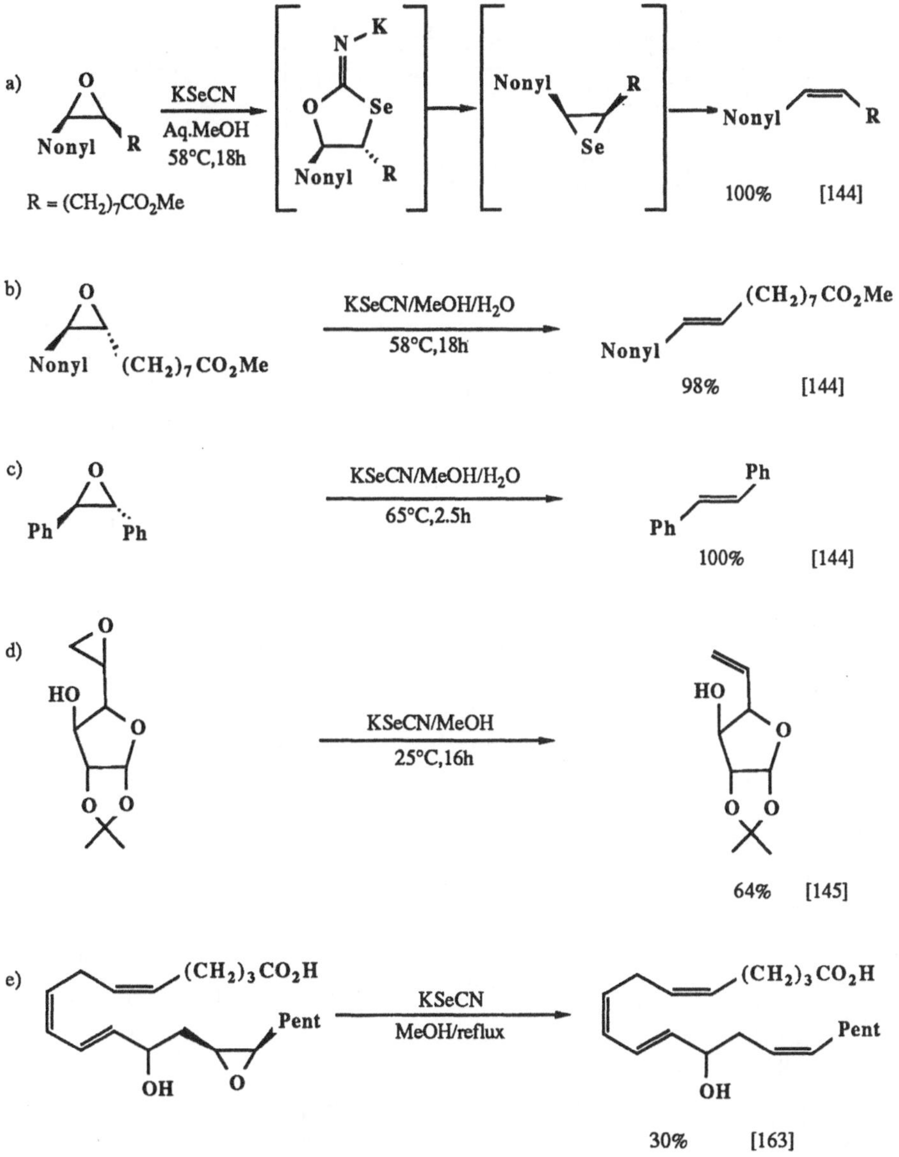

Scheme 30

trifluoroacetic acid is monitored [147] by ^{1}H NMR (Scheme 32). These reactions have been successfully applied to the synthesis of terminal [145, 147, 169] and α, β-disubstituted olefins [144, 146, 147, 169] including stilbenes [144, 146, 147, 167, 169], cyclohexenes [144, 146, 164, 169] (Schemes 30, 31) and unsaturated sugars [145] (Scheme 30, Entry d) from the corresponding epoxides. A syn deoxygenation (100% ss) was reported with 1,2-disubstituted epoxides [144–147, 167, 169]

29

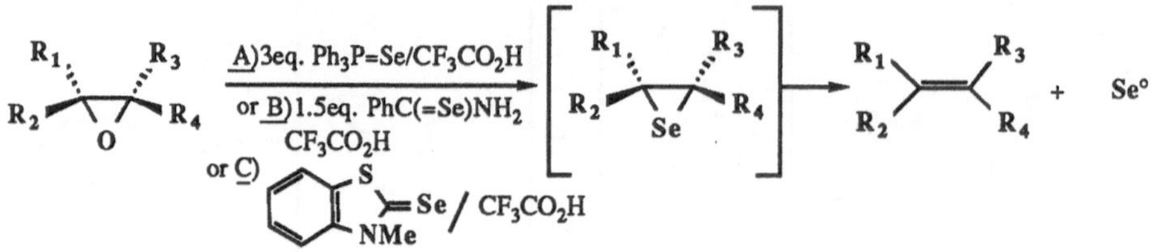

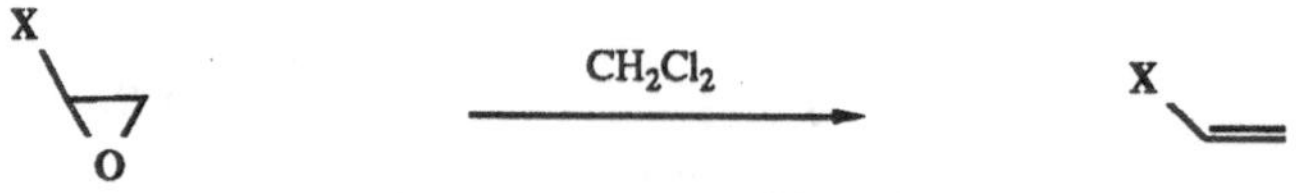

2. Reactions Involving Hydrogen Selenide, Selenols and Related Compounds

Reaction conditions:
A) 3eq. Ph$_3$P=Se/CF$_3$CO$_2$H
or B) 1.5eq. PhC(=Se)NH$_2$ / CF$_3$CO$_2$H
or C) (benzothiazole-2-selenone derivative with NMe) =Se / CF$_3$CO$_2$H

Epoxide (R$_1$, R$_2$, R$_3$, R$_4$) → episelenide intermediate (Se) → alkene (R$_1$, R$_2$, R$_3$, R$_4$) + Se°

	X	reagent	CH$_2$Cl$_2$	yield	ref
a)	X=Hex	B	0°C, 0.1h	89%	[169]
b)	X=Ph	B	0°C, 0.5h	54%	[169]
c)		C	-10°C	90%	[167]
d)	X=Cl	C	-10°C	100%	[167]

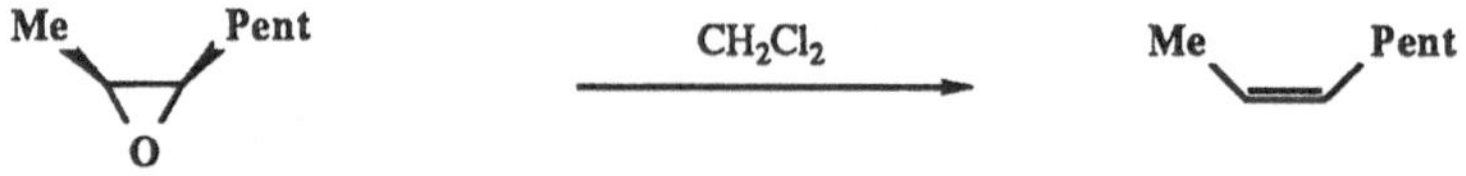

Me, Pent epoxide (cis) → Me, Pent alkene

	reagent	CH$_2$Cl$_2$	yield	ref
e)	A	---	71% (Z/E:100/0)	[146]
f)	B	0°C, 0.2h	75% (Z/E:100/0)	[169]

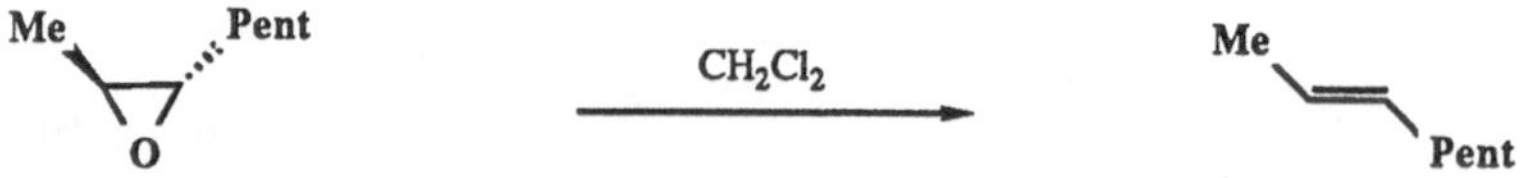

Me, Pent epoxide (trans) → Me, Pent alkene

	reagent	CH$_2$Cl$_2$	yield	ref
g)	A	---	68% (E/Z:100/0)	[146]
h)	B	0°C, 0.2h	74% (E/Z:100/0)	[169]

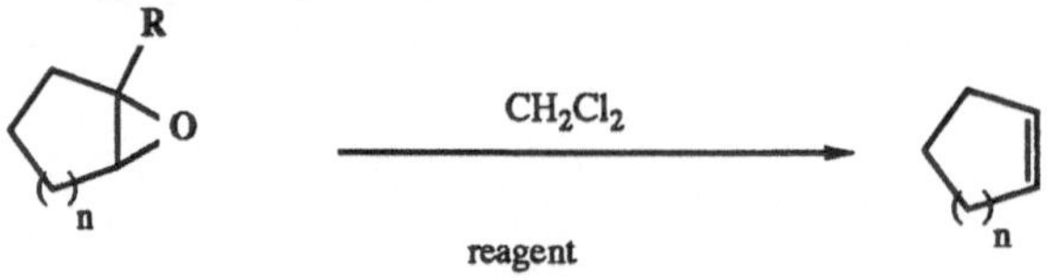

Bicyclic epoxide (R, n) → cyclic alkene (n)

			reagent	CH$_2$Cl$_2$	yield	ref
i)	n = 2	R=H	B	0°C, 0.1h	84%	[169]
j)	n = 2	R=H	C	-10°C, 5h	100% (NMR)	[167]
k)	n = 2	R=Me	B	0°C, 0.1h	51%	[169]
l)	n = 1	R=H	B	0°C, 30h	Sluggish	[169]
m)	n = 1	R=H	B	20°C, 30h	85%	[169]

Scheme 31

30

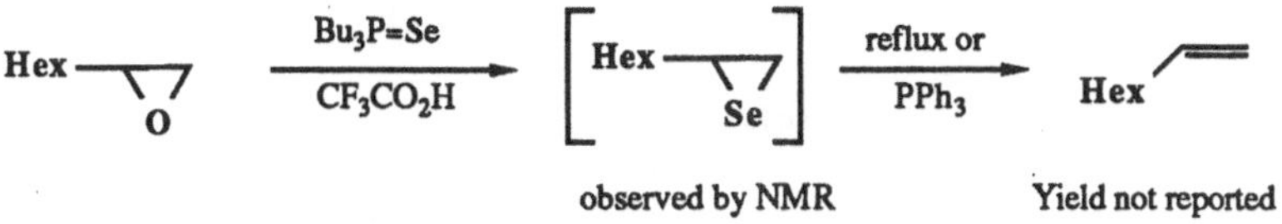

observed by NMR   Yield not reported

Scheme 32 [147]

and this feature was used by Corey [163] in his synthesis of 12-HETE (Scheme 30 Entry e).

Potassium selenocyanate also allows the synthesis of terminal olefins from a *vic*-ditosylate (Scheme 27) and 3-methyl-2-selenoxobenzothiazole proved valuable for the desulfuration of episulfides [167] (Scheme 33).

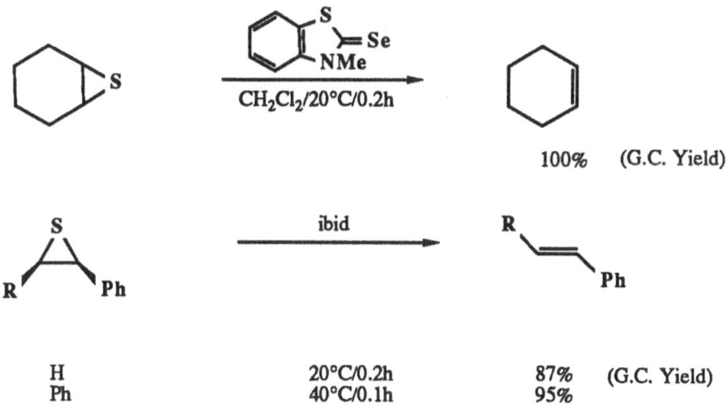

100%   (G.C. Yield)

| H | 20°C/0.2h | 87% | (G.C. Yield) |
| Ph | 40°C/0.1h | 95% | |

Scheme 33 [167]

2.2.3 Reduction of Disulfides to Thiolates

Sodium hydrogen selenide [149] (safely generated from sodium borohydride and elemental selenium in ethanol) as well as selenols [132] are effective reagents for the mild reduction of various disulfides to thiols. Dialkyl-, diaryl-, aryl alkyl disulfides and those bearing other functional groups such as cysteine dimer have been successfully transformed to the corresponding thiols [149]. The reaction however does not take place in the case of the hindered di-t-butyl disulfide (Scheme 34). Organic thiosulfates (Bunte salts) also react [149] with sodium hydrogen selenide and give disulfides or thiols according to the relative sodium hydrogen selenide/Bunte salt ratios used [149] (Scheme 35). Thus benzyl thiosulfate is exclusively reduced to dibenzyl disulfide when 0,5 molar equiv. of NaHSe is used whereas benzylthiol is exclusively formed with 1.6 molar equiv. of NaHSe (Scheme 35).

Metallic selenium is easily recovered from all these reactions [149] and since sodium hydrogen selenide can be prepared from sodium borohydride and

2. Reactions Involving Hydrogen Selenide, Selenols and Related Compounds

$$RSSR \quad + \quad 2HSeNa \quad \longrightarrow \quad 2RSH \quad + \quad Se° \quad + \quad H_2Se$$

R	Yield (%)
Ph	96
PhCH$_2$	100
nDec	95
HO$_2$CCH(NH$_2$)CH$_2$	97
tBu	0

Scheme 34 [149]

$$PhCH_2SSO_3^- \quad + \quad n \text{ eq. HSeNa} \quad \longrightarrow \quad PhCH_2SH \quad + \quad (PhCH_2S)_2 \quad + \quad Se°$$

n=0.55	97%	0	100
n=1.6	94%	100	0

Scheme 35 [149]

$$H_2NCH_2CH_2SSO_3H \quad + \quad n \text{ eq. HSeNa} \quad \xrightarrow[H_2O]{KOH} \quad H_2NCH_2CH_2SH \quad + \quad Se° \quad + \quad HSO_3Na$$

n=2	28%
n=4	100%

Scheme 36 [149]

selenium, the reaction can be performed with catalytic amounts of selenium and sodium borohydride which is the only material consumed. Sodium borohydride alone is also able to perform similar reactions [@149] but is not as selective when other functional groups are present. It also requires higher temperatures and longer reaction times to cleave the S-S bond and is unable [149] to reduce 2-aminoethanethiosulfuric acid to mercaptoethanolamine. NaHSe (4 equiv.) however permits this transformation in quantitative yield [149] (Scheme 36).

2.2.4 Reduction of Sulfoxides, Selenoxides and Telluroxides to Sulfides, Selenides and Tellurides Respectively

Selenols smoothly reduce sulfoxides to sulfides [132, 171] (Scheme 37, Entries a, c, d). The first report from Gunther [132] showed that the reaction occurred using hypophosphorous acid in the presence of catalytic amounts of diselenides. This mixture has been shown [132] to produce selenols. Benzeneselenol in CH$_2$Cl$_2$/EtOH or p-phenyl benzeneselenol in ethanol [171] at reflux of the solvent

Me$_2$SO $\longrightarrow$ Me$_2$S

a)　　　2.5 10^{-4}eq. (Me$_2$NHCH$_2$CH$_2$Se)$_2$Cl /1eq. H$_3$PO$_2$/20°C　　　[132]

b)　　　1eq. (EtO)$_2$P(O)SeH/CDCl$_3$/41°C/0.5h　　　[173]

R$_1$CONH—[β-lactam-cephem ring system with S, O, N, Me, CO$_2$R$_2$]　$\longrightarrow$　R$_1$CONH—[β-lactam-cephem ring system with S, O, N, Me, CO$_2$R$_2$]

c)	R$_1$=PhCH$_2$	R$_2$=CH$_2$CCl$_3$	PhSeH/CH$_2$Cl$_2$-MeOH/reflux/72h	85%	[171]
d)	R$_1$=PhCH$_2$	R$_2$=CH$_2$CCl$_3$	4-PhPhSeH/EtOH/80°C/2h	91%	[171]
e)	R$_1$=PhOCH$_2$	R$_2$=Me	(BuB)$_2$Se$_3$/CHCl$_3$/reflux/24h	25%	[172]

RS(O)tBu $\longrightarrow$ RStBu

f)	R=Me	(EtO)$_2$P(O)SeH/CH$_2$Cl$_2$/reflux/2.5h	79%	[173]
g)	R=tBu	(EtO)$_2$P(O)SeH/CH$_2$Cl$_2$/reflux/14h	14%	[173]

Scheme 37

deoxygenate deacetoxycephalosporin S-oxide in more than 85% yield
(Scheme 37 Entries c, d). Interestingly no other reactions such as addition of the
selenol across the carbon-carbon double bond of the α, β-unsaturated esters (see
Volume II), or the cleavage of the β-lactame ring, occur during this transformation.

　　Other reducing agents such as selenoboranes [172] [B(SePh)$_3$, B(SeMe)$_3$]
(Scheme 38a, b, d), (BuB)$_2$Se$_3$ (Schemes 37e, 38c) and O,O-diethyl hydrogen
phosphoroselenolate [173] (Scheme 37b, f, g) also reduce sulfoxides to sulfides.

(PhCH$_2$)$_2$SO $\longrightarrow$ (PhCH$_2$)$_2$S

a)	1.17eq. B(SePh)$_3$/-30°C/1h	91%
b)	1.15eq. B(SeMe)$_3$/0°C/0.2h	88%
c)	1.28eq. (BuB)$_2$Se$_3$/-30°C/1.5h	74%

d) C$_{15}$H$_{31}$COCH$_2$S(O)Me　$\xrightarrow[\text{0°C,1h; 20°C,1h}]{\text{1.1eq. B(SePh)}_3\text{/CHCl}_3}$　C$_{15}$H$_{31}$COCH$_2$SMe

90%

Scheme 38 [172]

33

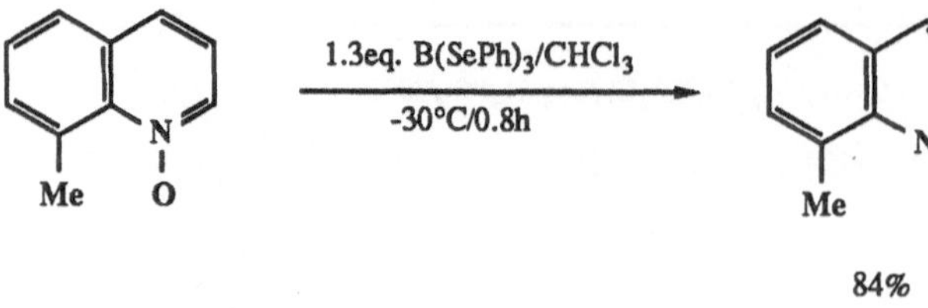

2. Reactions Involving Hydrogen Selenide, Selenols and Related Compounds

Scheme 39 [172]

(BuB)$_2$Se$_3$ seems to be the least efficient among the series of reagents (Schemes 37, 38), whereas tris(phenylseleno)borane is particularly powerful (Scheme 38). The latter is the only reagent which allows [172] the reduction of an N-oxide to the corresponding amine (Scheme 39), but it is unable, as are the other reagents in the series, to reduce phosphinoxides [172] and sulfones [172]. (Phenylseleno)trimethylsilane [148] (Scheme 40) and bis(trimethylsilyl) selenide [174] (Scheme 41) have a closely related reactivity. They reduce sulfoxides to sulfides [@148] as well as selenoxides [148] and telluroxides [148] to selenides and tellurides respectively (Schemes 40, 41). The reaction requires two equivalents of PhSeSiMe$_3$, most probably involves onium intermediates [148] and occurs in the presence of a variety of other functional groups such as ketones, phenols, alcohols, sulfones, nitro groups and olefins which remain untouched during the process (Scheme 40). Moreover bis(trimethylsilyl) selenide [174] (Scheme 41) offers the advantage over the other reagents of easy removal of the by-products (Se and Me$_3$SiOSiMe$_3$) formed during the reduction reaction.

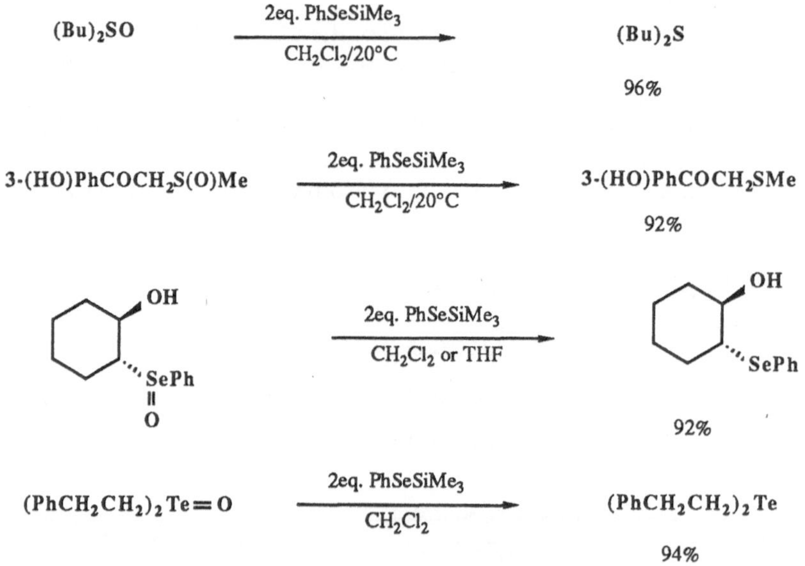

Scheme 40 [148]

34

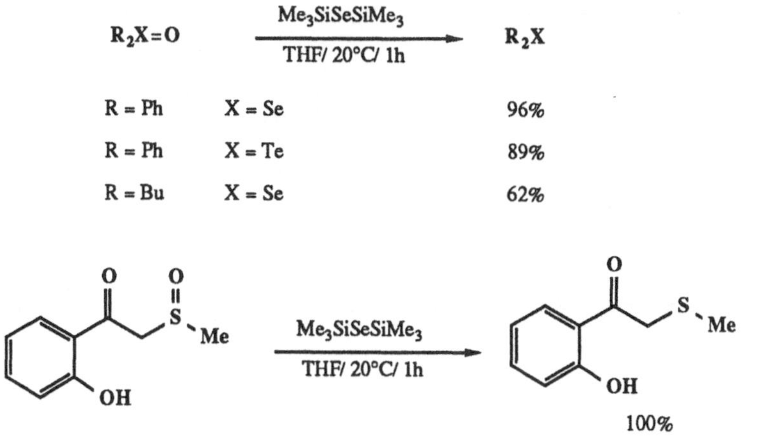
$R_2X{=}O$
$Me_3SiSeSiMe_3$
THF/ 20°C/ 1h
R_2X
R = Ph X = Se 96%
R = Ph X = Te 89%
R = Bu X = Se 62%
O
O
S
Me
OH
$Me_3SiSeSiMe_3$
THF/ 20°C/ 1h
O
S
Me
OH
100%

Scheme 41 [174]

2.2.5 Reduction of Nitro-, Nitroso-, Hydroxylamino-, Azo- and Hydrazo-aromatic Compounds to Aromatic Amines and Reduction of Aryldiazonium Salts to Hydrazinium Salts

Hydrogen selenide reduces nitrobenzene to aniline quantitatively [156] (Scheme 42a). Similar results are observed when nitrobenzene, or other title compounds are reacted with carbon monoxide, water and triethylamine in the presence of catalytic amounts of selenium [156] (Scheme 42) (see also Sect. 3.4.5). Hydrogen selenide is thought to be the effective reagent in this transformation [156, 175] (Scheme 43). Similar reductions can be performed with selenols [155] which are able to reduce aromatic nitroso- [155], hydroxylamino- [155], azo- [132, 155] and hydrazo- [155] compounds to aromatic amines [132, 155] (Scheme 44). Selenols also permit the reduction of arene diazonium fluoroborates to aryl hydrazinium fluorobarates [157] (Scheme 45) and leave untouched a nitro group when present in the molecule [157]. It must be recalled at this occasion that other reagents such as stannous chloride [@176], which also

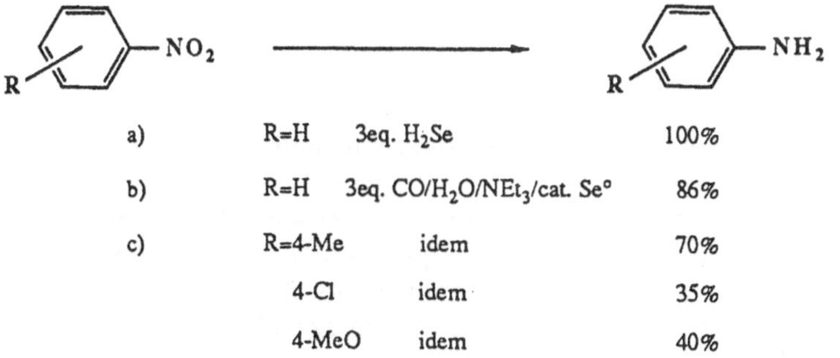
NO_2
R
NH_2
R
a) R=H 3eq. H_2Se 100%
b) R=H 3eq. $CO/H_2O/NEt_3$/cat. Se° 86%
c) R=4-Me idem 70%
4-Cl idem 35%
4-MeO idem 40%

Scheme 42 [156]

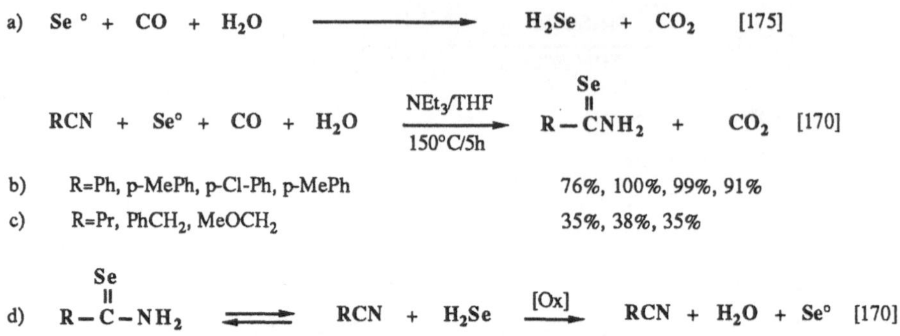

2. Reactions Involving Hydrogen Selenide, Selenols and Related Compounds

a) $\quad Se\degree + CO + H_2O \xrightarrow{\hspace{4cm}} H_2Se + CO_2 \quad$ [175]

$$RCN + Se\degree + CO + H_2O \xrightarrow[150\degree C/5h]{NEt_3/THF} \overset{\overset{\displaystyle Se}{\displaystyle \|}}{R-CNH_2} + CO_2 \quad [170]$$

b) R=Ph, p-MePh, p-Cl-Ph, p-MePh $\qquad\qquad$ 76%, 100%, 99%, 91%

c) R=Pr, PhCH$_2$, MeOCH$_2$ $\qquad\qquad\qquad$ 35%, 38%, 35%

d) $\quad \overset{\overset{\displaystyle Se}{\displaystyle \|}}{R-C-NH_2} \rightleftharpoons RCN + H_2Se \xrightarrow{[Ox]} RCN + H_2O + Se\degree \quad [170]$

Scheme 43

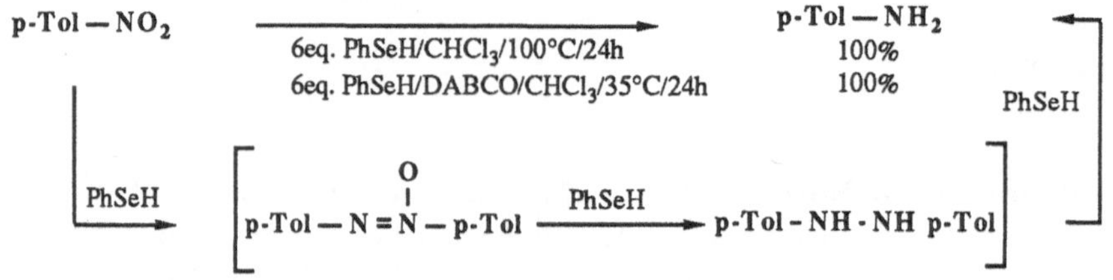

$$p\text{-Tol}-NO_2 \xrightarrow[\substack{6eq.\ PhSeH/CHCl_3/100\degree C/24h \\ 6eq.\ PhSeH/DABCO/CHCl_3/35\degree C/24h}]{} \substack{p\text{-Tol}-NH_2 \\ 100\% \\ 100\%}$$

$$\xrightarrow{PhSeH} \left[\overset{\overset{\displaystyle O}{\displaystyle |}}{p\text{-Tol}-N=N-p\text{-Tol}} \xrightarrow{PhSeH} p\text{-Tol}-NH\cdot NH\ p\text{-Tol} \right] \xrightarrow{PhSeH}$$

Scheme 44 [155]

$$X-\!\!\!\bigcirc\!\!\!-N_2^+,BF_4^- \xrightarrow[40\degree C/2h]{PhSeH/CH_2Cl_2} X-\!\!\!\bigcirc\!\!\!-N\overset{+}{H}NH_3,\ BF_4^- + (PhSe)_2$$

X=H, Cl, NO$_2$ $\qquad\qquad\qquad\qquad\qquad\qquad\qquad$ 80%

$$\text{(2-benzamide-N-Me-N-Ph, } N_2^+, BF_4^-) \xrightarrow[40\degree C/2h]{PhSeH/CH_2Cl_2} \text{(phthalazinedione NH-NH)} + PhNHMe$$

$\qquad\qquad\qquad\qquad\qquad\qquad\qquad\qquad\qquad$ 70% $\qquad\qquad$ 70%

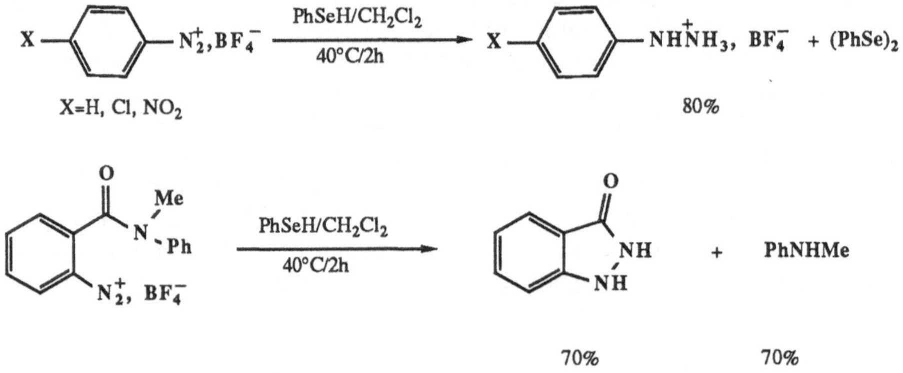

Scheme 45 [157]

reduce diazonium salts to hydrazinium salts would concomitantly reduce the nitro group.

The reaction between p-nitrotoluene and benzeneselenol (6 equiv.) takes place at 100 °C for 24 h and affords p-toluidine quantitatively [155] (Scheme 44a). If the reaction is conducted with one or two equivalents of benzeneselenol, 4,4'-dimethylazoxybenzene or N-(p-tolyl)hydroxylamine are obtained respectively (Scheme 44). These reactions probably involve an electron transfer and are greatly accelerated if benzeneselenolate anion (which is a much better electron donor than its conjugate acid) is used [155]. For example the reduction of *p*-nitrotoluene to toluidine proceeds at 35 °C instead of 100 °C when performed in the presence of DABCO (diazabicyclo-[2.2.2]-octane) (Scheme 44). The relative reactivity of oxygenated amino compounds with selenolate ions was reported [155] to follow the order: RN = O ≫ RN = NR > RNH-OH > RNO = NR ~ RNO$_2$ ≫ RNH-NHR.

2.2.6 Reduction of Schiff's Bases to Amines: Application to the One Pot Reductive Amination of the Carbonyl Group

Benzeneselenol is also an excellent reagent for the reduction of Schiff's bases to amines [150, 151] (Scheme 46). The reaction proceeds in good yield at room temperature and leaves other functional groups such as nitriles, olefins, amides and ester groups untouched when these are also present in the molecule [150]. It is carried out by simply mixing the reagent and the substrate without any pH control, and the amines are easily separated from the diselenide formed concomitantly. Some of these reductions have been performed under irradiation by sunlight [152] (Scheme 47). Reduction of imines with benzeneselenol offers several advantages over existing methods [@150] such as catalytic hydrogenation [*177].

A simple modification of this reaction [150, 151] permits the N-alkylation of primary and secondary amines by reacting them with carbonyl compounds and benzeneselenol (Scheme 48). Interestingly two different alkyl groups can be introduced on primary alkyl amines by successive addition of two different aldehydes in the presence of benzeneselenol (4.5 equiv.). This affords thus

$$\underset{\underset{R_1}{|}}{Ph-C=N-R_2} \xrightarrow{\text{2.5eq. PhSeH/20°C/0.2h}} \underset{\underset{R_1}{|}}{Ph-CHNHR_2}$$

R$_1$	R$_2$	Solvent	Yield (%)
H	p-Tol	MeCN	91
Ph	p-Tol	CHCl$_3$	77
H	Prop	Ether	89
H	Allyl	CHCl$_3$	83
H	(CH$_2$)$_2$CN	CHCl$_3$	80

Scheme 46 [150]

$$PhCH=N-X \xrightarrow{\text{PhSeH/h}\nu\text{ /20°C/2h}} PhCH_2-NHX$$

X=OH	70%
X=Ph	60%

Scheme 47 [152]

2. Reactions Involving Hydrogen Selenide, Selenols and Related Compounds

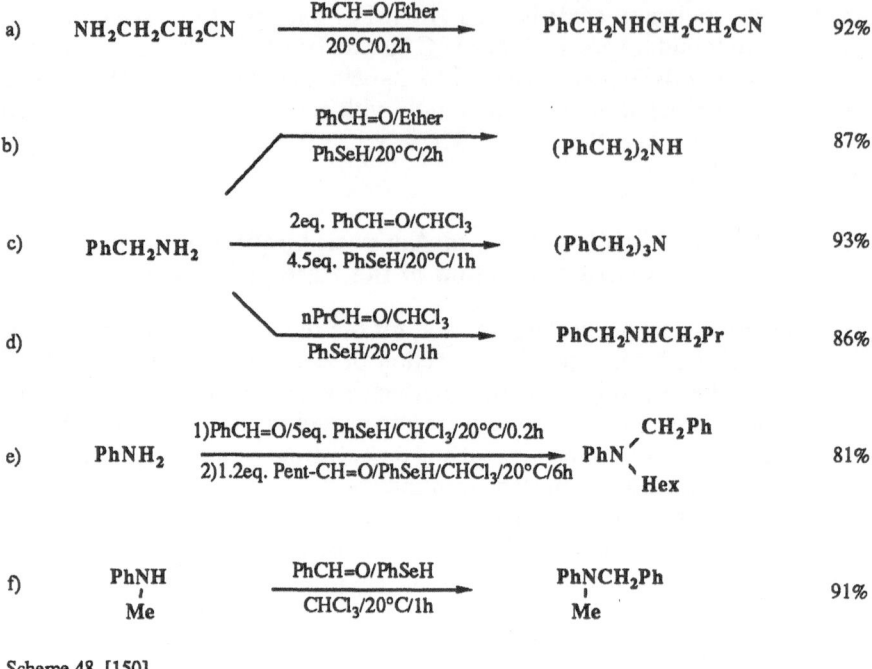

a) NH$_2$CH$_2$CH$_2$CN $\xrightarrow[\text{20°C/0.2h}]{\text{PhCH=O/Ether}}$ PhCH$_2$NHCH$_2$CH$_2$CN 92%

b) $\xrightarrow[\text{PhSeH/20°C/2h}]{\text{PhCH=O/Ether}}$ (PhCH$_2$)$_2$NH 87%

c) PhCH$_2$NH$_2$ $\xrightarrow[\text{4.5eq. PhSeH/20°C/1h}]{\text{2eq. PhCH=O/CHCl}_3}$ (PhCH$_2$)$_3$N 93%

d) $\xrightarrow[\text{PhSeH/20°C/1h}]{\text{nPrCH=O/CHCl}_3}$ PhCH$_2$NHCH$_2$Pr 86%

e) PhNH$_2$ $\xrightarrow[\text{2)1.2eq. Pent-CH=O/PhSeH/CHCl}_3\text{/20°C/6h}]{\text{1)PhCH=O/5eq. PhSeH/CHCl}_3\text{/20°C/0.2h}}$ PhN$\underset{\text{Hex}}{\overset{\text{CH}_2\text{Ph}}{<}}$ 81%

f) PhNH$\underset{\text{Me}}{|}$ $\xrightarrow[\text{CHCl}_3\text{/20°C/1h}]{\text{PhCH=O/PhSeH}}$ PhNCH$_2$Ph$\underset{\text{Me}}{|}$ 91%

Scheme 48 [150]

tertiary amines bearing three different substituents on nitrogen [150, 151] (Scheme 48e).

2.2.7 Reduction of Aldehydes and Ketones to Alcohols

Hydrogen selenide [153, 178] and bromomagnesium benzeneselenolate [154] reduce the carbonyl group of aldehydes and ketones and produce the corresponding alcohols.

In 1966 Mayer reported [178] that α-diketones are reduced to α-hydroxyketones on reaction with hydrogen selenide and pyridine in DMF (Scheme 49a, b). Cyclohexanone however is not reduced to cyclohexanol [178] but leads instead to a compound (probably cyclohexyl selenol) which on reaction with oxalyl chloride gives dicyclohexyl selenooxalate [178] (Scheme 49c). If the reactions are performed with H$_2$Se under irradiation with high pressure mercury lamp [153] aldehydes and ketones are reduced to alcohols in very high yields (Scheme 50). Interestingly, under these conditions cyclohexanone is also reduced to cyclohexanol (Scheme 50d). The reaction is much faster with aromatic compounds [153] and in the case of acetophenone it is slowed down if performed in the presence of biphenyl as triplet quencher [153]. Bromomagnesium benzeneselenolate reduces aromatic aldehydes to benzyl alcohols [154] in modest yield. This reaction has been rationalized as shown in Scheme 51. It is limited since aldehydes and ketones which possess active hydrogens are not reduced but afford condensed products instead [154] (Scheme 52).

38

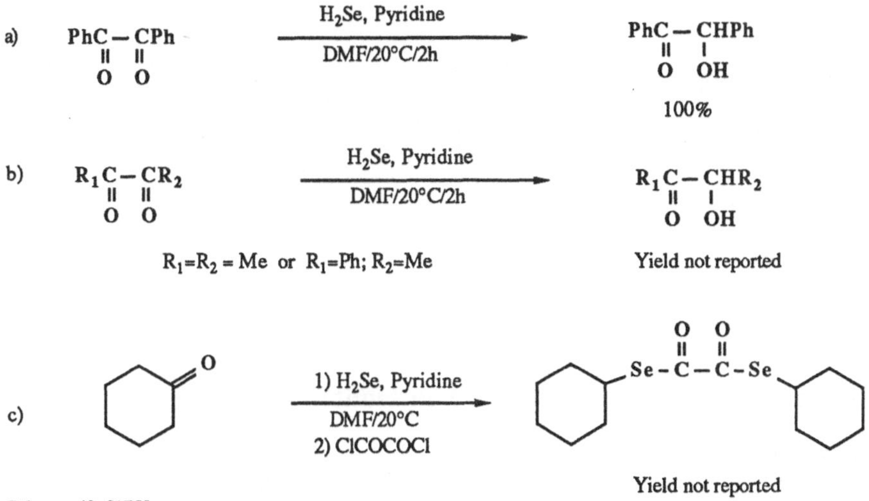

a)
$$PhC-CPh \xrightarrow[\text{DMF/20°C/2h}]{\text{H}_2\text{Se, Pyridine}} PhC-CHPh$$
with both carbonyls (O O) on the left, and O OH on the right

100%

b)
$$R_1C-CR_2 \xrightarrow[\text{DMF/20°C/2h}]{\text{H}_2\text{Se, Pyridine}} R_1C-CHR_2$$
(O O on left; O OH on right)

$R_1=R_2 = $ Me or $R_1=$Ph; $R_2=$Me

Yield not reported

c)

cyclohexanone $\xrightarrow[\substack{\text{DMF/20°C} \\ \text{2) ClCOCOCl}}]{\text{1) H}_2\text{Se, Pyridine}}$ cyclohexyl-Se-C(=O)-C(=O)-Se-cyclohexyl

Yield not reported

Scheme 49 [178]

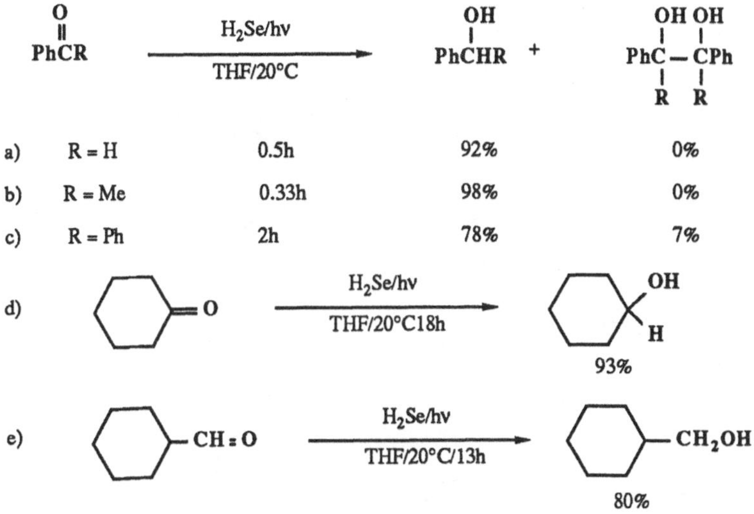

$$\underset{\text{PhCR}}{\overset{O}{\|}} \xrightarrow[\text{THF/20°C}]{\text{H}_2\text{Se/hv}} \underset{\text{PhCHR}}{\overset{OH}{|}} + \underset{\substack{R \quad R}}{\overset{OH\ OH}{PhC-CPh}}$$

a)	R = H	0.5h	92%	0%
b)	R = Me	0.33h	98%	0%
c)	R = Ph	2h	78%	7%

d)

cyclohexanone $\xrightarrow[\text{THF/20°C18h}]{\text{H}_2\text{Se/hv}}$ cyclohexane with OH and H

93%

e)

cyclohexyl-CH=O $\xrightarrow[\text{THF/20°C/13h}]{\text{H}_2\text{Se/hv}}$ cyclohexyl-CH$_2$OH

80%

Scheme 50 [153]

$$ArCHO \xrightarrow{\text{PhSeMgBr}} \left[\underset{\text{ArCHSePh}}{\overset{OMgBr}{|}}\right] \xrightarrow[\text{-PhSeSePh}]{\text{PhSeMgBr}} \left[\underset{\text{ArCHMgBr}}{\overset{OMgBr}{|}}\right] \xrightarrow{\text{H}_2\text{O}} ArCH_2OH$$

Ar = Ph, 4-MeOPh, 3,4-(MeO)$_2$Ph

37, 48, 30%

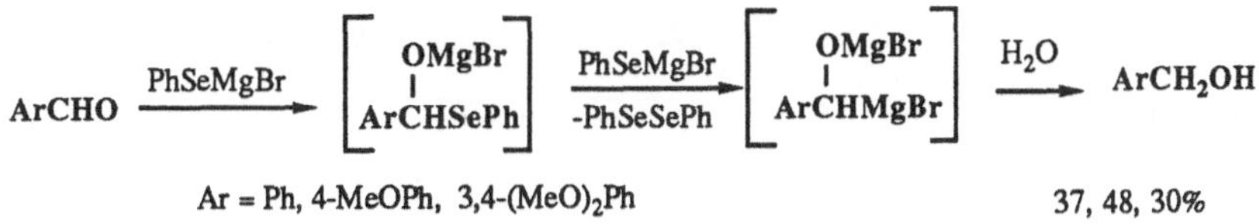

Scheme 51 [154]

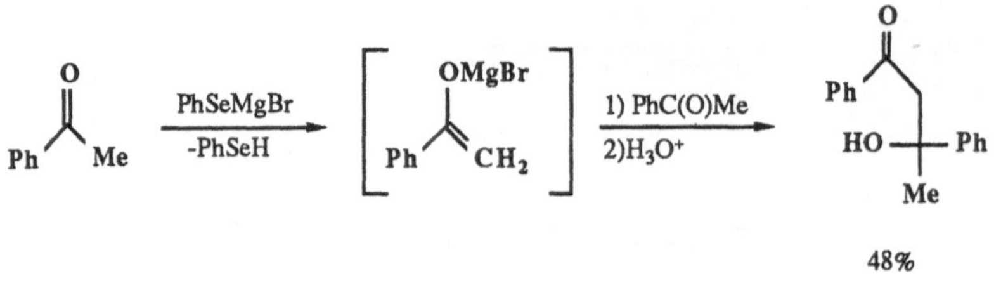
O
Ph
Me
PhSeMgBr
-PhSeH
OMgBr
Ph
CH2
1) PhC(O)Me
2)H3O+
O
Ph
HO
Ph
Me
48%

Scheme 52 [154]

2.2.8 Reduction of the Carbon-Carbon Double Bond of Enones

Benzeneselenol is also able to reduce the carbon-carbon double bond of stilbene [152] and of α,β-unsaturated carbonyl compounds [152, 180, 181] especially those bearing a β-aryl group [152]. The reactions are particularly slow in the dark [152] but the rate is greatly enhanced (20 °C, 2 h) if performed under sunlamp irradiation [152] (Scheme 53).

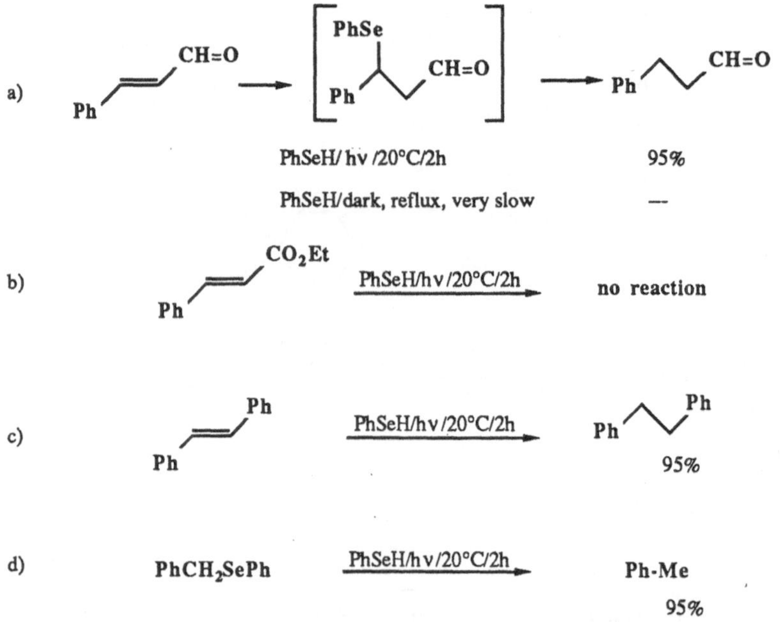
a)
CH=O
Ph
PhSe
Ph
CH=O
Ph
CH=O
PhSeH/ hv /20°C/2h
95%
PhSeH/dark, reflux, very slow
—
b)
CO2Et
Ph
PhSeH/hv /20°C/2h
no reaction
c)
Ph
Ph
PhSeH/hv /20°C/2h
Ph
Ph
95%
d)
PhCH2SePh
PhSeH/hv /20°C/2h
Ph-Me
95%

Scheme 53 [152]

2.2.9 Reduction of Benzyl Selenides to Aryl Alkanes and of Methylselenoacetals Derived From Aromatic Carbonyl Compounds to Benzyl Methyl Selenides

Under irradiation benzeneselenol reduces benzyl phenyl selenide to toluene [152] (Scheme 53d) and cinnamaldehyde to 3-phenyl propanal (Scheme 53a).

It is well known [182, 183] that selenols, including methaneselenol, react with aldehydes and ketones, in the presence of an acid catalyst such as zinc chloride (0.5 equiv.) or titanium tetrachloride (0.3 equiv.) to produce the corresponding selenoacetals in high yield (see Volume II). It was however found that the reaction takes another course in the case of hindered (such as diisopropyl ketone) or of aromatic ketones for which appreciable amounts of selenides, resulting from the formal reduction of the selenoacetal, is formed beside the expected selenoacetal [182, 183]. This side reaction is particularly observed with methaneselenol and ketones if an excess (5 equiv. instead of 0.5 equiv.) of the Lewis acid is used [182, 183] (Scheme 54a, b). Aliphatic aldehydes are usually not reduced under these conditions and produce the corresponding selenoacetals in high yield (Scheme 54c).

It was also found [182], although the mechanism of these reactions is not well understood, that methylselenoacetals derived from aromatic ketones are reduced to the corresponding selenides in very high yield when reacted with methaneselenol and 5 equivalents of an acid catalyst such as zinc chloride (Scheme 54d).

Similarly it was found [184–187] that selenides or ethers are formed besides

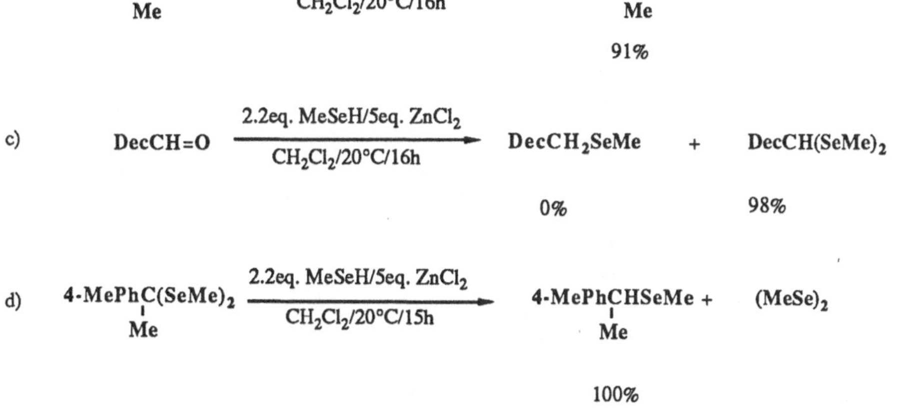

a)   4-MePhC=O $\xrightarrow[\text{CH}_2\text{Cl}_2/20°\text{C}/17\text{h}]{\text{2.2eq. MeSeH/5eq. ZnCl}_2}$ 4-MePhCHSeMe
       | |
       Me   Me

95%

b)   NonC=O $\xrightarrow[\text{CH}_2\text{Cl}_2/20°\text{C}/16\text{h}]{\text{2.2eq. MeSeH/5eq. ZnCl}_2}$ NonCHSeMe
         | |
         Me   Me

91%

c)   DecCH=O $\xrightarrow[\text{CH}_2\text{Cl}_2/20°\text{C}/16\text{h}]{\text{2.2eq. MeSeH/5eq. ZnCl}_2}$ DecCH$_2$SeMe   +   DecCH(SeMe)$_2$

                                                      0%   98%

d)   4-MePhC(SeMe)$_2$ $\xrightarrow[\text{CH}_2\text{Cl}_2/20°\text{C}/15\text{h}]{\text{2.2eq. MeSeH/5eq. ZnCl}_2}$ 4-MePhCHSeMe +   (MeSe)$_2$
           | |
           Me   Me

100%

Scheme 54 [182,183]

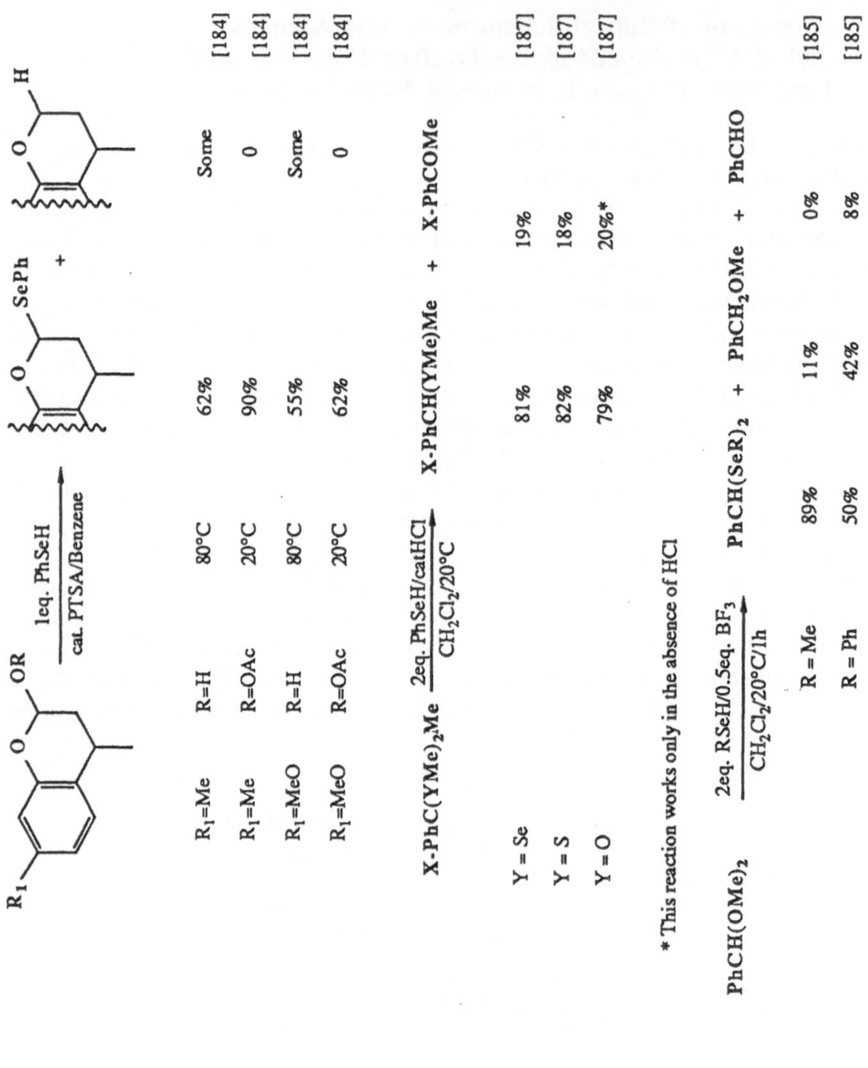

2. Reactions Involving Hydrogen Selenide, Selenols and Related Compounds

a)

$$R_1 \overset{O}{\underset{OR}{\bigcirc}} \xrightarrow[\text{cat. PTSA/Benzene}]{1\text{eq. PhSeH}} \overset{O}{\underset{SePh}{\bigcirc}} + \overset{O}{\underset{H}{\bigcirc}}$$

R_1=Me	R=H	80°C	62%	Some	[184]
R_1=Me	R=OAc	20°C	90%	0	[184]
R_1=MeO	R=H	80°C	55%	Some	[184]
R_1=MeO	R=OAc	20°C	62%	0	[184]

b)

$$\text{X-PhC(YMe)}_2\text{Me} \xrightarrow[\text{CH}_2\text{Cl}_2/20°\text{C}]{2\text{eq. PhSeH/catHCl}} \text{X-PhCH(YMe)Me} + \text{X-PhCOMe}$$

Y = Se	81%	19%	[187]
Y = S	82%	18%	[187]
Y = O	79%	20%*	[187]

* This reaction works only in the absence of HCl

c)

$$\text{PhCH(OMe)}_2 \xrightarrow[\text{CH}_2\text{Cl}_2/20°\text{C}/1\text{h}]{2\text{eq. RSeH/0.5eq. BF}_3} \text{PhCH(SeR)}_2 + \text{PhCH}_2\text{OMe} + \text{PhCHO}$$

R = Me	89%	11%	0%	[185]
R = Ph	50%	42%	8%	[185]

Scheme 55

the expected selenoacetal or the mixed (O,Se) acetal when (O,O) acetals or related derivatives are reacted with selenols in the presence of an acid catalyst (Scheme 55). This implies a reduction by the selenol at one stage of the reaction.

2.3 Use of Copper(I) Benzeneselenolate

2.3.1 As the Precursor of Mixed Alkyl Phenylselenocuprates

Although mixed alkyl phenylthiocuprates have found wide application in organic synthesis in the past ten years [@188], it is only recently that preliminary results concerning their seleno analogues appeared in the literature [188]. It was found that methyl-, *n*-butyl-, sec-butyl- and tertiary butyl-lithiums (RLi) react at 0 °C with copper(I) benzeneselenolate, [readily available by refluxing cuprous oxide with benzeneselenol (EtOH, 24 h)], to produce after 0,2 to 0,5 h, reagents, of general structure RCu(SePh)Li which possess a reactivity different from that of the corresponding organolithium reagents (RLi). These reagents are capable of cleanly and selectively transferring their alkyl groups to beta phenylselenovinylsulfones [188] leading to β-alkyl vinylsulfones in very high yield (Scheme 56). With respect to this transformation [188] these reagents were found to be far superior to other cuprates (Scheme 57) such as alkyl phenylthiocuprates.

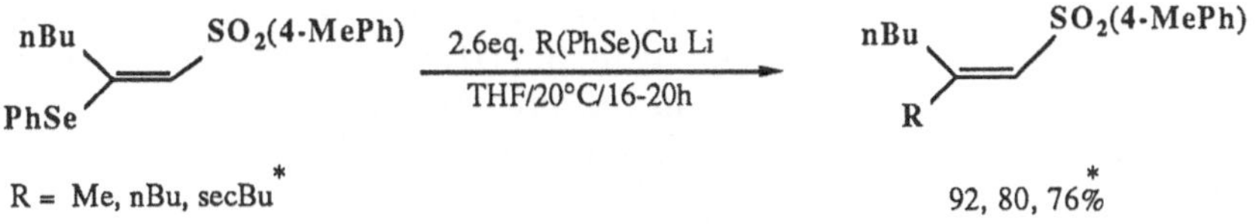
nBu
SO2(4-MePh)
PhSe
2.6eq. R(PhSe)Cu Li
THF/20°C/16-20h
nBu
SO2(4-MePh)
R
R = Me, nBu, secBu*
92, 80, 76%*

* This specific experiment has been performed in the presence of HMPA

Scheme 56 [188]

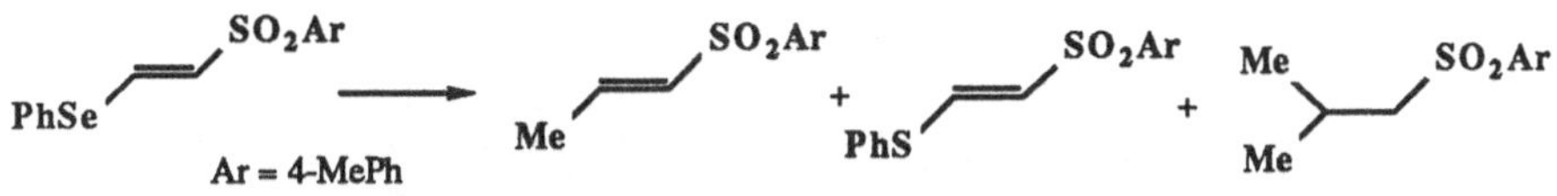
SO2Ar
PhSe
Ar = 4-MePh
SO2Ar
Me
+
SO2Ar
PhS
+
Me
SO2Ar
Me

1.1eq. Me_2Cu/0°C/Et_2O/2h	68%	...	20%
Me(CuCN)Li/-23°C/THF/2.5h	74%	...	...
Me(CuSPh)Li/-23°C/THF/1h	31%	55%	...
Me(CuSePh)Li/0°C/THF/2h	90%	...	...

Scheme 57 [188]

2.3.2 As a Catalyst in the Synthesis of α-Selenoketones from Selenolesters and Diazomethane

Copper(I) benzeneselenolate (PhSeCu) proved, with copper iodide and copper powder, one of the most effective catalyst [189] for the reaction between diazomethane and selenoesters which produces α-methylselenoketones (Scheme 58). Although the yields are identical whatever the catalyst used the shorter reaction times have been observed with PhSeCu.

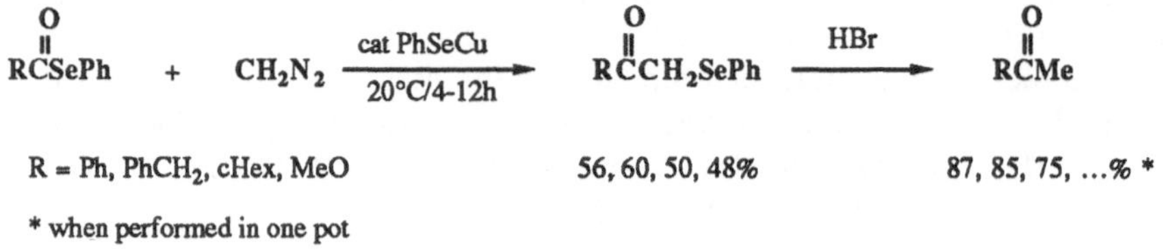

$$\underset{RCSePh}{\overset{O}{\underset{\|}{}}} \quad + \quad CH_2N_2 \quad \xrightarrow[20°C/4\text{-}12h]{cat\ PhSeCu} \quad \underset{RCCH_2SePh}{\overset{O}{\underset{\|}{}}} \quad \xrightarrow{HBr} \quad \underset{RCMe}{\overset{O}{\underset{\|}{}}}$$

R = Ph, PhCH₂, cHex, MeO   56, 60, 50, 48%   87, 85, 75, ...% *

* when performed in one pot

Scheme 58 [189]

2.4 Reduction of Dienes to Olefins with Dichloro Bis(diphenyl Selenide) Platinum (II)

Selenides are particularly efficient complexing agents for transition metal ions [54, 60]. Dichloro bis(diphenyl selenide) platinum (II) catalyses [190] the hydrogenation of non aromatic polyolefins in the presence of SnCl₂ (Scheme 59). Hydrogenation proceeds via stepwise migration of the double bonds to conjugation [190] and hydrogenation to the monoolefin [190]. Treatment of the monoole-

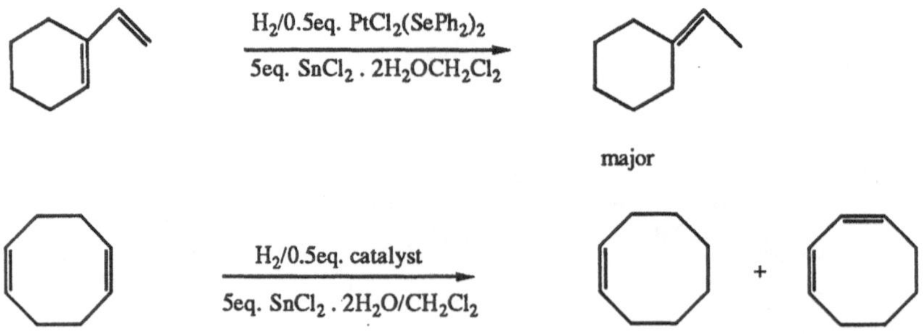

H₂/0.5eq. PtCl₂(SePh₂)₂

5eq. SnCl₂ . 2H₂OCH₂Cl₂

major

H₂/0.5eq. catalyst

5eq. SnCl₂ . 2H₂O/CH₂Cl₂

catalyst			
PtCl₂(SePh)₂	10h	25%	75%
PtCl₂(PPh₃)₂	10h	15%	85%

Scheme 59 [190]

44

2.4 Reduction of Dienes to Olefins with Dichloro Bis(Diphenyl Selenide) Platinum (II)

fin with this catalytic system leads only to isomerisation through stepwise migration of the double bond. The similar behaviour of dichloro bis(diphenyl sulfide) platinum (II) and dichloro bis(triphenyl phosphine) platinum (II) were simultaneously reported [190].

Chapter 3

Reactions Involving Metallic or Amorphous Selenium with Organic Molecules

Metallic selenium has been used inter alias for:

A the cis/trans isomerisation of olefins,
B the oxidation and the aromatization of cyclic and polycyclic hydrocarbons,
C the oxido-reduction reaction of some hydrocarbons,
D the oxidation of carbon monoxide,
E the oxidation of formates and formamides,
F the oxidation of hydrazine to diimide.

3.1 Transformation of (Z) Alkenes to Their (E) Isomers

Metallic selenium catalyses the Z to E isomerisation of alkenes. The reaction requires about 200 °C and the solubilisation of selenium in the medium. The interconversion of both olefinic stereoisomers to give equilibrium mixtures seems to involve the fast formation of a (cis-π) complex between the olefin and diradical selenium species, followed by σ complex formation and release of selenium and of the olefin (Scheme 60). Selenium was found superior to other catalysts such as methanesulfonic acid [191a] or potassium t-butoxide [191a] in promoting the (Z to E) isomerisation of stilbenes [191] (relative rates at 190 °C: Se/MeSO$_3$H/t-BuOK 734/20.6/1). Metallic selenium has also been used for the (Z-E) isomerisa-

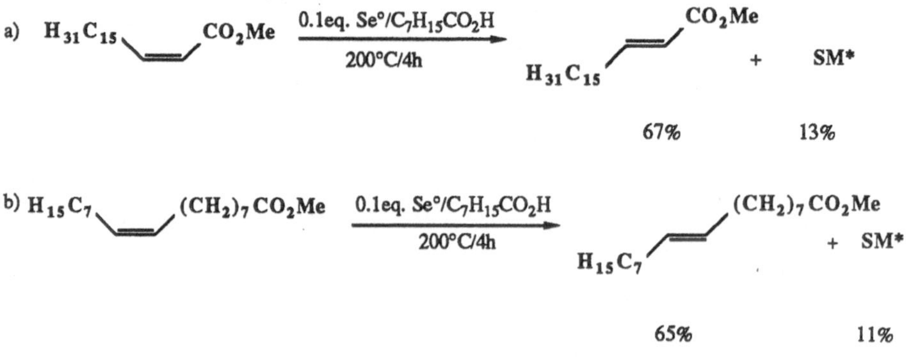

a) $H_{31}C_{15}$—CO$_2$Me $\quad \xrightarrow[200°C/4h]{\text{0.1eq. Se°/C}_7\text{H}_{15}\text{CO}_2\text{H}}$ $\quad H_{31}C_{15}$—CO$_2$Me $\quad + \quad$ SM*

67%   13%

b) $H_{15}C_7$—(CH$_2$)$_7$CO$_2$Me $\quad \xrightarrow[200°C/4h]{\text{0.1eq. Se°/C}_7\text{H}_{15}\text{CO}_2\text{H}}$ $\quad H_{15}C_7$—(CH$_2$)$_7$CO$_2$Me $\quad + \quad$ SM*

65%   11%

* Other regioisomers are formed in about 20% yield

Scheme 60 [194]

46

tion of unsaturated fatty acids [192, 193] and their esters [194] (Scheme 60). Cis-trans linoleic acid [195] or esters have been isomerized to their trans-trans configurated isomers, which have been directly trapped with dienophiles [195]. In all the cases when an allylic methylene group is present, organoselenium derivatives ($\sim$ 1%, difficult to remove from the reaction medium) [195] have been observed besides the desired trans olefin derivatives. A careful study of the reaction has been carried out [195] on the whole series of possible regioisomeric (Δ^2 to Δ^{16}) methyl cis-octadecenoates. After equilibration at 200 °C for 4 h the mixtures contain mainly the (E) isomers (E/Z $\sim$ 4/1) and also appreciable amounts (27 to 8%) of products [195] resulting from the migration of the carbon-carbon double bond away from the position originally occupied (Scheme 60).

3.2 Oxidation of Cyclic and Polycyclic Hydrocarbons and Heterocycles to Aromatic Compounds Using Elemental Selenium

Metallic and amorphous selenium, used in large amounts (1 to 3 equiv. weight) promote at high temperatures ($\sim$ 300 °C) the oxidation of cyclic and polycyclic hydrocarbons to aromatic compounds [196–209] in low to moderate yields (Schemes 61–65, 66b–e). Selenium is usually reduced [199] to hydrogen selenide (Scheme 62) during the process.

Depending upon the nature of the starting material, the aromatisation requires the departure of hydrogen atoms, or a hydrogen atom and an alkyl group or a hydrogen and a heteroatom. As a general trend the ethyl group linked to the same carbon atom as the hydrogen has a greater tendency to be removed (Schemes 61d, e, 62) and is incorporated in ethaneselenol [199] (Scheme 62). Compounds bearing quaternary centres are also aromatized by selenium at around 340 °C by loss of one of the two groups (Schemes 63a, b, d, 64) [197, 199–201, 210]. With this respect ethyl groups seem to be more prone to removal than methyl groups (Scheme 64) [202]. Angular alkyl groups present in polycyclic derivatives are also lost and this occurs preferentially with respect to the cleavage of one of the rings (Schemes 63d, 65b, c) [204, 208, 209]. Again an angular ethyl group is more prone to removal than a methyl group. In some cases a rearrangement takes place and this is particularly true when the reaction is carried out at very high temperatures (450 °C). Thus under theses conditions methyl hydrindenes are transformed to naphthalene [205] (Scheme 66b, c) and cholic acid (Scheme 65c) [205, 209] and cholesterol [205, 208] (Scheme 65b) are both aromatized to chrysene.

The selenium oxidation of hydrocarbons [211], although it occurs at higher temperature than sulfur oxidation [@212], proved to be more powerful and reliable [212]. It has been widely used during the last sixty years for the determination of the carbon framework of several products such as cholic acid (Scheme 65c) [205, 209], cholesterol (Scheme 65b) [205, 208], vaginatin [213], trametenolic acid [214], phytosterols [215], oestrone [216] and other natural products [179, 209, 210, 217–224].

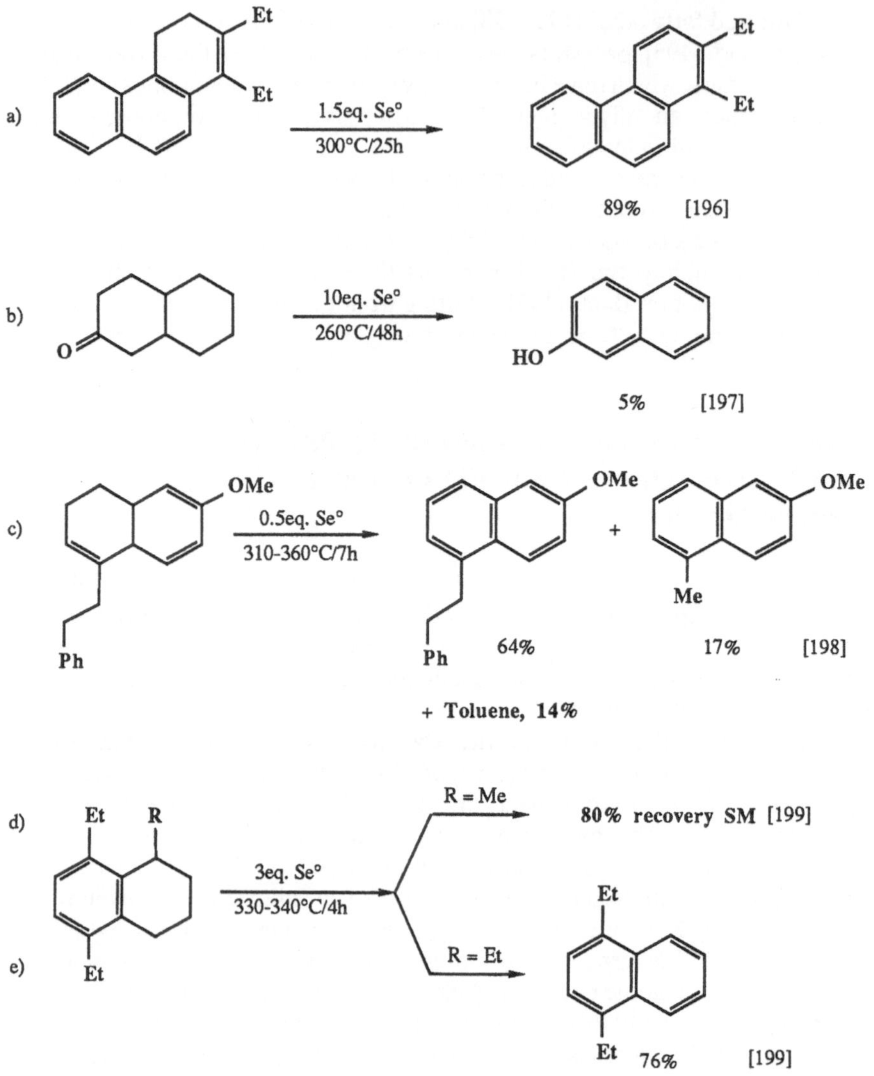

3. Reaction of Metallic or Amorphous Selenium with Organic Molecules

a)

1.5eq. Se°
300°C/25h

89%   [196]

b)

10eq. Se°
260°C/48h

5%   [197]

c)

0.5eq. Se°
310-360°C/7h

64%   17%   [198]

+ Toluene, 14%

d)

3eq. Se°
330-340°C/4h

R = Me   80% recovery SM [199]

R = Et   76%   [199]

e)

Scheme 61

3eq. Se°
330-340°C

14%   25%   +   SM   +   3/2eq. H₂Se   +   EtSeH
                70%   60%

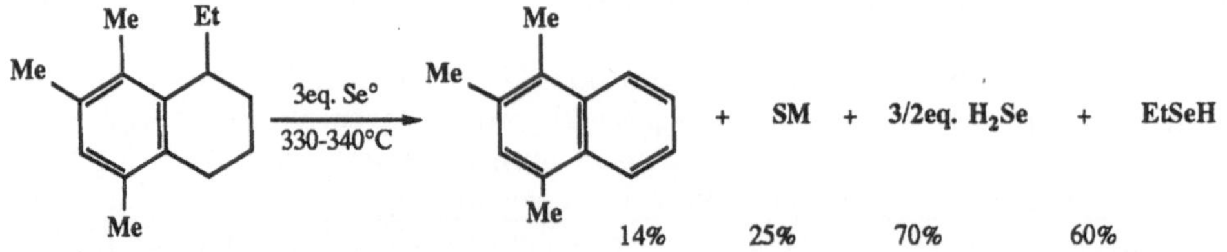

Scheme 62 [199]

48

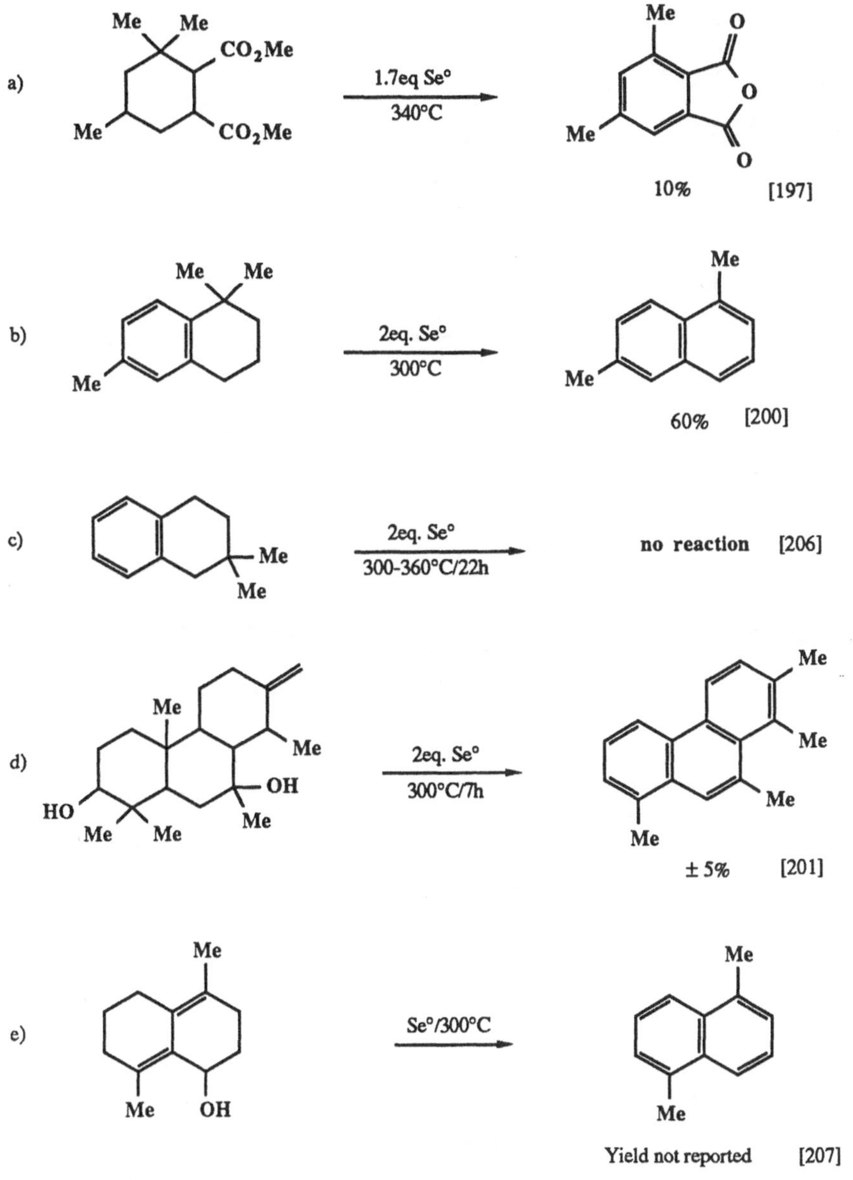

a)

1.7eq Se°
340°C

10%   [197]

b)

2eq. Se°
300°C

60%   [200]

c)

2eq. Se°
300-360°C/22h

no reaction   [206]

d)

2eq. Se°
300°C/7h

± 5%   [201]

e)

Se°/300°C

Yield not reported   [207]

Scheme 63

Aromatization also takes place:

A by formation of a new ring as for example 1,2-diethyl cyclohex-1-ene which is transformed to naphthalene on heating at 450 °C with elemental selenium [205] (Scheme 66e);

B by ring contraction [225]: bicyclo [5.1.0] octa-2,5-diene leads (in very low yield) to ethyl benzene and o-xylene when it is heated for 168h with selenium

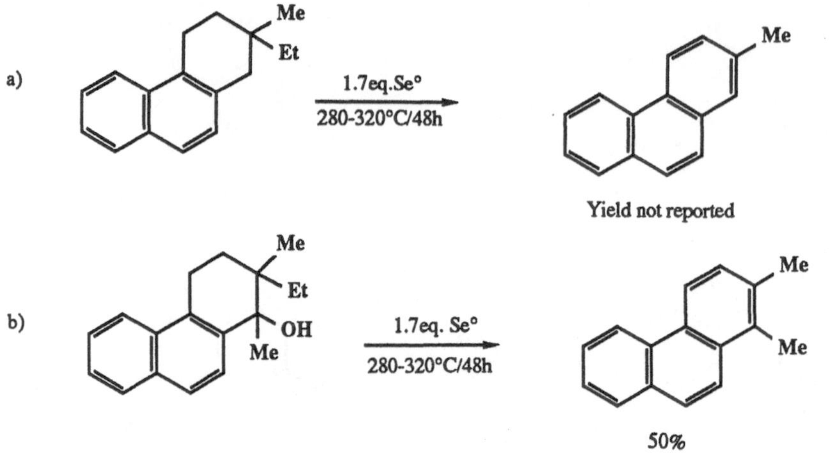

3. Reaction of Metallic or Amorphous Selenium with Organic Molecules

a)

1.7eq.Se°
280-320°C/48h

Yield not reported

b)

1.7eq. Se°
280-320°C/48h

50%

Scheme 64 [202]

a)

Se°
280-330°C
R = Me, H

[203]

Yield not reported

b)

Cholesterol

Se°/ 360°C

[204, 209]

Se°/ 400°C

c)

Cholic acid

Se°/ 400°C

[204, 209]

Yields not reported

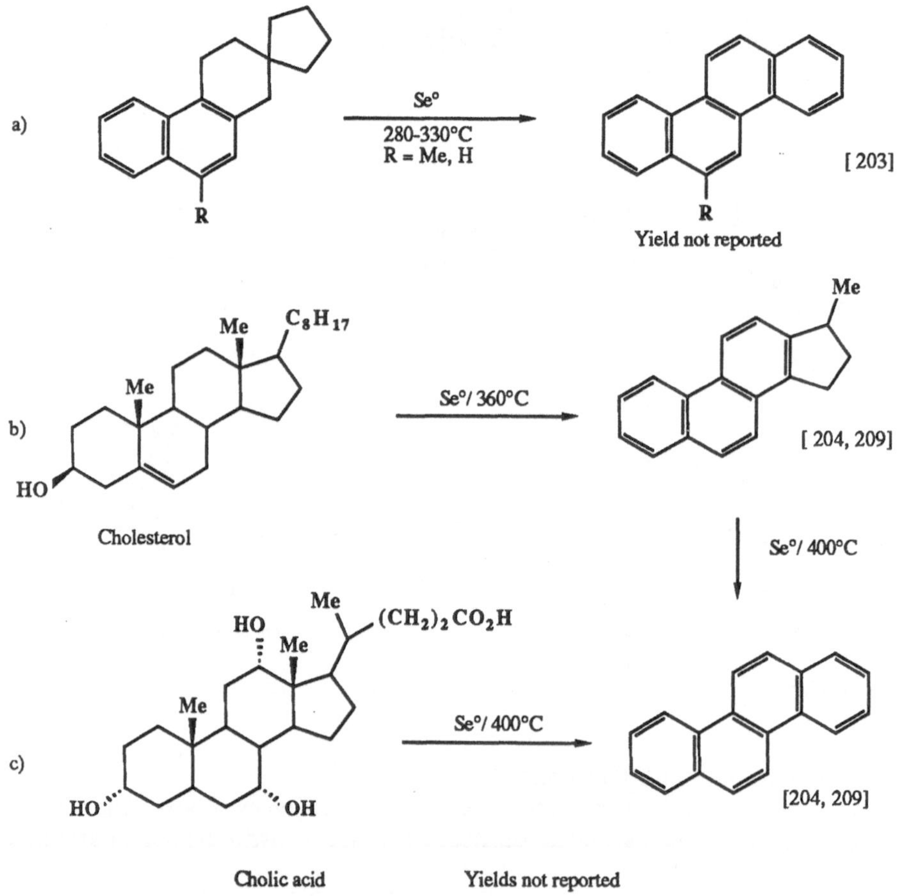

Scheme 65

50

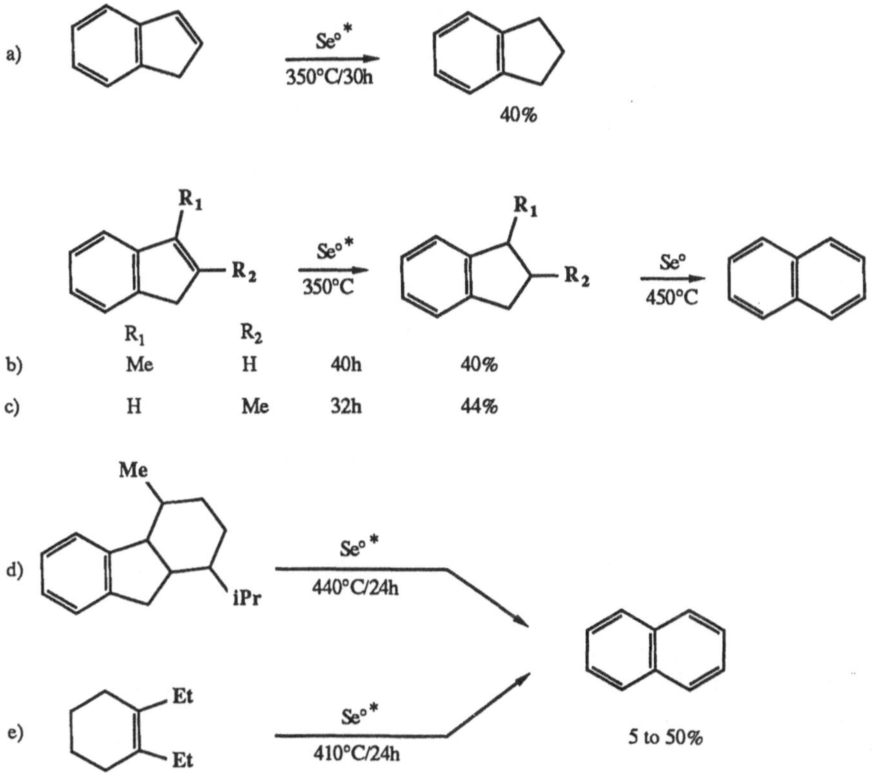

a)

$$\xrightarrow[350°C/30h]{Se°*}$$

40%

R$_1$

$$\xrightarrow[350°C]{Se°*}$$

R$_1$

$$\xrightarrow[450°C]{Se°}$$

	R$_1$	R$_2$		
b)	Me	H	40h	40%
c)	H	Me	32h	44%

d)

Me

iPr

$$\xrightarrow[440°C/24h]{Se°*}$$

e)

Et

Et

$$\xrightarrow[410°C/24h]{Se°*}$$

5 to 50%

* The same weights of Se° and starting material were used

Scheme 66 [205]

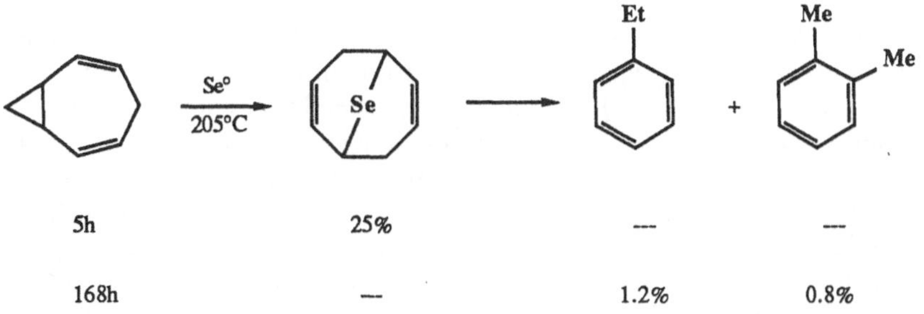

$$\xrightarrow[205°C]{Se°}$$

Se

5h

25%

168h

--

1.2%

0.8%

Et

Me

Me

Scheme 67 [225]

(Scheme 67) (see also Scheme 66d) [205]. Selenium powder in triphenyl-methane has also been used [226] for the dehydrogenation of 2-aryl-2-imidazolines to 2-aryl imidazoles (Scheme 68).

3. Reaction of Metallic or Amorphous Selenium with Organic Molecules

Ar	(Mole ratio Se°/S.M.)/T°C/time	Yield(%)
Ph	(4)/275°C/5h	56
3-MePh	(2.8)/250°C/9h	64
3-ClPh	(4.5)/300°C/9h	40

Scheme 68 [226]

3.3 Oxido-reduction Reactions of Hydrocarbons

In few cases metallic selenium at moderate temperatures (230–350 °C) promotes the reduction of olefinic compounds to the corresponding alkanes [204, 227, 228, 229]. Thus, whereas indenes are reduced [205] to hydrindenes at 350 °C, they are transformed to naphthalenes at higher temperatures [205] (450 °C) (Scheme 66a–d) and oleic acid produces 14% stearic acid when heated [229] at 300 °C. Finally it is worthwile to mention, although the reaction does not fall in this class of redox reaction, that cholestanone is produced by heating [228] cholesterol and selenium at 250 °C. It was suggested [228] that the hydrogens required for the reduction result from the breakdown of a portion of the starting material. Metallic selenium facilitates the transfer of hydrogens.

3.4 Reactions Involving Carbon Monoxide and Catalytic Amounts of Selenium

Elemental (metallic or amorphous) selenium is readily reduced under mild conditions [170, 175] to hydrogen selenide by carbon monoxide and water in the presence of a base (Scheme 43a). This method is exceedingly convenient since it avoids isolation of air sensitive hydrogen selenide [@175]. This reactive system proved particularly useful *inter alias* for the synthesis of various reagents such as selenocarboxamides [which are able to reduce epoxides to olefins [169] (Sect. 2.2.2.2)] and selenocarbamate salts. The latters, on reaction with strong mineral acids, led to high yields of carbonyl selenide which gave ureas when reacted with amines [233, 234], (Scheme 69a). Selenocarbamate salts also proved valuable intermediates for the synthesis of various carbonic acid derivatives, (see below) [230–238]. The last type of reaction when performed under oxygen pressure permits the reoxidation of selenium containing intermediates to elemental selenium which, can in turn be recycled [230].

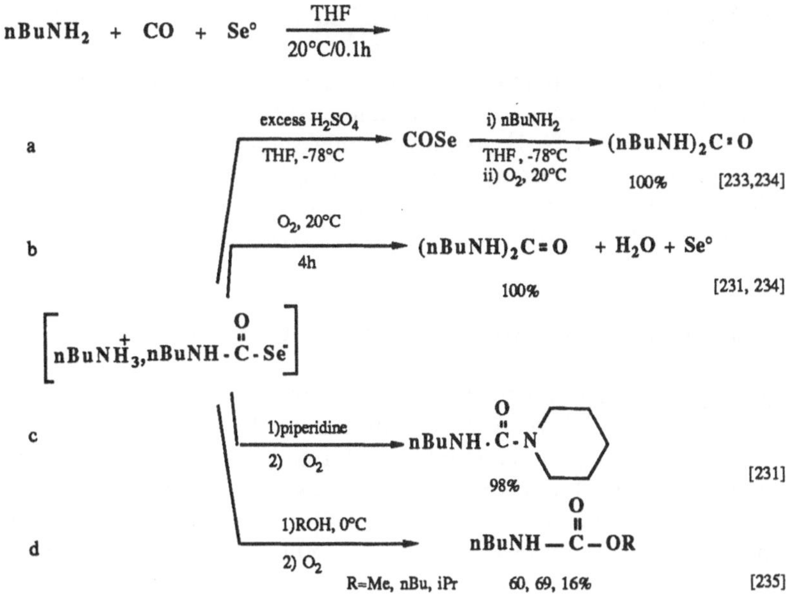

$$nBuNH_2 \;+\; CO \;+\; Se° \xrightarrow[20°C/0.1h]{THF}$$

a $\xrightarrow[\text{THF, -78°C}]{\text{excess } H_2SO_4}$ COSe $\xrightarrow[\text{ii) } O_2, 20°C]{\text{i) nBuNH}_2 \text{ THF, -78°C}}$ $(nBuNH)_2C \cdot O$
        100%   [233,234]

b $\xrightarrow[\text{4h}]{O_2, 20°C}$ $(nBuNH)_2C=O$ $+ H_2O + Se°$
        100%   [231, 234]

$$\left[nBu\overset{+}{N}H_3, nBuNH\text{-}\overset{\overset{O}{\|}}{C}\text{-}Se^- \right]$$

c $\xrightarrow[\text{2) } O_2]{\text{1)piperidine}}$ $nBuNH\text{-}\overset{\overset{O}{\|}}{C}\text{-}N\bigcirc$
        98%   [231]

d $\xrightarrow[\text{2) } O_2]{\text{1)ROH, 0°C}}$ $nBuNH-\overset{\overset{O}{\|}}{C}-OR$
        R=Me, nBu, iPr   60, 69, 16%   [235]

Scheme 69

3.4.1 Synthesis of Acyclic Derivatives of Carbonic Acid

3.4.1.1 Synthesis of Ureas
3.4.1.1.1 Synthesis of Symmetrical N,N-Dialkyl Ureas

Ammonia [231, 232] and amines [231–234, 237, 238] react with carbon monoxide and amorphus selenium to produce after air-oxidation the corresponding ureas. Isolation of selenocarbamate salts as inermediates provides strong evidence that the process takes place in at least two steps and therefore allows the synthesis of unsymmetrical ureas by reacting a different amine on the selenocarbamate salt [Scheme 69]. The reaction occurs under very mild conditions [THF, 20 °C] with primary alkyl amines [231–234], it can be performed with stoichiometric or even catalytic amounts of Se and leads to N,N' dialkyl ureas in almost quantitative yields.

3. Reaction of Metallic or Amorphous Selenium with Organic Molecules

3.4.1.1.2 Synthesis of Unsymmetrical N,N'-Dialkyl and Triakyl Ureas

Stoichiometric amounts of selenium are necessary if unsymmetrical ureas are needed since the selenocarbamate salt must be produced before the addition of the second amine [231]. Although only selenocarbamate salts derived from primary alkyl or aryl amines have been described [231–236], they can either react with a different primary alkyl or a secondary alkyl amine allowing thus the synthesis of di- and trisubstituted ureas [231] (Scheme 69c).

3.4.1.1.3 Synthesis of Symmetrical N,N'-Diaryl Ureas

Primary aryl amines fail to produce the corresponding ureas under the conditions where the N,N'-dialkyl ureas were formed stoichiometrically. Diaryl ureas can be however formed if the reaction is performed in the presence of a base such as triethyl amine [234, 237] (Scheme 70). This reaction involves ammonium hydrogen selenide as the effective catalyst [238]. Its course is altered when performed under forced conditions [NEt_3, $100\,°C$, $50\ kg/cm^2$] since formanilides are produced instead of the ureas [238] (Scheme 70c).

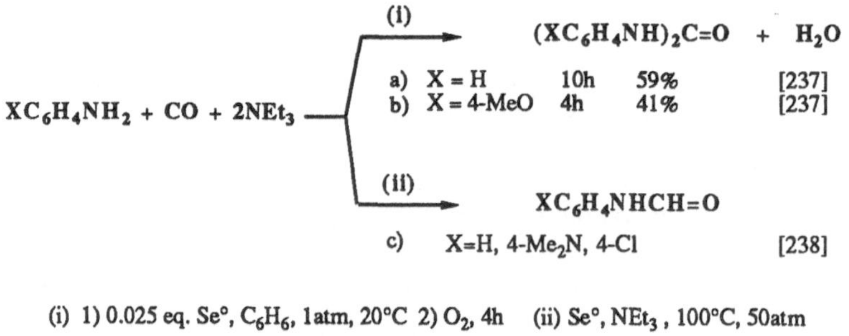

$$XC_6H_4NH_2 + CO + 2NEt_3$$

(i) ⟶ $(XC_6H_4NH)_2C=O$ + H_2O

a) X = H	10h	59%	[237]
b) X = 4-MeO	4h	41%	[237]

(ii) ⟶ $XC_6H_4NHCH=O$

c)   X=H, 4-Me$_2$N, 4-Cl   [238]

(i) 1) 0.025 eq. Se°, C_6H_6, 1atm, 20°C 2) O_2, 4h   (ii) Se°, NEt_3, 100°C, 50atm

Scheme 70

3.4.1.1.4 Synthesis of Tetraalkyl Ureas

The process of the formation of ureas from dialkyl amines is drastically changed from that described with primary ones. The amine salts of selenocarbamates are still produced in the first stage of the reaction. They are stable under nitrogen at $25\,°C$ but are susceptible to molecular oxygen already at $0\,°C$ and lead to bis (N,N-dialkylcarbamoyl)diselenides in almost quantitative yields [232] (Scheme 71). These have been successfully transformed to tetraalkyl ureas when pyrolised at $200\,°C$ (Scheme 71) whereas at lower temperatures ($100–150\,°C$), bis (N,N-dialkylcarbamoyl)selenide can be isolated (Scheme 71) and further transformed at $200\,°C$ to tetraalkyl ureas.

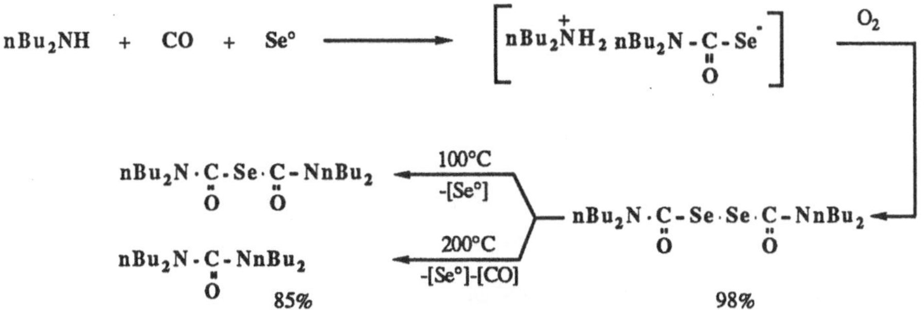

$$nBu_2NH + CO + Se° \longrightarrow \left[nBu_2\overset{+}{N}H_2 \ nBu_2N-\overset{\overset{\displaystyle ..}{C}}{\underset{\displaystyle O}{}}-Se^- \right] \xrightarrow{\ O_2\ }$$

$$nBu_2N\cdot\underset{\overset{\|}{O}}{C}\cdot Se\cdot\underset{\overset{\|}{O}}{C}-NnBu_2 \xleftarrow[\ -[Se°]\ ]{100°C}$$

$$\longrightarrow nBu_2N\cdot\underset{\overset{\|}{O}}{C}-Se\cdot Se\cdot\underset{\overset{\|}{O}}{C}-NnBu_2 \longleftarrow$$

$$nBu_2N\cdot\underset{\overset{\|}{O}}{C}-NnBu_2 \xleftarrow[\ -[Se°]-[CO]\ ]{200°C}$$

85%

98%

Scheme 71 [232]

3.4.1.2 Synthesis of Carbamates and Thiocarbamates

Selenocarbamate salts derived from primary alkyl or aryl amines proved also excellent precursors of carbamates [235] (Scheme 69d) and of thiocarbamates [236] (Scheme 72). The formers are produced on reaction with alcohols when carried out in the presence of molecular oxygen. Thiocarbamates are obtained by simple addition of disulfides to the selenocarbamates salts. This reaction does not require the presence of oxygen and proceeds more rapidly with aromatic than with aliphatic disulfides.

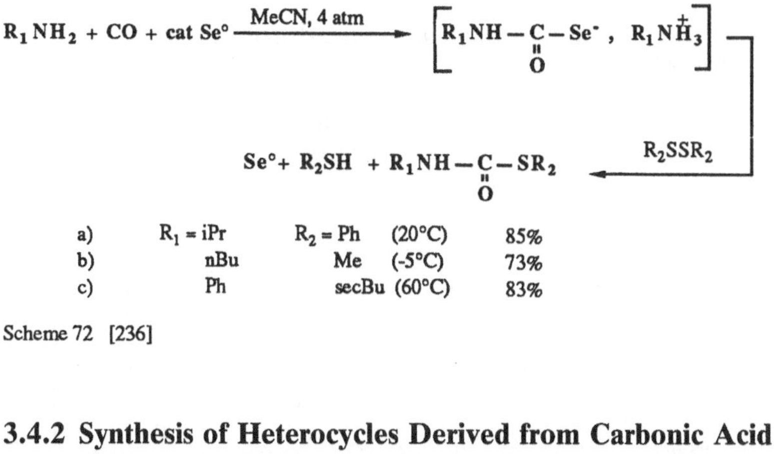

$$R_1NH_2 + CO + cat\ Se° \xrightarrow{\ MeCN,\ 4\ atm\ } \left[R_1NH-\underset{\overset{\|}{O}}{C}-Se^- , \ R_1\overset{+}{N}H_3 \right] \longrightarrow$$

$$Se° + R_2SH + R_1NH-\underset{\overset{\|}{O}}{C}-SR_2 \xleftarrow{\ R_2SSR_2\ }$$

	R_1 =	R_2 =		
a)	iPr	Ph	(20°C)	85%
b)	nBu	Me	(-5°C)	73%
c)	Ph	secBu	(60°C)	83%

Scheme 72 [236]

3.4.2 Synthesis of Heterocycles Derived from Carbonic Acid

Primary β-aminoalcohols [233, 234, 239, 240], β-aminothiols [239, 240], β-aminosulfides [239, 240] and other bifunctional compounds such as β-diamino-derivatives [231, 234, 240], β-glycols [240], β-dithiols [239], β,β'-di(hydroxyalkyl)-disulfides [239] provide a convenient and easy route to five membered heterocycles by simple treatment with a CO/O_2 mixture (4:1 to 10:1, vol:vol) and catalytic amounts of selenium (Scheme 73). The presence of a base [230, 239] such as triethylamine [239] proved to be essential when an amino function is not present (Scheme 73a, f).

55

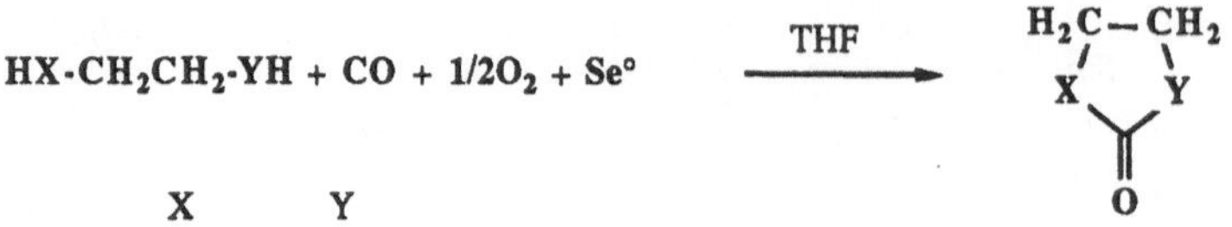

3. Reaction of Metallic or Amorphous Selenium with Organic Molecules

$$HX\cdot CH_2CH_2\cdot YH + CO + 1/2O_2 + Se^{\circ} \xrightarrow{\text{THF}}$$

(product: five-membered ring H_2C-CH_2 with X, Y and C=O)

	X	Y			
a)	S	S	NEt₃/25°C/10h	90%	[239]
b)	O	NH	60°C/2h	95%	[239]
c)	S	NH	60°C/2h	95%	[239]
d)	NH	NH	60°C/50atm	98%	[231]

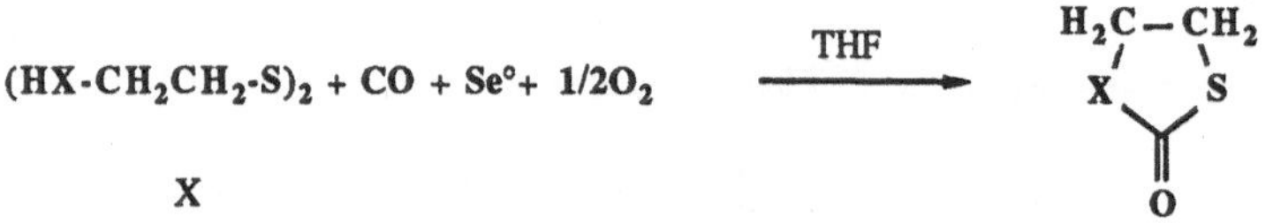

$$(HX\cdot CH_2CH_2\cdot S)_2 + CO + Se^{\circ} + 1/2O_2 \xrightarrow{\text{THF}}$$

(product: five-membered ring H_2C-CH_2 with X, S and C=O)

	X			
e)	NH	60°C/2h	90%	[239]
f)	OH	NEt₃/60°C/20h	60%*	[239]

* Selenium was not recovered and CO_2 was formed during this reaction

Scheme 73

3.4.3 Synthesis of Carbonohydrazides, Semicarbazides, Carbazates and Carbonates

N,N-dialkyl hydrazines react [241] with carbon monoxide in the presence of catalytic amounts of selenium to produce selenocarbazic acid salts which give on further oxidation (i) carbonohydrazides, (ii) semicarbazides [241] if an amine is

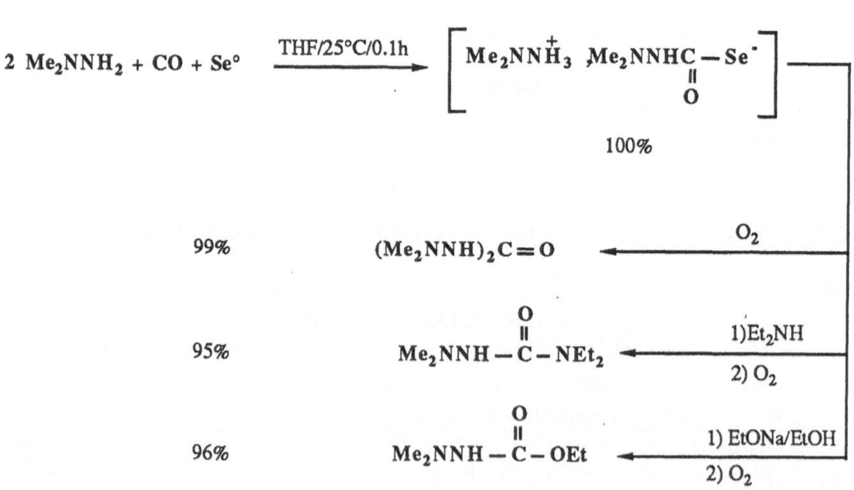

$$2\ Me_2NNH_2 + CO + Se^{\circ} \xrightarrow{\text{THF/25°C/0.1h}} \left[Me_2NN\overset{+}{H}_3 \ , Me_2NNHC\overset{\|}{\underset{O}{-}}Se^- \right]$$

100%

99% $(Me_2NNH)_2C=O$ $\xleftarrow{O_2}$

95% $Me_2NNH-\overset{O}{\overset{\|}{C}}-NEt_2$ $\xleftarrow{\text{1)Et}_2\text{NH}}{\text{2) O}_2}$

96% $Me_2NNH-\overset{O}{\overset{\|}{C}}-OEt$ $\xleftarrow{\text{1) EtONa/EtOH}}{\text{2) O}_2}$

Scheme 74 [241]

introduced to the medium and (iii) carbazates if an alcohol is used instead (Scheme 74).

Carbonates are formed in very high yield if primary alcohols are directly reacted with carbon monoxide and stoichiometric [230] or catalytic [230, 242] amounts of amorphous selenium (Scheme 75). Disappointingly low yields are however observed [230] with secondary and tertiary alcohols (6–16%) and the reaction does not proceed with phenols.

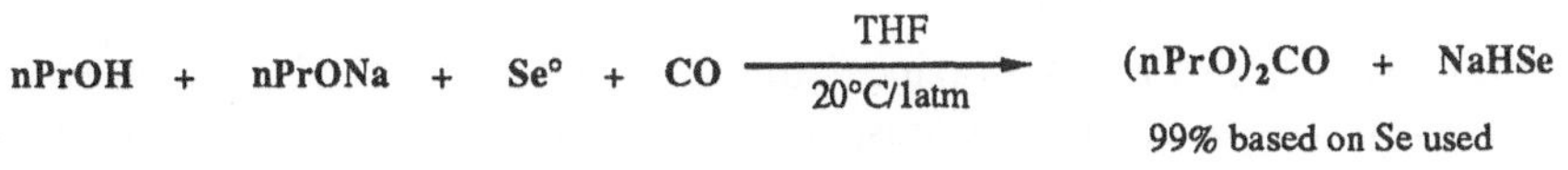

$$nPrOH \; + \; nPrONa \; + \; Se° \; + \; CO \xrightarrow[20°C/1atm]{THF} (nPrO)_2CO \; + \; NaHSe$$

99% based on Se used

Scheme 75 [230, 242]

3.4.4 Oxidation of Formates and Formamides to Carbonates and Carbamates

Catalytic amounts of elemental selenium promote, in the presence of alkoxide ions and under mild conditions, the oxidation of formates [243] and of N,N-dialkyl formamides [243]. Carbonates are formed in high yield in the first case whereas N,N-dialkyl carbamates are obtained in modest yield in the latter (Scheme 76). NaHSe, formed in the process can be oxidized back to elemental selenium if oxygen is introduced in the medium, therefore selenium can be recycled many times.

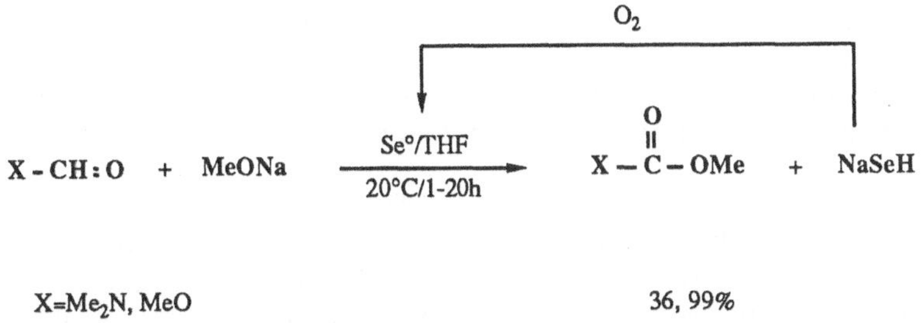

$$X-CH{:}O \; + \; MeONa \xrightarrow[20°C/1-20h]{Se°/THF} \overset{O}{\overset{\|}{X-C-OMe}} \; + \; NaSeH$$

O₂

X=Me₂N, MeO   36, 99%

Scheme 76 [243, 244]

3.4.5 Synthesis of Hydrogen Selenide and Some of Its Application

As we have already pointed out hydrogen selenide is smoothly formed on reaction of elemental selenium with water and carbon monoxide in the presence of a base (Scheme 43a).

This reaction has been used for the synthesis of selenocarboxamides [@170] from nitriles (Scheme 43b, c). Aryl derivatives give better results than aliphatic ones. Aliphatic selenocarboxamides are not so stable: they decompose to nitriles and elemental selenium on exposure to air, suggesting that selenoamides are in equilibrium with the corresponding nitriles and hydrogen selenide [170] (Scheme 43d). The same reagent reduces nitroaromatic compounds to aryl amines [156] (Scheme 42b, c) (see Sect. 2.2.5) and nitroalkanes, such as nitrocyclohexane and 2-nitropropane to the corresponding oximes [245] and/or to ketones [245]. Ketone formation is dependent [245] especially on solvent and pH.

3.4.6 Oxidation of Hydrazine to Diimide: Application to the *Cis* Hydrogenation of Olefins

Elemental selenium oxidizes hydrazine to diimide [246]. The latter compound has been successfully used for the stereospecific *cis* hydrogenation of 1,2-dimethyl cyclohexene to *cis* 1,2-dimethyl cyclohexane, of diphenyl acetylene to cis stilbene or dibenzyl depending upon the amount of hydrazine used [246] and of styrene to ethylbenzene (Scheme 77). H_2Se_2, believed to be formed in the process, can be oxidized back to elemental selenium (Scheme 77).

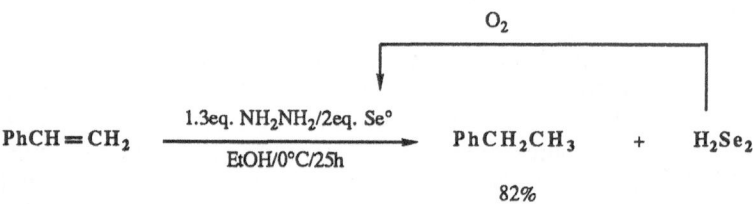

Scheme 77 [246]

3.4.7 Carbonylation of Alkyl Aryl Ketones to 1,3-Dicarbonyl Compounds: Application to the Synthesis of 4-Hydroxycoumarins

Ortho-hydroxy acetophenones react [247, 248] with carbon monoxide in the presence of bases and equimolecular amounts of selenium to produce [247, 248] 4-hydroxycoumarins and hydrogen selenide (Scheme 78). Therefore the carbonyl group must have been derived from carbon monoxide through a C-C bond formation with the ketone enolates. The nature of the base has a marked effect on this reaction. Thus whereas 1,5-diazabicyclo-[5.4.0]-undec-5-ene (DBU) and 1,5-diazabicyclo-[4.3.0]-non-5-ene (DBN) are most effective for the carbonylation reaction [CO (30 kg/cm^2), 100 °C, 30h] it does not take place with tertiary amines such as triethylamine, N-methyl pyrrolidine or 1,4-diazabicyclo-[2.2.2]-octane (DABCO).

As already mentionned, selenium is reduced to hydrogen selenide during the process. It was therefore expected that if the last compound could be in situ oxidized back by an appropriate oxidizing agent, the carbonylation could proceed with a catalytic amount of selenium. This proved to be feasible only if the reaction was performed with nitrobenzene as the oxidant [catal. Se. 4 equiv. DBU, CO (30 kg/cm^2), 90 °C, 48h] (Scheme 78f).

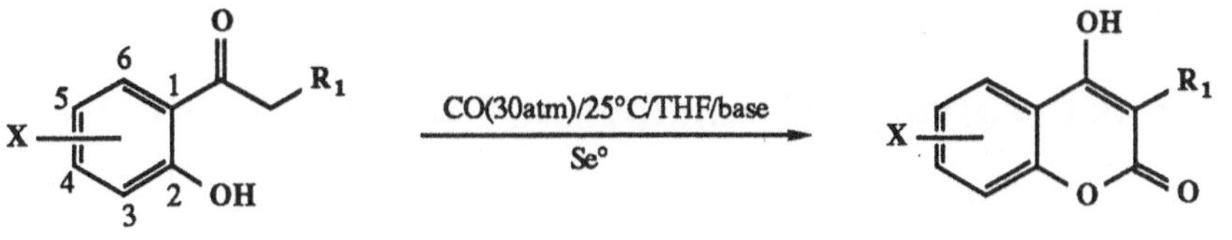

	X	R_1	Conditions	Yield (%)
a)	H	H	4eq. DBU/1eq. Se°/100°C/30h	100
b)	5-Me	H	4eq. DBU/1eq. Se°/100°C/30h	35
c)	5-Me	H	8eq. DBU/1eq. Se°/100°C/30h	77
d)	4-MeO	H	8eq. DBU/1eq. Se°/100°C/30h	36
e)	H	Me	4eq. DBU/1eq. Se°/130°C/30h	42
f)	H	H	4eq. DBU/0.2eq.Se°/0.5 eq. $PhNO_2$ 90°C/30h	68

Scheme 78 [247]

Chapter 4

Reactions Involving Selenoxides and Related Derivatives

4.1 Reactions Involving Selenoxides

Selenoxides have been used as mild oxidizing agents resembling hydroperoxides more than sulfoxides [249]. They are readily reduced (especially the vinyl alkyl selenoxides) for example with stannous chloride [249], phosphorus triiodide and diphosphorus tetraiodide [*250, 251, 252]. Those, such as dibenzyl or diaryl selenoxides, which have no hydrogen on the β-carbon atom and which therefore are not prone to the *syn* β-elimination reaction leading to selenenic acids and olefins [*61, *62, *63, *64, *88], have been used for the oxidation of several functional groups. Dibenzyl- or dimethyl selenoxides were found to be stronger oxidizing agents than the diphenyl analogue [249]. In several instances the presence of an acid was essential for the success of the reaction.

4.1.1 Oxidation of Sulfides, Amines and Acyl Hydrazines to Sulfoxides, Aminoxides, and to Symmetrical Diacyl Hydrazines, Respectively

Several selenoxides such as dimethyl selenoxide, dibenzyl selenoxide, diphenyl selenoxide and bis(p-methoxyphenyl) selenoxide have been used much more than their sulfur counterparts, as mild reagents for the oxidation of various functional groups. The milder oxidizing properties of bis(p-methoxyphenyl) selenoxide have been demonstrated [253]. Thus, whereas tertiary amines have been oxidized to aminoxides [254] (Scheme 79), anilines to azo derivatives [255] on reaction with dimethyl, diphenyl or dibenzyl selenoxides, they do not react with bis(p-methoxyphenyl) selenoxide [253].

On the other hand diphenyl selenoxide in acetic acid efficiently oxidizes acyl hydrazines to sym-diacyl hydrazines [256] (Scheme 80a). Further oxidation was reported to take place in the case of phthaloylhydrazine [257] (Scheme 80b).

All the selenoxides cited, including bis(p-metoxyphenyl) selenoxide, are however able to oxidize sulfides, including the hindered ones, to sulfoxides [249, 253] (Scheme 81). In addition bis (p-methoxyphenyl) selenoxide oxidizes thiols to disulfides under very mild conditions [253] (Scheme 82). However sulfoxides are unable to perform the above transformations [253] except the last one which occurs under forced conditions (160 °C/5–19h) [@258].

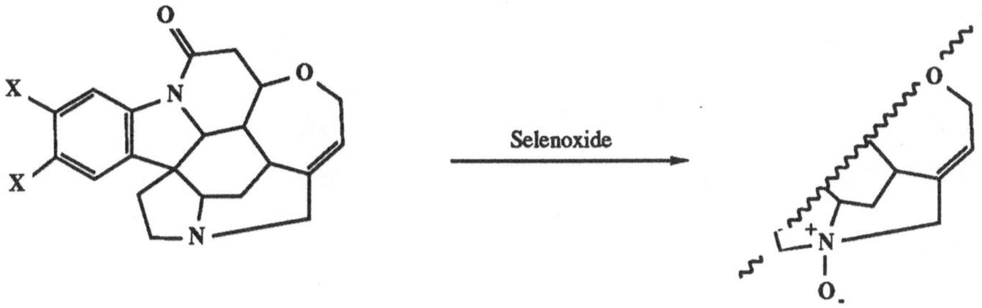

Starting Material	Selenoxide	Conditions	Yield (%)
X=MeO Brucine	Ph$_2$SeO	(AcOH/100°C/10h)	trace (+84% S.M.)
	NCH$_2$CH$_2$Se(O)Me	(AcOH/60°C/2h)	40%
X=H Strychnine	Ph$_2$SeO	(AcOH/100°C/10h)	4%
	NCH$_2$CH$_2$Se(O)Me	(AcOH/60°C/2h)	60%

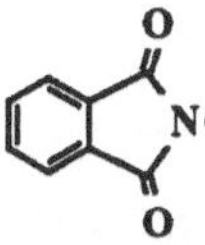

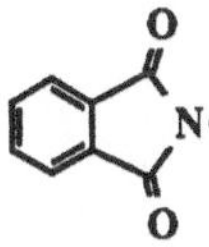

Scheme 79 [254]

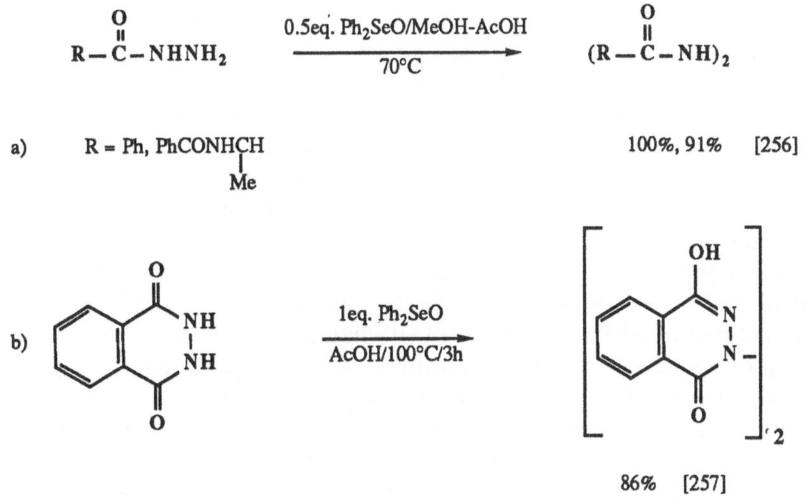

a)   R = Ph, PhCONHCH
                           |
                          Me

100%, 91%   [256]

b)

1eq. Ph$_2$SeO
AcOH/100°C/3h

86%   [257]

Scheme 80

$$R_2S \quad + \quad R'_2SeO \quad \xrightarrow[\text{20°C/1 to 2h}]{\text{AcOH}} \quad R_2SO \quad + \quad R'_2Se$$

| R=Bu, tBu | R'=PhCH$_2$ | 98, 68% | [249] |
| R=PhCH$_2$, C$_{12}$H$_{25}$ | R'=4-MeOPh | 89, 97% | [253] |

Scheme 81

a) $\quad$ RSH $\quad \xrightarrow[\text{20°C/0.2 - 0.5h}]{\text{(4-MeOPh)}_2\text{SeO/CH}_2\text{Cl}_2} \quad$ RSSR

$\quad$ R = Ph, PhCH$_2$, C$_6$H$_{13}$ $\qquad\qquad\qquad$ 90, 94, 95%

b) $\quad$ XCH$_2$CH$_2$SH $\quad \xrightarrow[\text{20°C/0.2 - 0.5h}]{\text{(4-MeOPh)}_2\text{SeO/CH}_2\text{Cl}_2} \quad$ (XCH$_2$CH$_2$S)$_2$

$\quad$ X=NH$_2$, OH $\qquad\qquad\qquad\qquad\qquad$ 82, 92%

Scheme 82 [253]

4.1.2 Oxidation of Enediols to Dicarbonyl Compounds

Selenoxides efficiently oxidize compounds bearing the enediol moiety to α-dicarbonyl compounds. The reaction has been used for the transformation of L-ascorbic acid to L-dehydroascorbic acid (50% yield) [259] and for the oxidation of pyrocatechols to o-quinones [253, 260–262] (Schemes 83–85) (see also Sect. 6.1). The reaction works particularly well with diaryl selenoxides in methanol [*253, 262]. It permits the oxidation, at 0 °C and under neutral conditions [*253, 262], of t-butyl pyrocatechols (Scheme 83) and proceeds cleanly even in the presence of other functional groups such as phenolic hydroxyl [262] or amino [260] groups which are also susceptible to be oxidized but remain untouched under these conditions [262]. In several instances the o-quinones have been trapped by a nucleophile such as an amine (Scheme 85) [250, 261] or another activated aromatic group [262] (Scheme 84) present in the molecule in the right position to effect an intramolecular reaction. In some cases the resulting intermediates are further oxidized to novel o-quinones [260, 261, 262] (Scheme 85). The reaction allows the synthesis of adrenochrome from adrenaline (Scheme 85) [260] and the construction of aporphine and homoaporphine skeletons [262] from benzyliso-quinoline and phenethylisoquinoline systems respectively (Scheme 84). Of particular interest is the latter reaction which proceeds in less than 10% yield when o-chloranil, a common catechol oxidant [@262] is used. Finally, whereas phenols which do not bear an additional hydroxyl group are inert under the reported conditions [262] (Scheme 84), hydroquinone is cleanly oxidized to p-quinone [253, 262] (Scheme 83c, d).

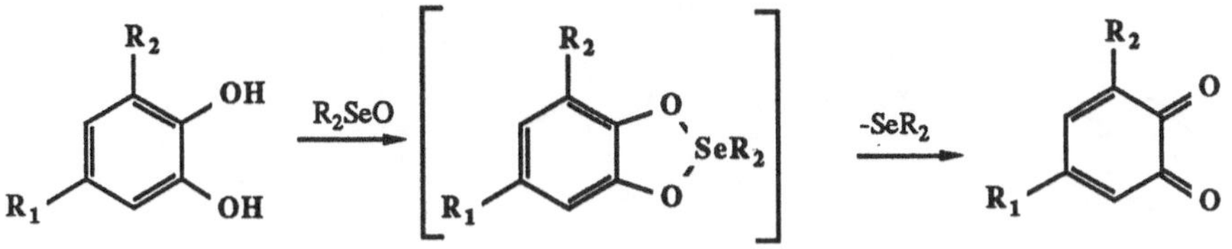

a) R=Ph, R_2=R_1=tBu or R_2=H, R_1=tBu   (MeOH/0°C/0.5h)   100%   [262]

b) R=4-MeOPh, R_1=R_2=Bu   (EtOH-Et$_2$O/20°C/1.7h)   97%   [253]

c)   Ph$_2$SeO/MeOH   95%   [262]

d)   (4-MeOPh)$_2$SeO/MeOH/20°C/0.7h   87%   [253]

Scheme 83

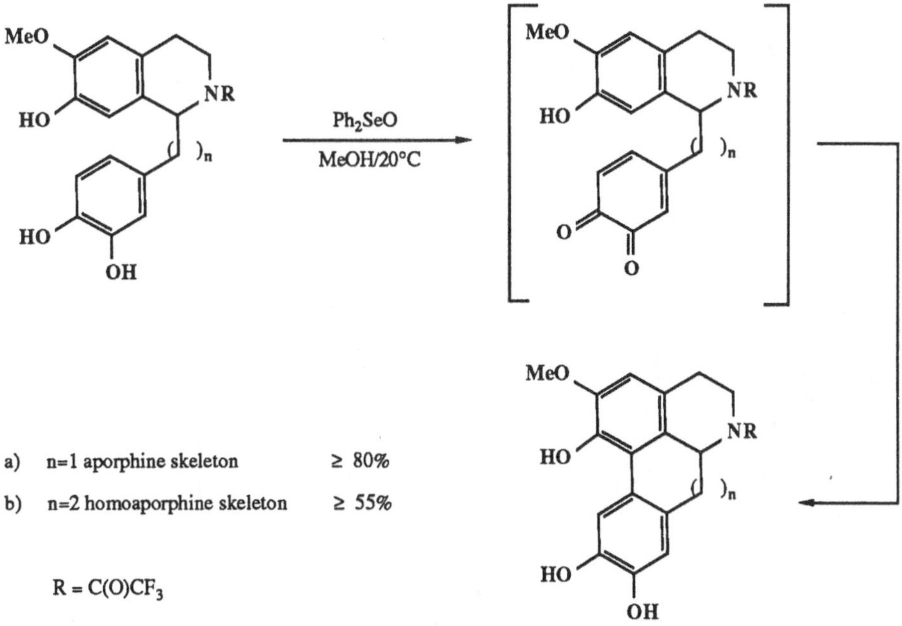

a) n=1 aporphine skeleton   ≥ 80%

b) n=2 homoaporphine skeleton   ≥ 55%

R = C(O)CF$_3$

Scheme 84 [262]

63

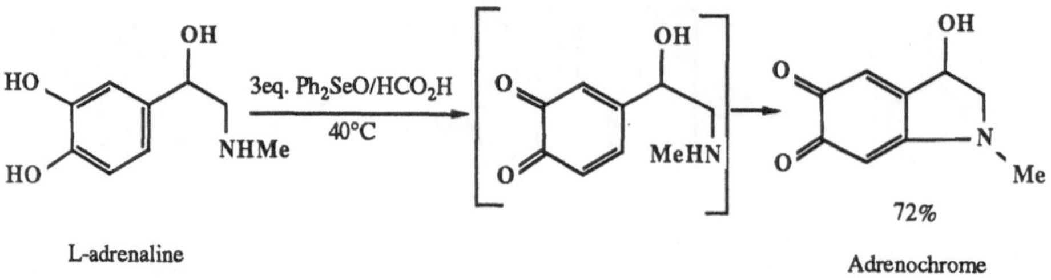

4. Reactions Involving Selenoxides and Related Derivatives

L-adrenaline

Adrenochrome

Scheme 85 [260]

4.1.3 Oxidative Conversion of Thiocarbonyl Compounds to Carbonyl Compounds

Dimethyl selenoxide [263], dibenzyl selenoxide [264] and diaryl selenoxides [264] react with thiocarbonyl compounds such as thioureas [263, 264] and thioamides [263, 264], including cyclic ones [264], to produce ureas (Scheme 86) and amides respectively. Thiouracyls are converted to uracyls [263] (Scheme 87a). This reaction was successfully applied [263] to the synthesis of uridylyl-(3',5')-4-

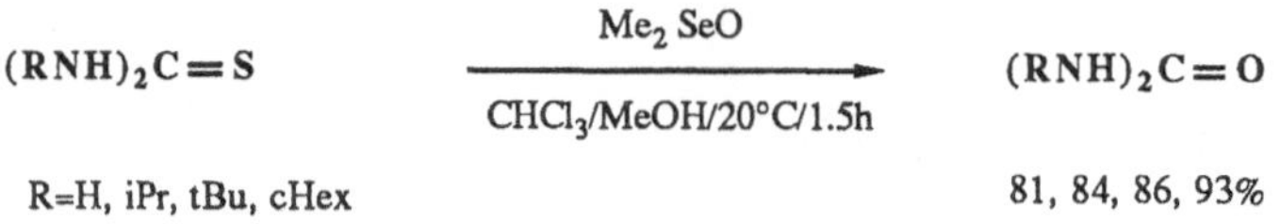

$$(RNH)_2C=S \xrightarrow[\text{CHCl}_3/\text{MeOH}/20°C/1.5h]{\text{Me}_2\,SeO} (RNH)_2C=O$$

R=H, iPr, tBu, cHex

81, 84, 86, 93%

Scheme 86 [263]

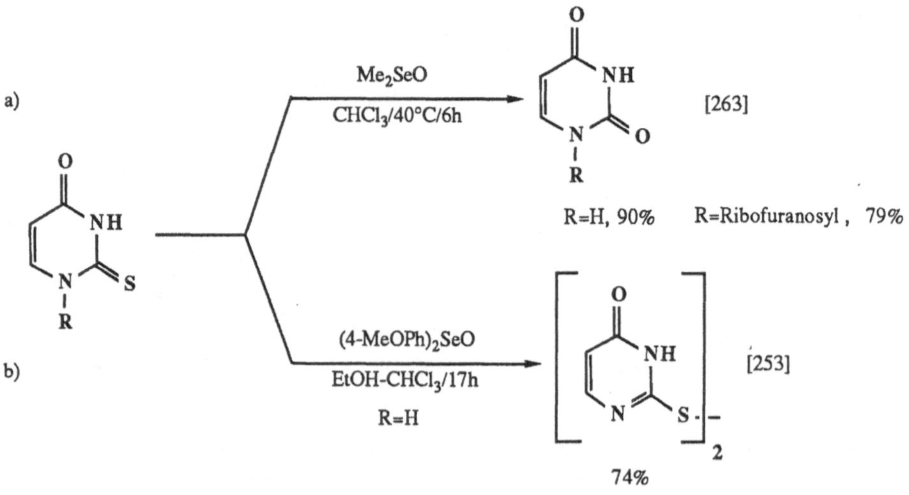

Scheme 87

64

uridine from uridylyl-(3′,5′)-4-thiouridine and occurs without degradation of the sugar moiety and of the internucleotide bond [263] (Scheme 88). It takes place cleanly on the unprotected thionucleotide. Bis(p-methoxyphenyl) selenoxide however behaves differently [253]. It acts as a milder oxidizing agent and oxidative dimerisation of the starting materials [253] often occurs instead of the S to O exchange [263, 264] reported above. Thus thioamides and thioureas possessing a free amino group are cleanly converted to thiadiazoles [253] (Schema 89a). However thiouracil leads to 2,2′-dithio-di-4(3H)-pyrimidone (Scheme 87 compare b to a) and N,N′-diphenyl thiourea is smoothly oxidized to 2,4-diphenyl-3,5-bis-(phenylimino)-1,2,4-thiadiazolidine (Scheme 89b). The latter reaction is remarkable since it proceeds in high yield in contrast to the reaction with benzoyl peroxide which gives the same product in poor yields.

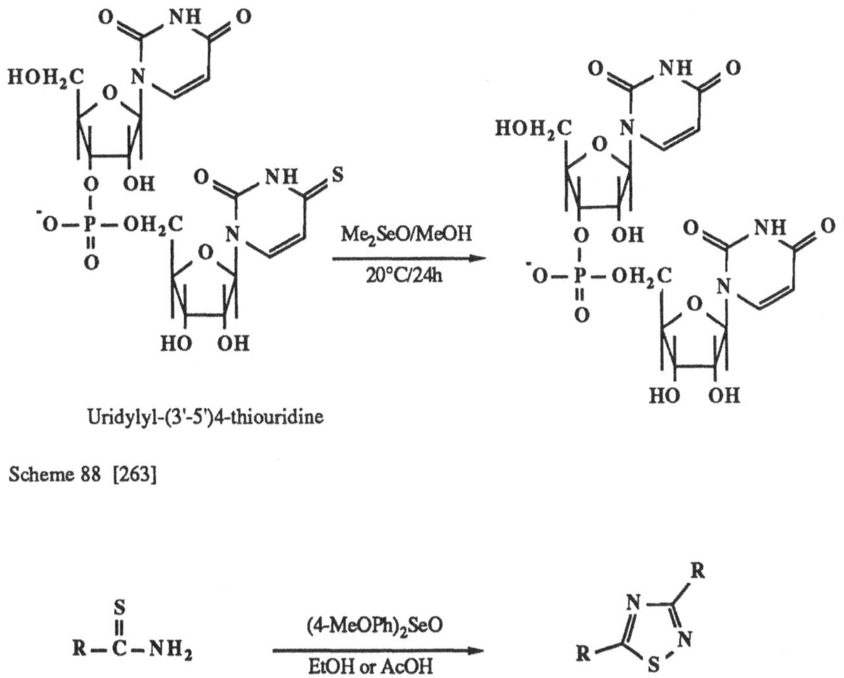

Uridylyl-(3'-5')4-thiouridine

Scheme 88 [263]

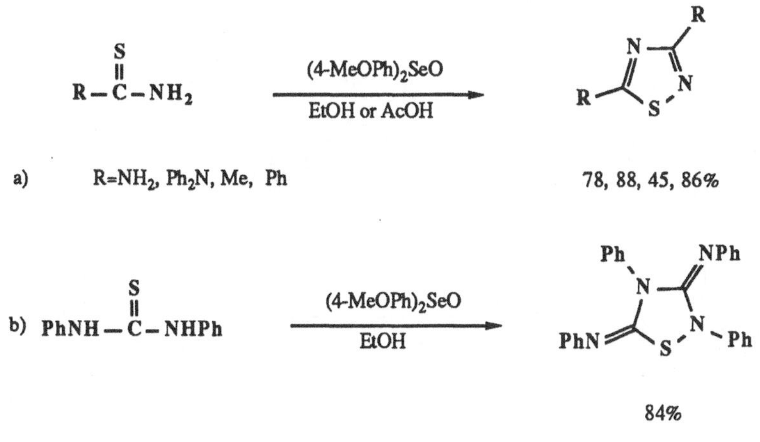

a)   R=NH$_2$, Ph$_2$N, Me, Ph   78, 88, 45, 86%

b) PhNH—C—NHPh   84%

Scheme 89 [253]

4.1.4 Oxidation of Trivalent Phosphorus Compounds to Their Oxides and of Thio- and Selenophosphorus Derivatives to Their Oxygenated Analogues

Trivalent phosphorus compounds have been readily converted to their oxides on reaction with dimethyl selenoxide [263] and with bis(p-methoxyphenyl) selenoxide [253] (Schemes 90–92). Dimethyl selenoxide has also been used to convert thio- and selenophosphoryl compounds to their oxygenated analogues [263]

$$R_3P \ + \ R'_2 \ SeO \longrightarrow R_3P{=}O \ + \ R'_2Se$$

R=Ph, EtO, Me$_2$N; R'=Me   CHCl$_3$/ 20°C/1h   100, 83, 84% [263]

R=Ph; R'=4-MeOPh   CH$_2$Cl$_2$/20°C/0.2h   99%   [253]

Scheme 90

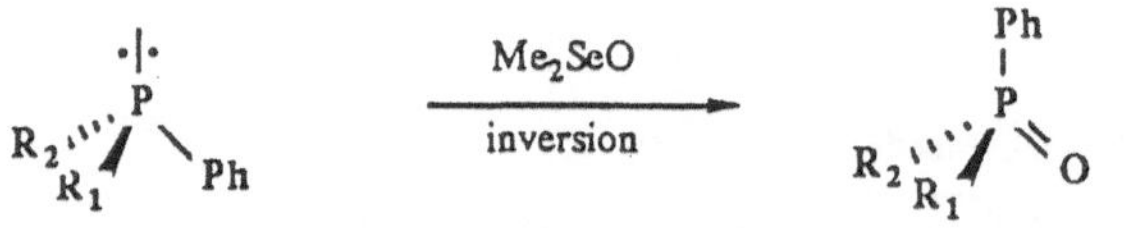

R$_1$	R$_2$	Yield %	ee %
Me	Pr	64	97
MeO	Et	84	86
EtS	Et	78	88

Scheme 91 [263]

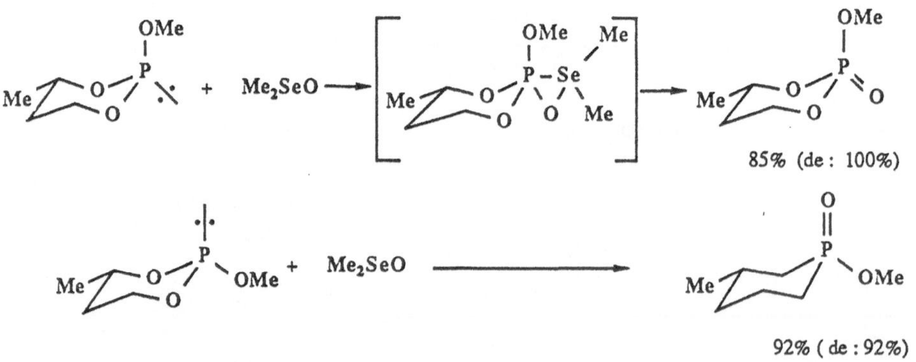

85% (de : 100%)

92% (de : 92%)

Scheme 92 [263]

(Scheme 93–95). The stereochemistry of these reactions has been investigated [263] (Schemes 91, 92, 94, 95). It generally takes place in acyclic derivatives [263] with inversion of configuration[1] (Schemes 91, 94), whereas almost full retention

$$R_3P{=}X \quad + \quad Me_2SeO \longrightarrow R_3P{=}O \quad + \quad Me_2Se$$

R	X	Yield (%)
Ph	Se	100
EtO	Se	92
Bu	S	76

Scheme 93 [263]

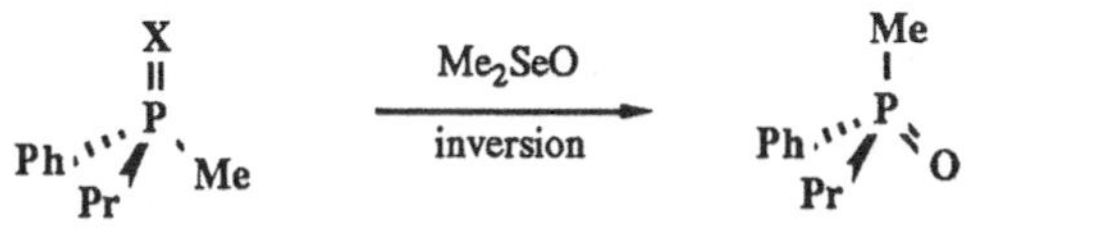

X	Yield(%)	ee(%)
S	76	75
Se	80	100

Scheme 94 [263]

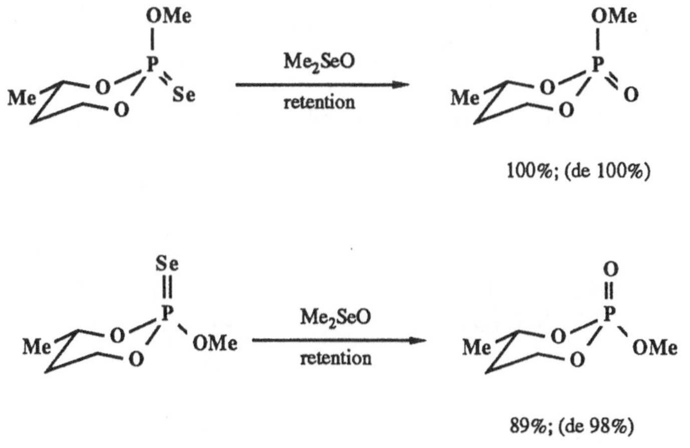

100%; (de 100%)

89%; (de 98%)

Scheme 95 [263]

---

[1] Retention of configuration at the phosphorous atom occurs when the oxidation is performed with perbenzoic acid [@263].

of configuration ($> 98\%$ ds) was observed when diastereoisomeric 2-methoxy-4-methyl-1,3,2-dioxaphosphorinanes (Scheme 92) as well as their 2-thioxo or 2-selenoxo derivatives [263] (Scheme 95) were reacted. The results have been rationalized by assuming nucleophilic attack of phosphorus on the selenium atom of the selenoxide and internal cyclisation leading to an intermediate (Scheme 92) in which the methoxy group and the three membered ring oxygen atom occupy apical positions and selenium occupies an equatorial position.

4.1.5 Oxidation of Olefins to vic-Glycols with Osmium Tetraoxide-Selenoxide Reagent

Diphenyl selenoxide and methyl phenyl selenoxide are able to oxidize [266, 267], in basic media, soluble osmium (VI) to osmium (VIII) derivatives. Selenoxides are cleanly reduced to selenides in this process (Scheme 96a). Selenoxides however fail to oxidize osmate esters in neutral organic solvents [266] and are stable over a period of hours in alkaline solutions containing catalytic amounts of osmium tetraoxide. These features were successfully used for the homogeneous liquid phase oxidation of olefins to glycols [@266]. Thus ethylene, 1,2-di-, tri- and tetrasubstituted olefins have been transformed to vic-glycols on reaction with selenoxides and catalytic amounts of osmium tetraoxide (Scheme 96b, c). Cis hydroxylation takes place with cyclohexene [266] and no further oxidation to α-ketols is observed [266]. Based on the yield of glycol recovered, at least 100 catalyst turnovers have been obtained and there is no evidence of irreversible catalyst deactivation [266]. The reaction also proceeds at a good rate with high molecular weight olefins since diorganyl selenoxides are soluble in aqueous organic solvents [266]. Another attractive feature of this process is the easy synthesis of selenoxides by singlet oxygen oxidation of selenides [268]. These can act in a catalytic manner as a singlet oxygen transfer agent. Thus cis-1,2-cyclohexane diol, contaminated by trace amounts of cyclohex-2-ene-1-ol and cyclohex-2-ene-1-one, has been prepared [266] on irradiation (200 watts incandescent light bulb) of cyclohexene, in the presence of methyl phenyl selenide, osmium tetraoxide, oxygen and catalytic amounts of rose bengal.

a) R_2SeO + $OsO_4^=$ + H_2O ———————→ R_2Se + $OsO_4(OH)_2^=$

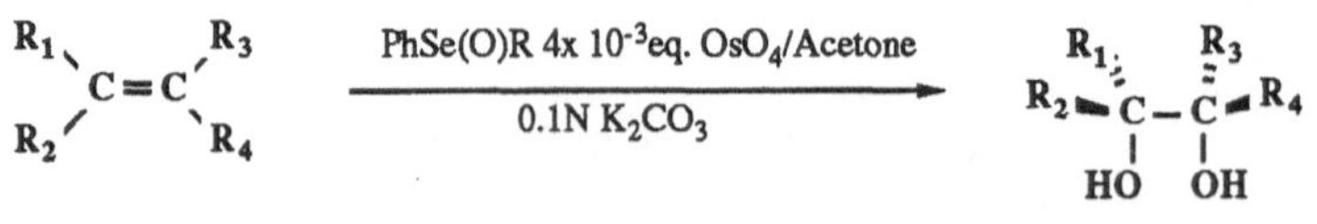

b) ethylene, propylene, 2,3-dimethyl-2-butene (*); R = Ph   80, 95, 54% (*)

c) ethylene, propylene;   R = Me   77, 85%

(*) 98% yield if the reaction performed at 50°C for 24h

Scheme 96 [266]

4.2 Oxidation of Alcohols to Carbonyl Compounds with Dimethyl Selenide/N-chlorosuccinimide

Dimethyl selenide/N-chlorosuccinimide (NCS) complex converts various alcohols in the presence of 1,8-diazabicyclo-[5.4.0]-undec-7-ene (DBU) to the corresponding carbonyl compounds [269] (Scheme 97). The reaction which has close analogy with the Corey-Kim method (Me₂S/NCS/NEt₃) [@269, 270] possesses different features since the selenium (IV) complex [269] is more stable than the corresponding sulfur (IV) complex, hence allyl alcohols are oxidized

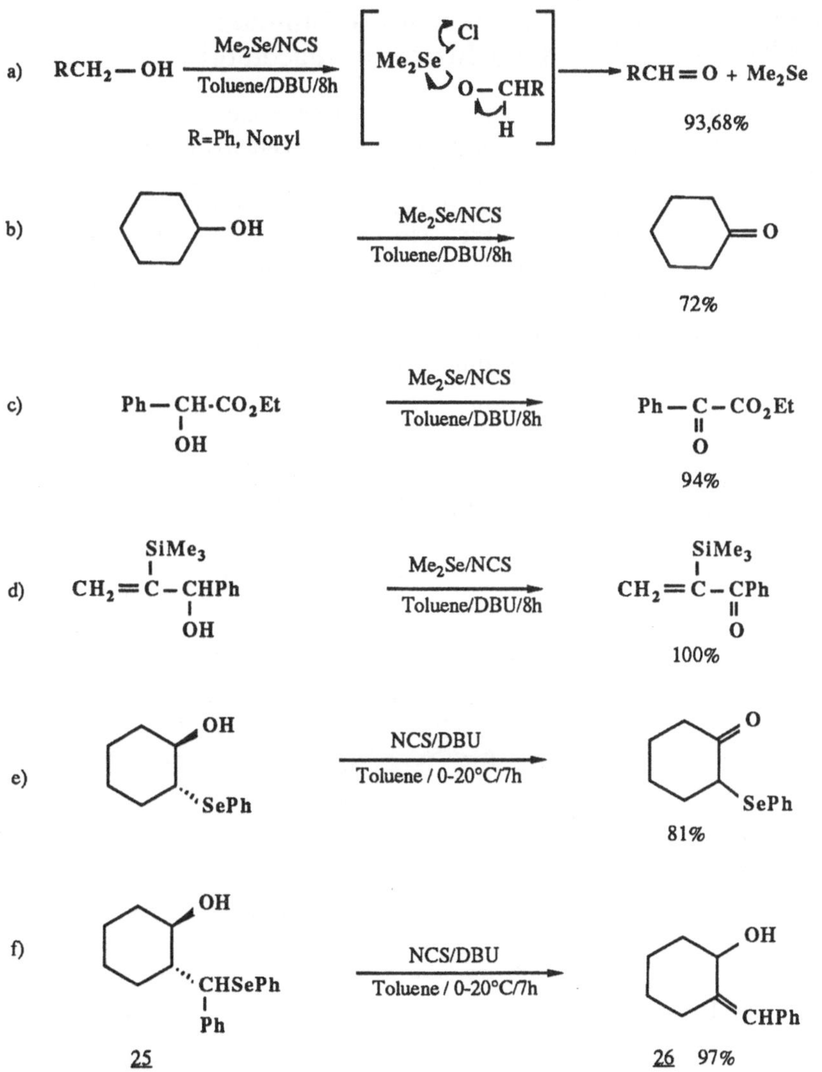

Scheme 97 [269]

69

[269] to the corresponding α, β-unsaturated carbonyl compounds without the concomitant formation of allylic chlorides [270].

β-Hydroxyalkyl phenyl selenides are efficiently oxidized to α-selenoketones [269] with NCS alone and interestingly the reaction does not require the presence of dimethyl selenide or sulfide [271, 272]. The formation of the allyl alcohol **26** [269] from the γ-hydroxyalkyl phenyl selenide **25** may be due to the presence in this specific case of a phenyl group on the carbon bearing the phenylseleno group (Scheme 97f).

4.3 Oxidative Conversion of sec-Benzylamines to Imines and Tertiary Benzylamines to Iminium Salts by Diphenylselenium Bis(trifluoroacetate)

Diphenylselenium bis(trifluoroacetate), prepared from diphenyl selenium dibromide and silver trifluoroacetate [273] or from diphenyl selenoxide and trifluoroacetic anhydride [274] is able to oxidize [274] at 20 °C the secondary and tertiary amino groups present in substituted tetrahydropyridine systems to the corresponding imines or iminium salts at 20 °C. The reactions are thought to proceed through the amino selenane derivative shown in Scheme 98. A series of 1-substituted 1,2,3,4-tetrahydroisoquinolines have been oxidized in high yields ($\sim$ 80%) to the corresponding 3,4-dihydroisoquinolines [274] which are often contaminated with small amounts of isoquinoline ($\sim$ 10%) (Scheme 98). Further

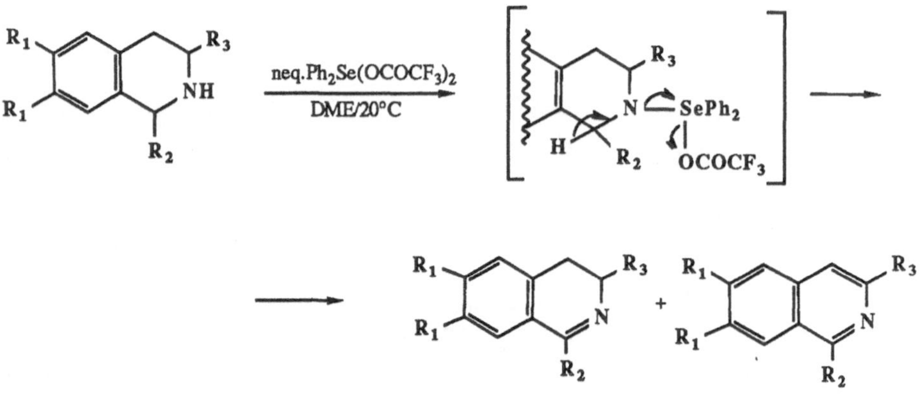

	R_1	R_2	R_3	n		
a)	MeO	H	H	2	80%	00%
b)	MeO	Me	H	3	81%	10%
c)	MeO	CO_2H	H	2	100%	00%
d)	H	H	CO_2Me	3	00%	100%

Scheme 98 [274]

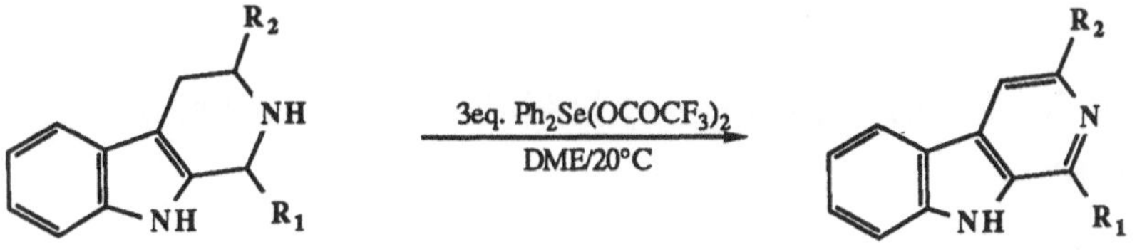

R_1	R_2	Yield (%)
H	H	61
Me	CO_2Me	70

Scheme 99 [274]

oxidation to isoquinoline does not usually occur and only those derivatives bearing an ester moiety in position 2 (Scheme 98 Entry d) are converted in high yield to isoquinoline when an excess of oxidant is used [274] (Scheme 98). The same reagent smoothly oxidizes 1,2,3,4-tetrahydrocarbolines (Scheme 99) to their 3,4-dihydro derivatives or to β-carbolines depending upon the equivalents of oxidant used (1.2 or 3 equiv. respectively). These reactions performed in the presence of potassium cyanide lead to α-aminonitriles also obtainable via the Reissert reaction [@274]. Accordingly N-methyl-1,2,3,4-tetrahydroisoquinolines are oxidized to the 1,2-iminium species which have been trapped with potassium cyanide to produce 1-cyano-tetrahydoquinolines (Scheme 100). Diphenylselenium dichloride, a closely related reagent efficiently converts [255] primary aryl amines to azobenzenes.

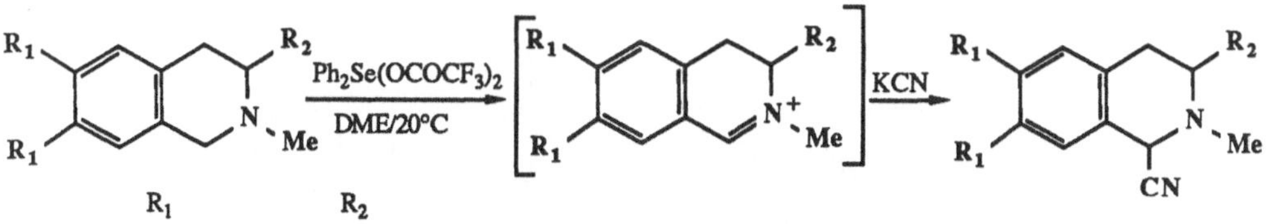

R_1	R_2	
MeO	H	90%
H	CO_2Me	78%

Scheme 100 [274]

Chapter 5

Reactions Involving Selenenyl Halides and Related Compounds

5.1 Allylic Halogenation of Olefins

Areneselenenyl chlorides and diaryl diselenides are effective catalysts in the allylic chlorination of olefins which implies N-chlorosuccinimide as the reagent [275]. The major product is a rearranged allylic chloride often contaminated with

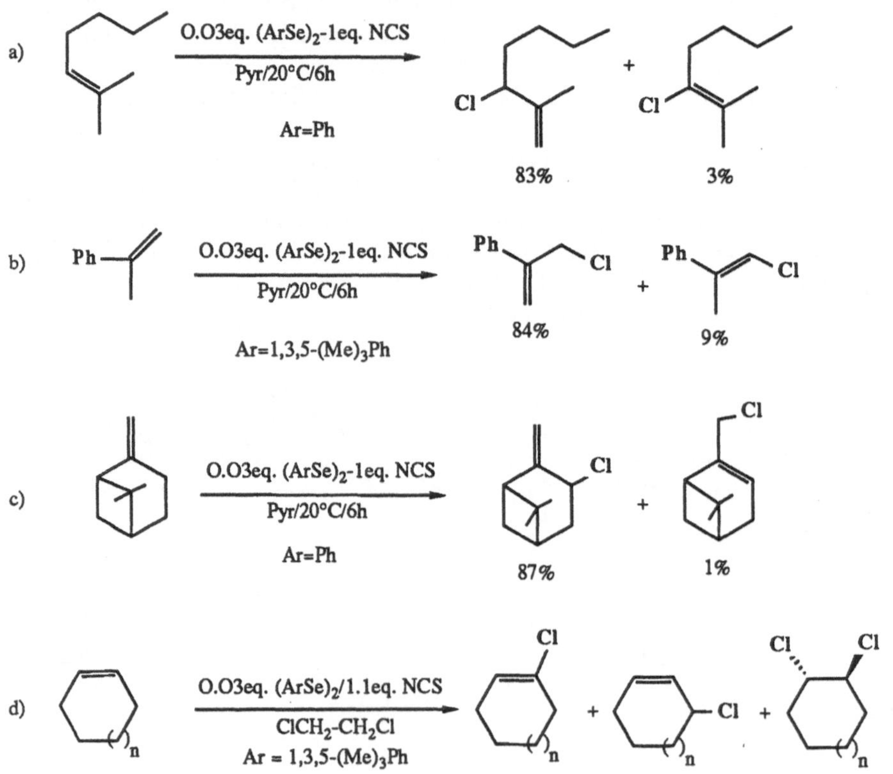

Scheme 101 [275]

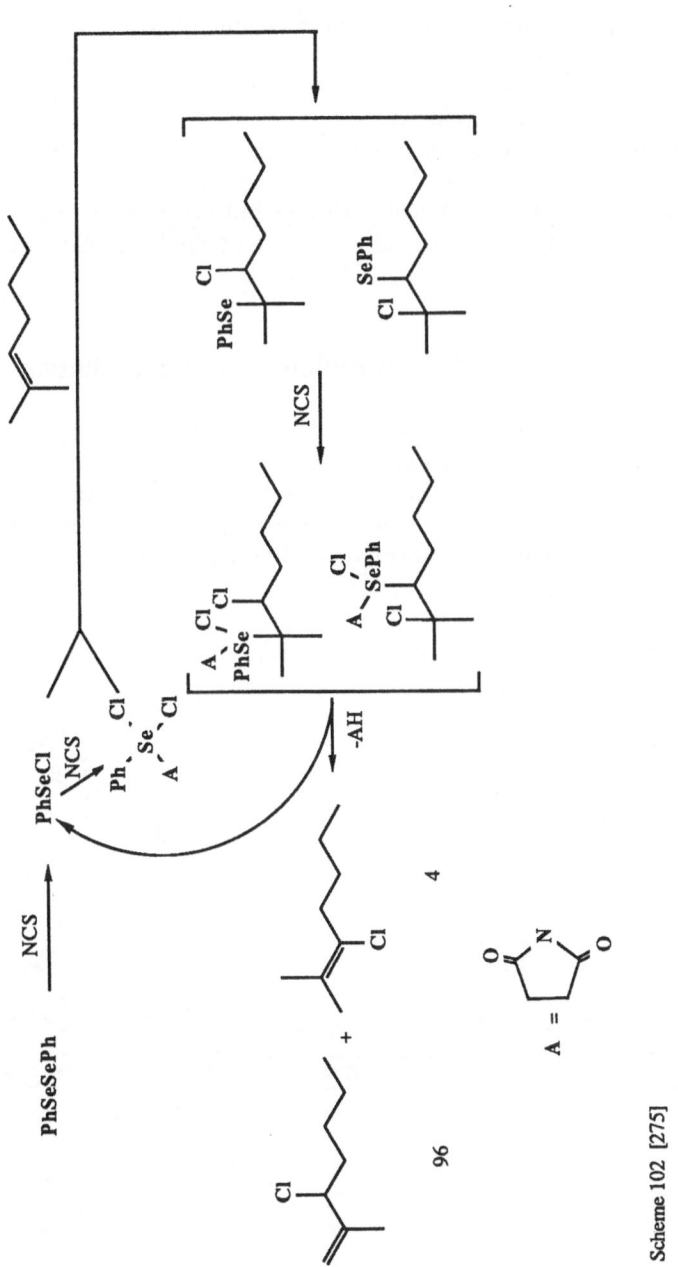

Scheme 102 [275]

a small amount of vinylic chloride (Scheme 101). The chlorination of β-pinene (Scheme 101c) produces almost exclusively the unrearranged allylic chloride [275] which is probably formed via the corresponding allylic phenylselenide, itself formed from β-pinene and N-phenylselenosuccinimide [276]. This method [275, 276] is particularly interesting as it does not proceed by free radical processes which usually occur in other direct allylic halogenation of olefins [@276].

The possible catalytic cycle yielding rearranged allylic chlorides and vinyl chlorides is shown in Scheme 102. Other halogenated selenium compounds [275] (e. g. O = SeCl₂, TsN = SeCl₂, PhCON = SeCl₂ and SeCl₄) also lead to allylic oxidation of olefins by non radical pathways but none of these selenium (IV) species proved as efficient (especially in catalytic applications) as the method reported above (Schemes 101 and 102).

5.2 Ring Expansion of 1,3-Dithiolans and 1,3-Dithians

Benzeneselenenyl chloride is a mild reagent for the ring expansion of 1,3-dithiolans and 1,3-dithians leading to dihydro-1,4-dithiins and dihydro-1,4-dithiepins respectively [@277] (Scheme 103). Best results are obtained when 2.1 equivalents of the reagent are used but in some cases the ketone corresponding to the thioacetal is concomitantly formed [277] (Scheme 103b).

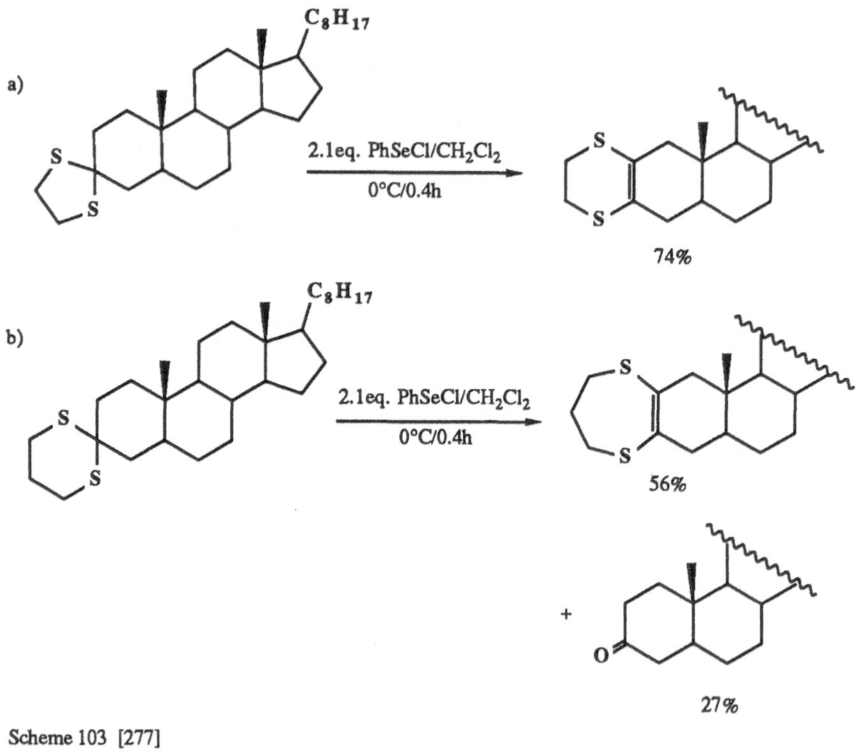

Scheme 103 [277]

5.3 Oxidation of Alcohols to Carbonyl Compounds

The oxidation of alcohols to carbonyl compounds by diphenyl diselenide/t-butyl hydroperoxide is presented in Sect. 6.4.2.

5.4 Synthesis of Oligonucleotides and Nucleoside Phosphoramidates

The title derivatives can be synthesized in high yields by a combined use of triphenylphosphite and 2,2'-dipyridyl diselenide as a coupling reagent for the internucleotidic bond formation [278] (Scheme 104). Pyridine was found a more suitable solvent than for example hexamethyl phosphotriamide (HMPA) or dimethyl formamide (DMF). This method avoids the formation of undesirable symmetrical pyrophosphates from the nucleoside components [@278]. It also offers definitive advantage over a related method involving triphenylphosphine/ 2,2'-dipyridyl disulfide, since it avoids the reaction of the sugar moiety of nucleosides with the coupling reagent observed with the latter method [@278].

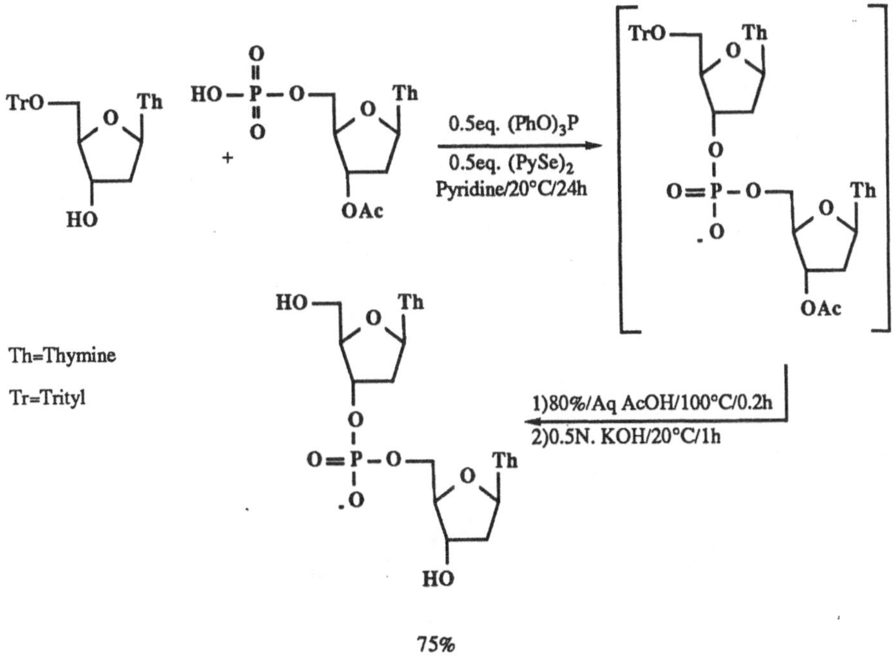

Th=Thymine

Tr=Trityl

Scheme 104 [278]

Chapter 6

Reactions Involving Benzeneseleninic Anhydride (BSA) and Related Reagents

6.1 Oxidation of Phenols, Pyrocatechols and Hydroquinones [*92]

Benzeneseleninic anhydride (BSA) and in some cases benzeneseleninic acid (which readily produces the anhydride) proved to be mild reagents able to oxidize in neutral or basic media pyrocatechols (1,2-dihydroxybenzenes) to orthoquinones [279] (Scheme 105) and hydroquinones to p-quinones [279] (Scheme 106).

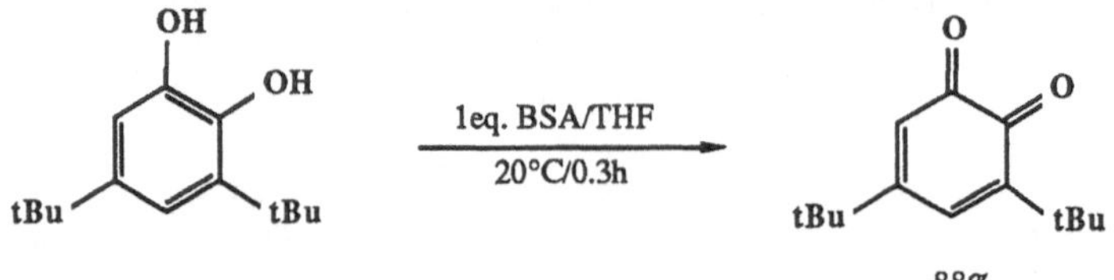

Scheme 105 [279]

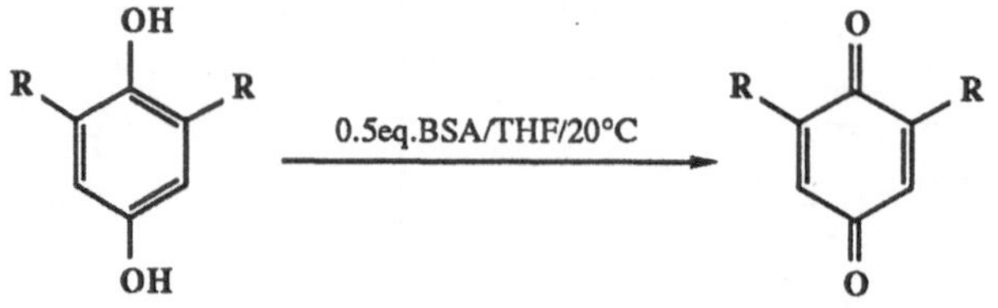

| a) | R=H | 84% | [279] |
| b) | R=Me | 88% | [279] |

| c) | R=MeO | 98% | [282] |
| d) | R=NH$_2$ | 85% | [282] |

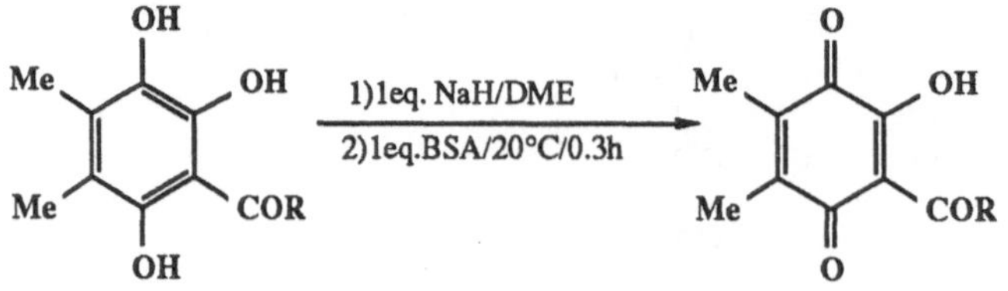

Scheme 106

76

More remarkably phenols are oxidized [@282] into quinones [279–282] (Scheme 107) or into α-hydroxydienones [282–285] depending upon the positions occupied by the substituents in the aromatic ring. When the reactions are carried out in neutral medium, phenyl aryl selenides are often observed as side products arising from electrophilic substitution on the starting phenols [279, 282] (Schemes 107a, b, 108c–f).

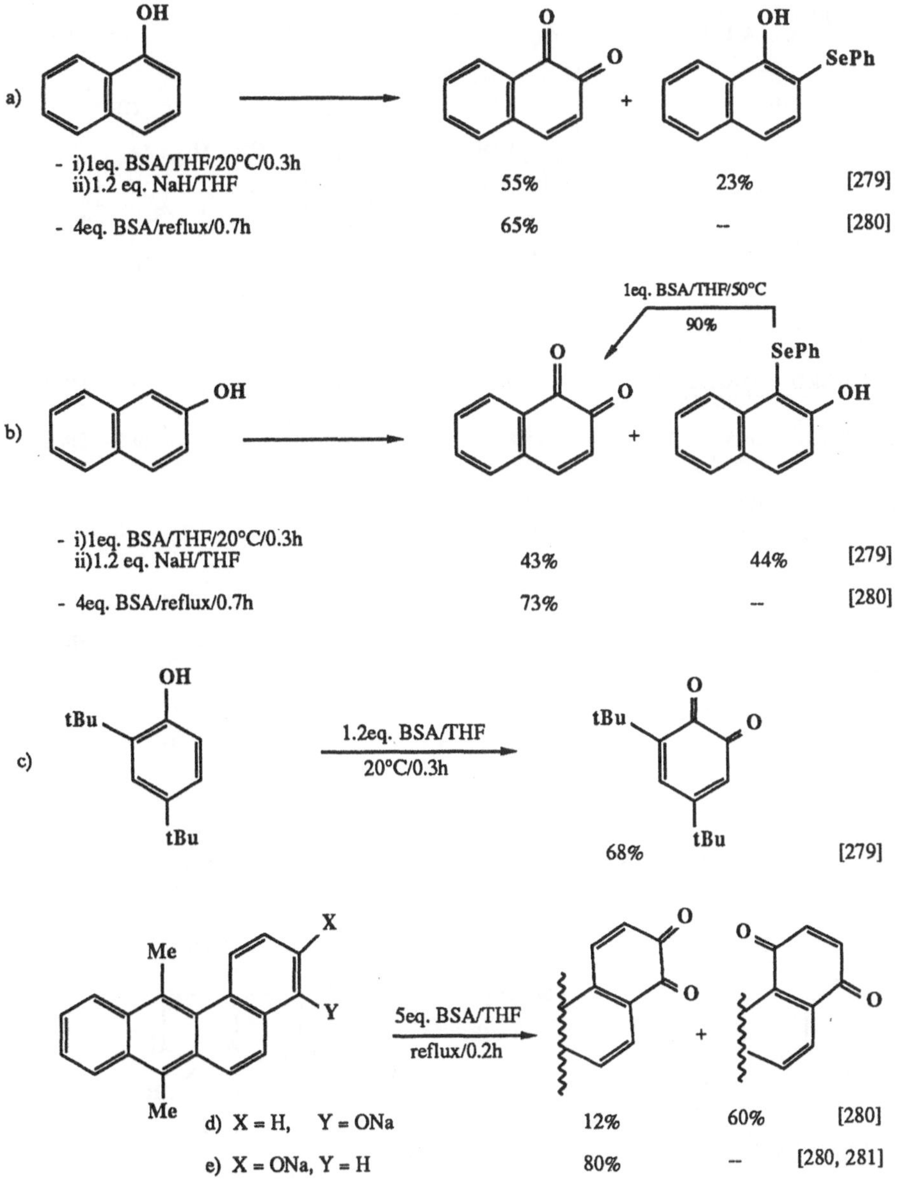
a)
OH
O
O
+
OH
SePh
- i)1eq. BSA/THF/20°C/0.3h
ii)1.2 eq. NaH/THF
55%
23%
[279]
- 4eq. BSA/reflux/0.7h
65%
--
[280]
b)
OH
1eq. BSA/THF/50°C
90%
O
O
+
SePh
OH
- i)1eq. BSA/THF/20°C/0.3h
ii)1.2 eq. NaH/THF
43%
44%
[279]
- 4eq. BSA/reflux/0.7h
73%
--
[280]
c)
OH
tBu
tBu
1.2eq. BSA/THF
20°C/0.3h
tBu
O
O
tBu
68%
[279]
Me
Me
X
Y
5eq. BSA/THF
reflux/0.2h
O
O
+
O
O
d) X = H, Y = ONa
12%
60%
[280]
e) X = ONa, Y = H
80%
--
[280, 281]

Scheme 107

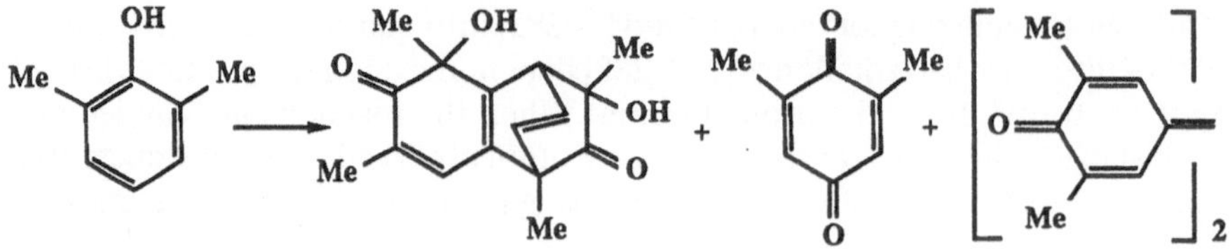

6. Reactions Involving Benzeneseleninic Anhydride (BSA) and Related Reagents

a)	1eq. BSA/DMF/20°C/2h	5%	25%	40%	[282]
b)	1) 1eq. NaH/DMF 2) 1eq. BSA/THF20°C/2h	44%	0%	0%	[282]

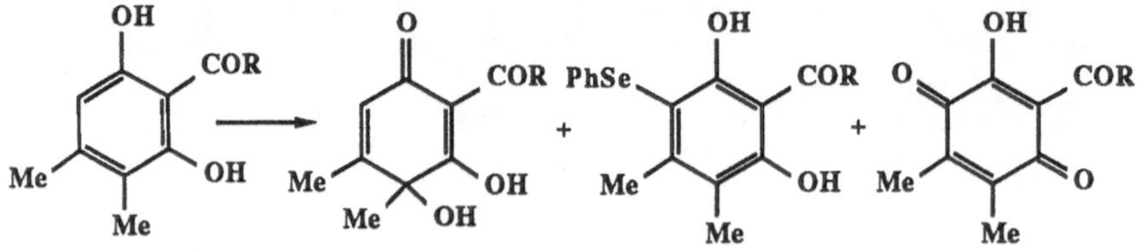

c)	R=MeO	1.08eq. BSA/CH₂Cl₂ 20°C/0.4h	35%	55%	--	[282]
d)	R=MeO	1) NaH/DME 2) BSA/20°C/0.5h	75%	17%	--	[282]
e)	R=NH₂	0.85eq. BSA/CHCl₃ 20°C/0.5h	25%	45%	20%	[282]
f)	R=NH₂	1) NaH/DME 2) BSA/20°C/12h	68%	trace	--	[282]

g)	1.1eq. BSA/CH₂Cl₂/20°C/3h	0%	[282]
h)	1)NaH/THF 2)BSA/55°C/2h	55% (based on S.M. used)	[282]

i)   BSA/CH₂Cl₂   reflux/3h   85%   [285]

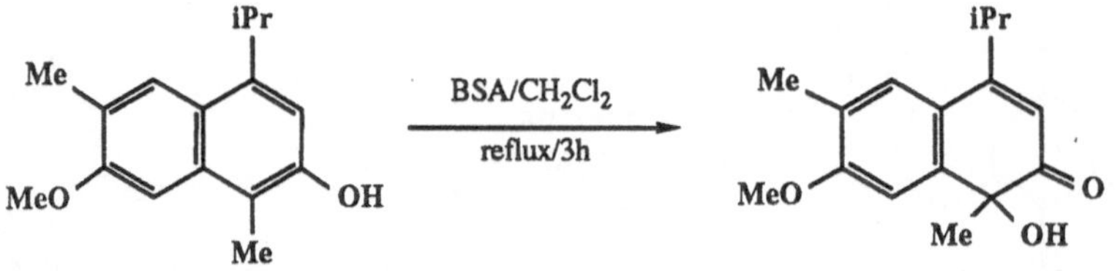

Scheme 108

78

The reactions are ortho directing even in some cases where the para position is unblocked [279–284]. These observations have been rationalized [282] by Barton in termes of the mechanism outlined in Scheme 109 and which involves the intermediary formation of a seleninate ester (**27**, Scheme 109) and its [2, 3] sigmatropic rearrangement into selenenate ester **28**. The latter species then breaks down into o-hydroxy dienone **29** or o-quinone **30**.

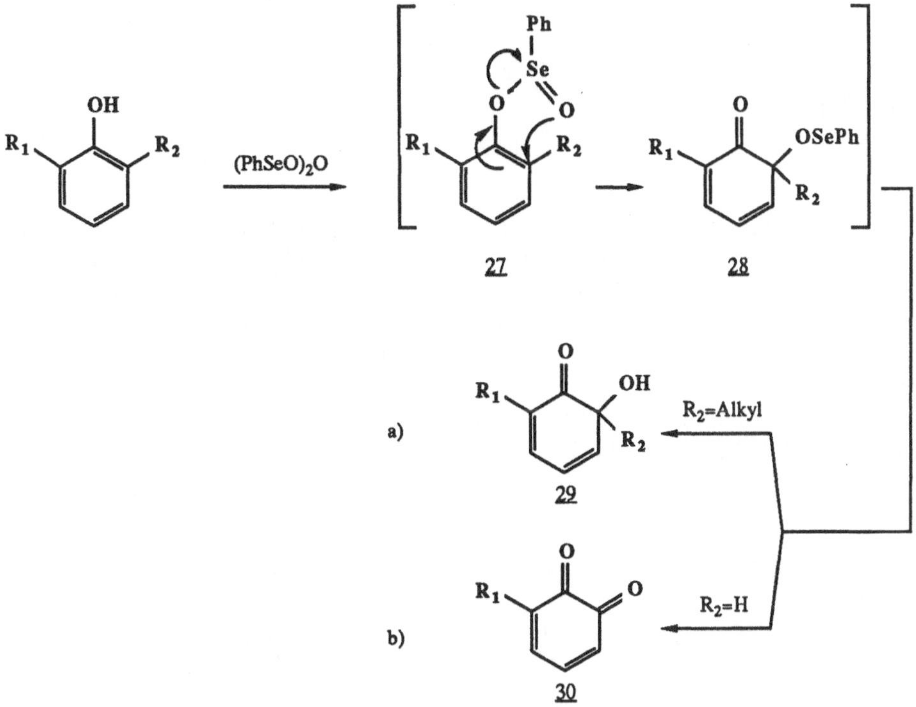

Scheme 109 [279, 282, 283]

Ortho-quinones have been successfully prepared from phenols [279, 280] naphthols [279, 280, 282], hydroxyphenanthrenes [280], hydroxyanthracenes [280], hydroxybenz[a]anthracenes [280, 281] (Scheme 107a–d) having unsubstituted ortho positions. In the case of fused aromatic compounds where two ortho positions are available, oxidation regioselectively occurs near to the ring fusion [279–282] (Scheme 107b, d).

These reactions were used by Barton [282] in the synthesis of hydroxy dienone models of the A ring of tetracyclines (Scheme 108d–f); by Jeffs [285] in the synthesis of the 1-hydroxynaphthalenone derivative isolated from cotton (Scheme 108i) and by Harvey [280, 281] in the synthesis of an o-quinone precursor of a highly carcinogenic dihydrobenzochrysene (Scheme 107d, e). In isolated cases the reactions do not follow the usual course: for example it was reported [280] that 4-hydroxy-7,12-dimethyl benz[a]anthracene, an α-type phenol [280], mainly provides a p-quinone derivative (Scheme 107e) and 3-hydroxyphenanthrene leads [280] to the ring contracted 2,2-dihydroxybenz[e]indane-1,3-dione (Scheme 110) instead of the expected o-quinone.

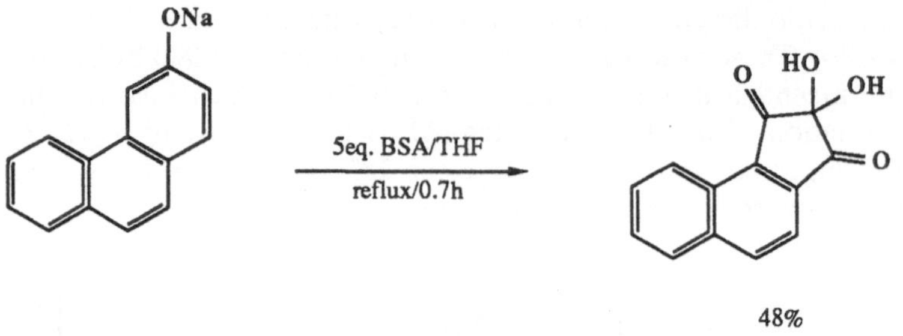

6. Reactions Involving Benzeneseleninic Anhydride (BSA) and Related Reagents

Scheme 110 [280]

The reactions proceed differently in the presence of hexamethyldisilazane when quinophenylselenoimines [286] (Scheme 111) are formed instead of o-quinones. Para substitution leading to p-quinophenylselenoimines occurs with phenols possessing two alkyl groups in 2 and 6 positions [286] (Scheme 111a) whereas ortho substitution is preferred with phenols with unsubstituted para positions [286] (Scheme 111b,c). The value of this reaction is enhanced since quinophenylselenoimines can be readily reduced to aminophenols under mild conditions [286] (Scheme 111).

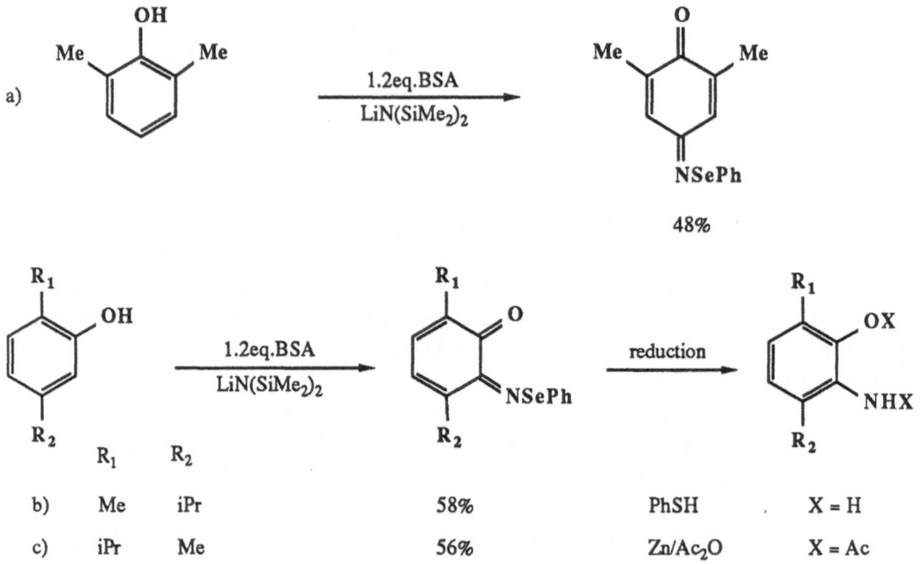

	R_1	R_2				
b)	Me	iPr		58%	PhSH	X = H
c)	iPr	Me		56%	Zn/Ac$_2$O	X = Ac

Scheme 111 [286]

80

6.2 Dehydrogenation of Ketones to Enones and Dienones

Benzeneseleninic anhydride has also been successfully used for the oxidation of ketones at the α-position [287–293, @294]. Steroids and triterpenoid ketones are dehydrogenated in chlorobenzene at 95 °C [287, 288]. The reaction allows the synthesis of cholesta-1,4-dien-3-one from cholestanone in high yield (Scheme 112a) and of cholesta-1,4,6-trien-3-one from cholesta-4,6-dien-3-one (Scheme

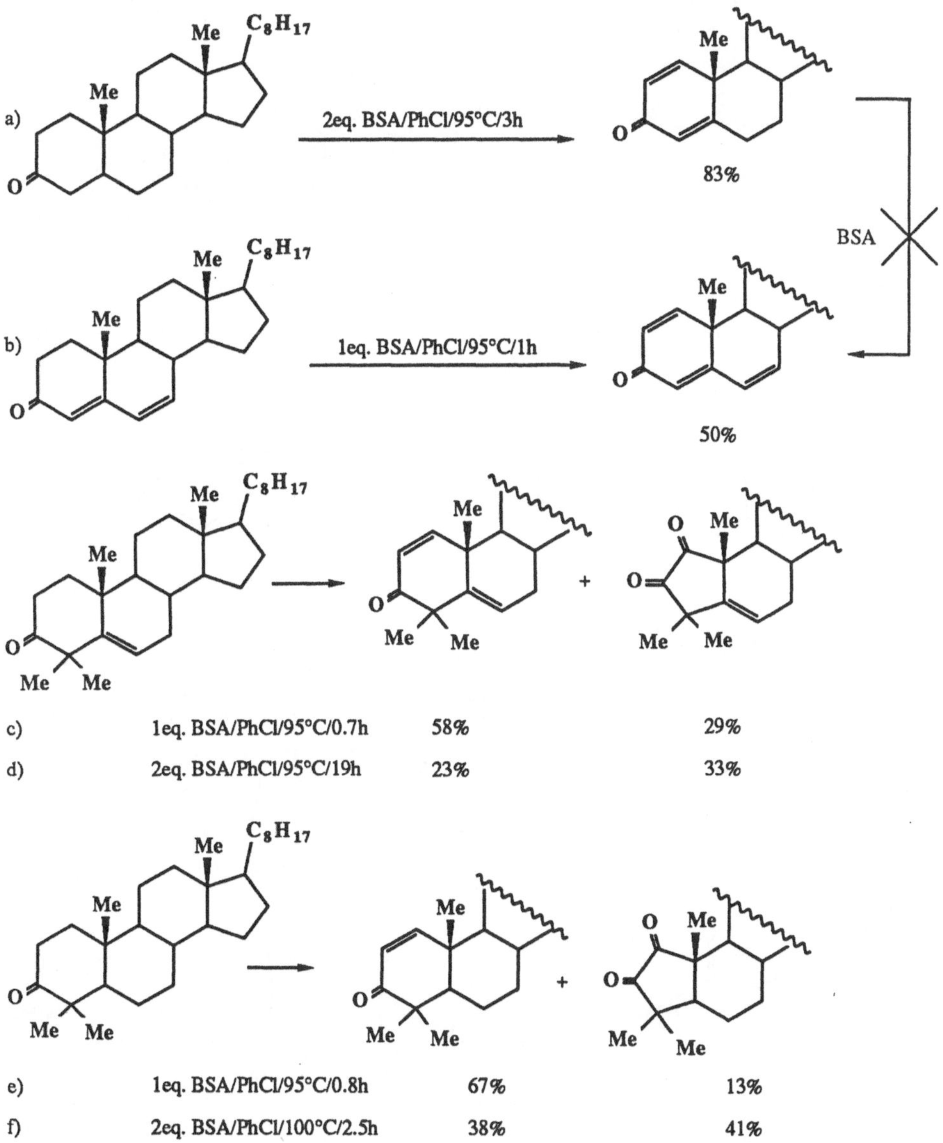

| a) | 2eq. BSA/PhCl/95°C/3h | | 83% |
| b) | 1eq. BSA/PhCl/95°C/1h | | 50% |

c)	1eq. BSA/PhCl/95°C/0.7h	58%	29%
d)	2eq. BSA/PhCl/95°C/19h	23%	33%

e)	1eq. BSA/PhCl/95°C/0.8h	67%	13%
f)	2eq. BSA/PhCl/100°C/2.5h	38%	41%

Scheme 112 [288]

6. Reactions Involving Benzeneseleninic Anhydride (BSA) and Related Reagents

112b). Interestingly the remote double bond present on 4,4-dimethyl cholest-5-en-3-one [288] (Scheme 112c, d) and in α- and β-amyrone [288] are not affected by BSA under the usual reaction conditions. This contrasts with the oxidation of the same compounds using SeO_2 [@295] (Sect. 9.3).

Side reactions leading to ring A contracted diketones, have been frequently observed with 4,4-dimethyl-3-steroidal ketones [288] (Scheme 112c–f). They are favoured [289] when an excess of BSA and/or longer reaction times are used (Scheme 112 compare e to f). It was also found that certain enolisable ketones [290, 291, 292] adjacent to a tertiary centre are oxidized to α-hydroxy ketones

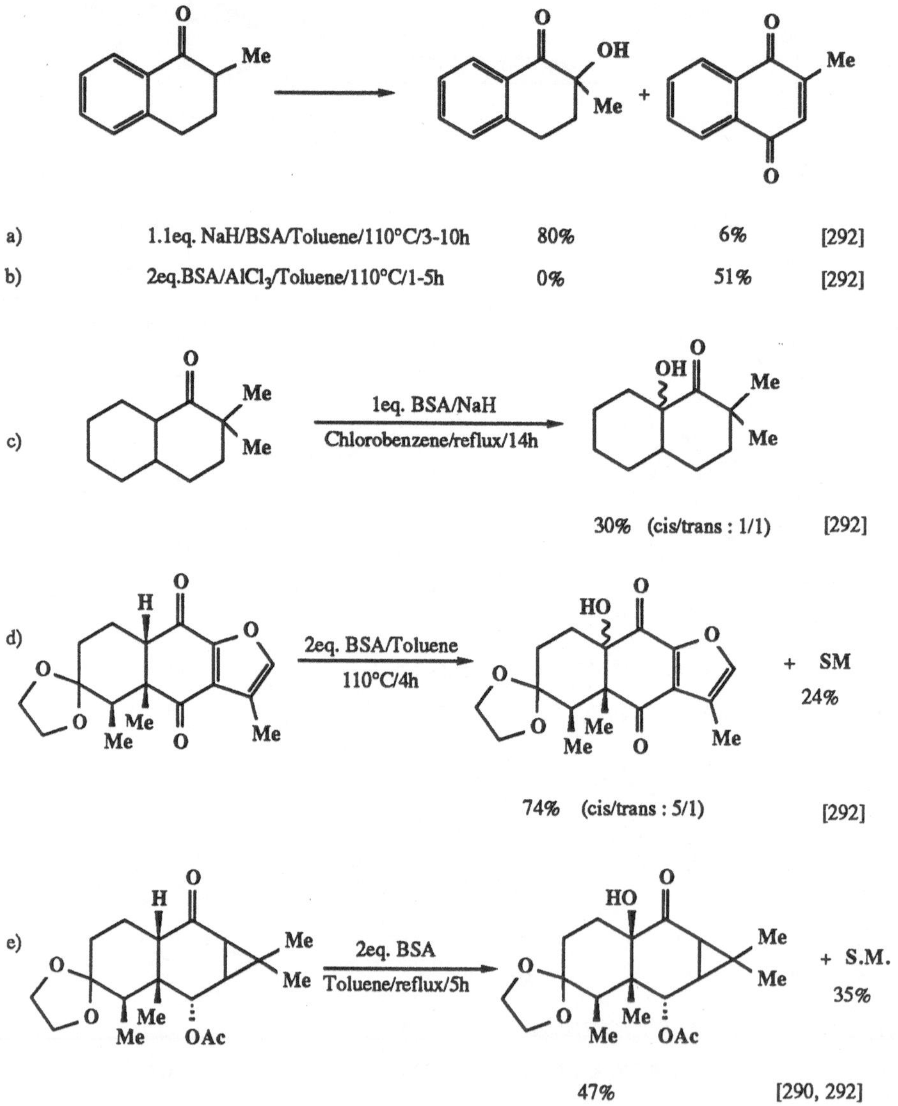

a)	1.1eq. NaH/BSA/Toluene/110°C/3-10h	80%	6%	[292]
b)	2eq.BSA/AlCl₃/Toluene/110°C/1-5h	0%	51%	[292]

30% (cis/trans : 1/1)   [292]

74% (cis/trans : 5/1)   [292]

47%   [290, 292]

Scheme 113

(Scheme 113a, c–e). This observation has been successfully aplied to the angular hydroxylation of polycyclic ketones especially in the eremophilane series [290–292] (Scheme 113d). In some cases better results have been achieved by performing the reactions under basic conditions such as in the presence of NaH for example [292] (Scheme 113a, c). The case of chaparrinone triacetate [293] is related to those just described since hydroxylation occurs at the allylic position of the enone system (Scheme 114). Other conditions have been tried with varying degree of success. For example, enones can be obtained if the reactions are performed with BSA in the presence of aluminum trichloride [292] (Scheme 113b). In the case of 3-cholestanone (**31**, Scheme 115) the α-phenylselenoenone (**32**, Scheme 115) is formed [295] when the reaction is performed at 20 °C with stoichiometric amounts of BSA in acetic acid. It is smoothly dehydrogenated to the corresponding 1,4-dienone **33** on further reaction with BSA in chlorobenzene at higher temperature [295] (Scheme 115).

A very important improvement of the method was achieved [289, 295] by performing the reaction with BSA generated in situ [by oxygen atom transfer from iodoxy benzene or m-iodoxy benzoic acid to catalytic amounts of diphenyl diselenide [289, 295], or of benzene seleninic anhydride [295]. Better results are

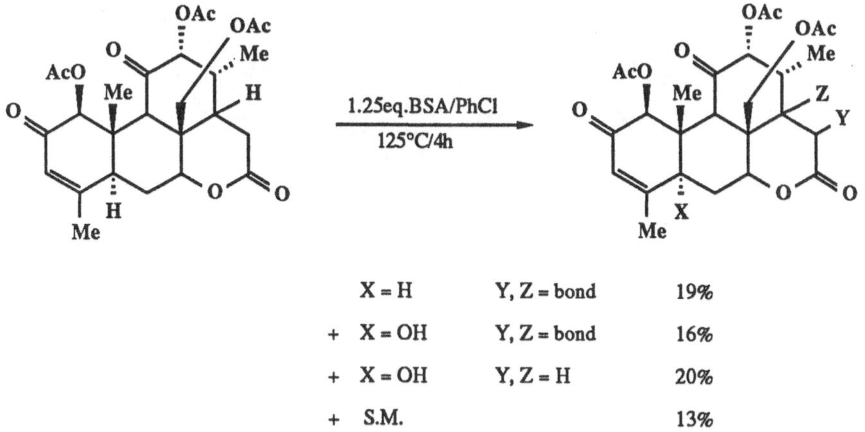

X = H	Y, Z = bond	19%
+ X = OH	Y, Z = bond	16%
+ X = OH	Y, Z = H	20%
+ S.M.		13%

Scheme 114 [293]

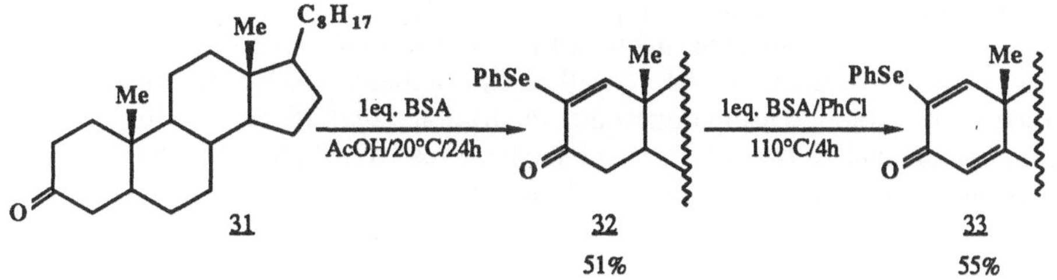

Scheme 115 [295]

83

6. Reactions Involving Benzeneseleninic Anhydride (BSA) and Related Reagents

a) X,Y=O	0.3eq. PhSeSePh/N$_2$O$_3$/PhCl/110°C/4 days	63%
b) X,Y=O	0.3eq. PhSeSePh/3.3eq. PhIO/PhCl/110°C/12h	65%
c) X=OH, Y=H	0.2eq. BSA/3.3eq.PhIO/PhCl/110°C/24h	81%
d) X=OH, Y=H	0.2eq. BSA/3.3eq. PhIO/p-TsOH, trace/PhCl/110°C/12h	77%
e) X=OH, Y=H	0.2eq. PhSeSePh/3.3eq. 3-OI-PhCO$_2$H Toluene/110°C/3h	88%
f) X,Y=O	0.5eq. SeO$_2$/3eq. PhIO/PhCl/110° C	58%

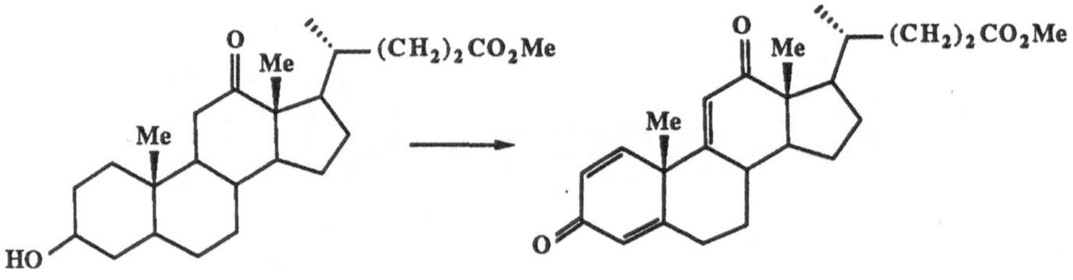

g) 0.1eq. PhSeSePh/5eq. 3-OI-PhCO$_2$H   89%
Toluene/110°C/96h

Scheme 116   [289, 295]

obtained if the reactions are performed with trace amounts of pTsOH [295]. The method permits the synthesis of dienones [295] especially from various 3-keto and 12-keto steroids (Scheme 116) and also from the corresponding alcohols (Scheme 116a, d, e), avoiding the unwanted side reactions [due to the selenenic acid generated in the elimination step [287–289, 292] and reducing the cost of the process [289, 295]. This is particularly true when m-iodoxy benzoic acid is used as the iodobenzoic acid formed as well as the diphenyl diselenide can easily be recovered without chromatographic separation [295] (Scheme 116g). Other compounds such as N-oxides which have been tested [295] for oxygen transfer to diphenyl diselenide proved unsuitable for BSA formation.

6.3 Dehydrogenation of Lactones and Lactams to α,β-Unsaturated Compounds

Some steroidal lactones [296, 297] (Scheme 117a–c) and lactams [298, 299] (Scheme 117d) have been smoothly dehydrogenated with BSA in refluxing chlorobenzene or diglyme. Over-oxidation leading to γ-hydroxy-α,β-unsaturated lactones (Scheme 117b) can be avoided by shorter reaction times [297] (Scheme 117c). On the basis of published results [297, 299] there appears no general trend.

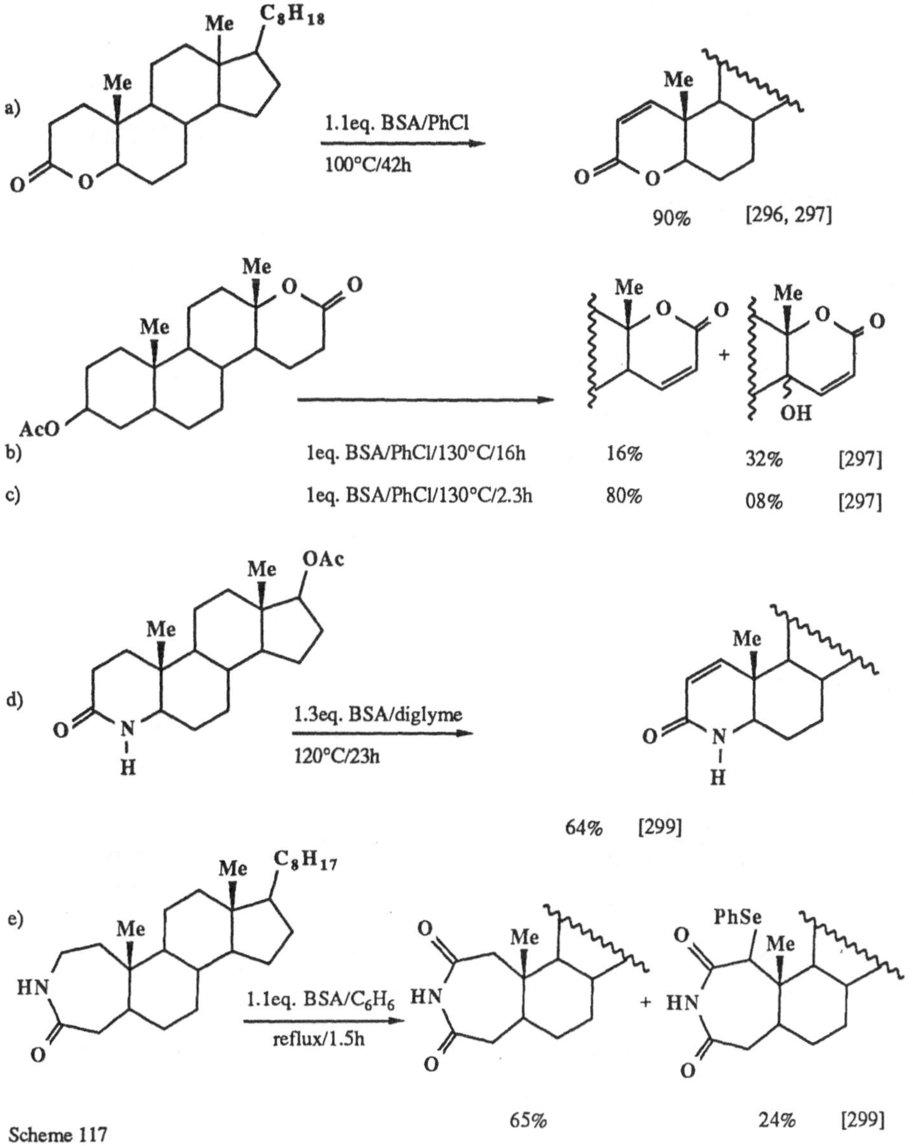

Scheme 117

6. Reactions Involving Benzeneseleninic Anhydride (BSA) and Related Reagents

For example five membered ring lactones [297], acyclic esters [296] and amides [299] do not react with BSA. However BSA reacts [283] with the ester enolates (LiNPr$_2$ + ester) to give α, β-unsaturated esters without the need for in situ oxidation which is often required when carbonyl compounds are reacted with other organoselenium reagents [302, 303]. The reaction takes a different course with aza-steroidal lactams [299] (Scheme 117e) and with 4-aza-steroidal enamides [300, 301].

6.4 Oxidation of Alcohols to Carbonyl Compounds or to Enones and Dienones

6.4.1 With BSA

A number of benzylic [304], allylic [304] and saturated alcohols [304] have been oxidized by BSA to the corresponding carbonyl compounds in high yields (Scheme 118). The reaction usually takes place in refluxing benzene, chloroben-

R-C$_6$H$_4$—CH$_2$OH $\longrightarrow$ R-C$_6$H$_4$·CH=O

a)	R=H	2eq. BSA/PhH/80°C/0.3h	99.5%	[304]
b)	R=4-NO$_2$	0.5eq. BSA/PhH/80°C/0.1h	97%	[304]
c)	R=4-MeO	0.5eq. BSA/PhH/80°C/0.2h	85%	[304]

d) 1.1eq. BSA/PhH/ 67°C/16h 94% [304]

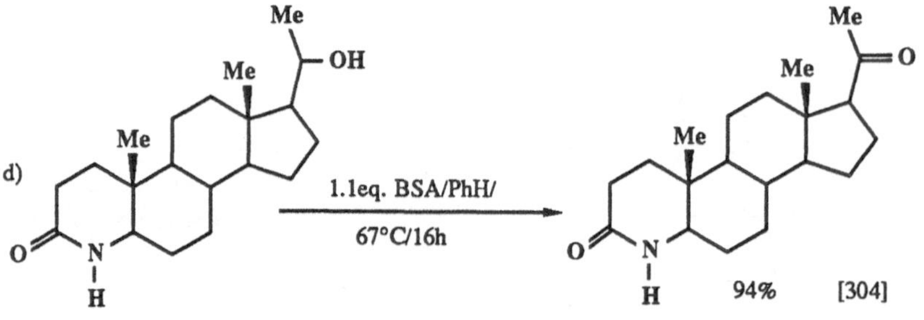

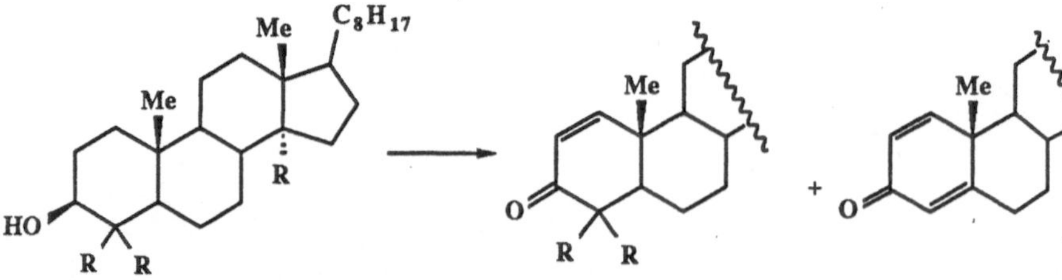

e)	R=Me	2eq. BSA/PhH/130°C/0.3h	65%	--	[304]
f)	R=H	4eq. BSA/PhH/130°C/0.2h	--	60%	[304]

Scheme 118

zene or THF and is much more rapid with benzylic (0,1 to 3h) than with allylic (10h) or saturated alcohols (16–20h) (Scheme 118a, d). The oxidation is particularly facile with steroidal ring A alcohols [304] (Scheme 118e, f) where further oxidation yields enones or commercially important 1,4-dienones in a one step process. These compounds otherwise require multistep processes to be prepared [304]. The smooth oxidation of ring A ketones, involved in these transformations, has been described previously [288] (Scheme 112). Also of interest is the selective oxidation of the secondary alcohol shown in Scheme 118d which occurs chemoselectively without touching the potentially reactive lactame site [304].

6.4.2 With Diphenyldiselenide and t-Butyl Hydroperoxide

The oxidation of alcohols to the corresponding aldehydes and ketones has also been achieved with *t*-butyl hydroperoxide in the presence of diaryldiselenides in refluxing benzene [305–307] (Scheme 119) or of polymer bounded phenylseleninic acid in refluxing carbon tetrachloride [308] (Scheme 120). Both conditions seem to involve similar intermediates.

In the former case, the effect of substituents on the diaryl diselenide has been examined using 3-phenyl-1-propanol as substrate. 3-Phenylpropanal was obtained in 4, 64, 98% yields when $(2\text{-}NO_2PhSe)_2$, $(PhSe)_2$ or $[2,4,6\text{-}(Me)_3PhSe]_2$ were used respectively. The reaction proceeds cleanly with 0.5 equiv. of bis (2,4,6-trimethylphenyl)diselenide and with a wide range of primary, secondary and allyl alcohols including geraniol (Scheme 119c) and with benzylalcohols (Scheme 120).

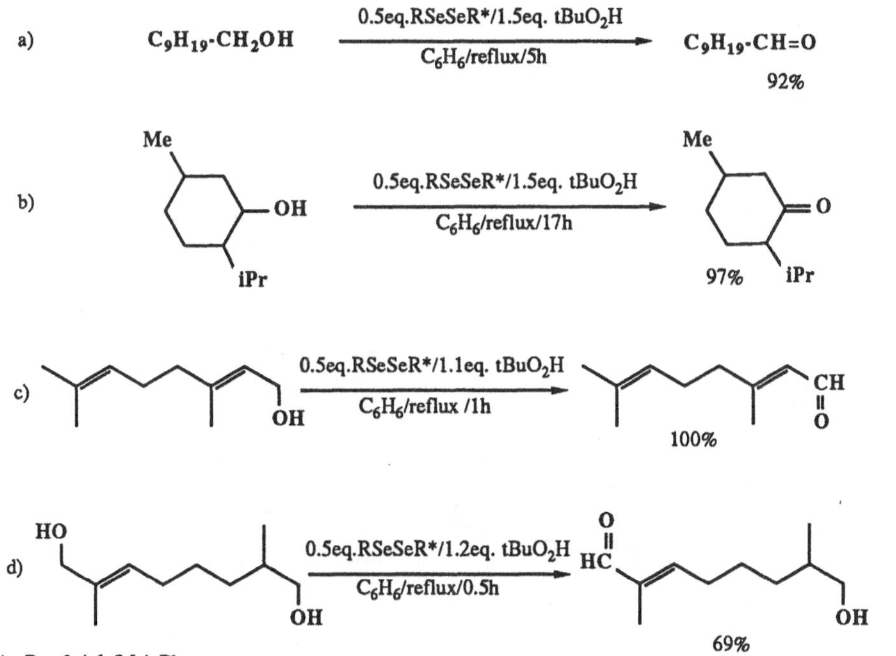

a)　　　　$C_9H_{19}\text{-}CH_2OH$ 　$\xrightarrow[C_6H_6/reflux/5h]{0.5eq.RSeSeR*/1.5eq.\ tBuO_2H}$　 $C_9H_{19}\text{-}CH{=}O$
　　　　　　　　　　　　　　　　　　　　　　　　　　　　　　92%

b) $\xrightarrow[C_6H_6/reflux/17h]{0.5eq.RSeSeR*/1.5eq.\ tBuO_2H}$ 97%

c) $\xrightarrow[C_6H_6/reflux\ /1h]{0.5eq.RSeSeR*/1.1eq.\ tBuO_2H}$ 100%

d) $\xrightarrow[C_6H_6/reflux/0.5h]{0.5eq.RSeSeR*/1.2eq.\ tBuO_2H}$ 69%

* R = 2,4,6-(Me)$_3$Ph

Scheme 119 [307]

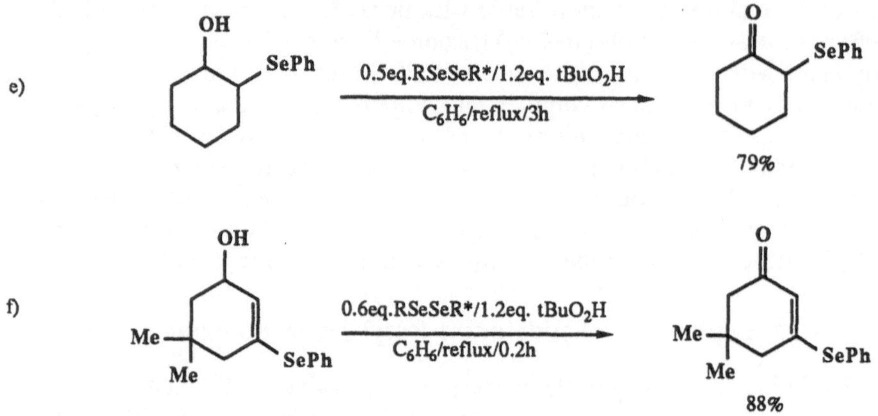

6. Reactions Involving Benzeneseleninic Anhydride (BSA) and Related Reagents

e)

0.5eq.RSeSeR*/1.2eq. tBuO₂H
$$\text{0.5eq.RSeSeR*/1.2eq. tBuO}_2\text{H}$$
C₆H₆/reflux/3h

79%

f)

0.6eq.RSeSeR*/1.2eq. tBuO₂H
C₆H₆/reflux/0.2h

88%

* R = 2,4,6-(Me)₃Ph

Scheme 119 (contd.) [307]

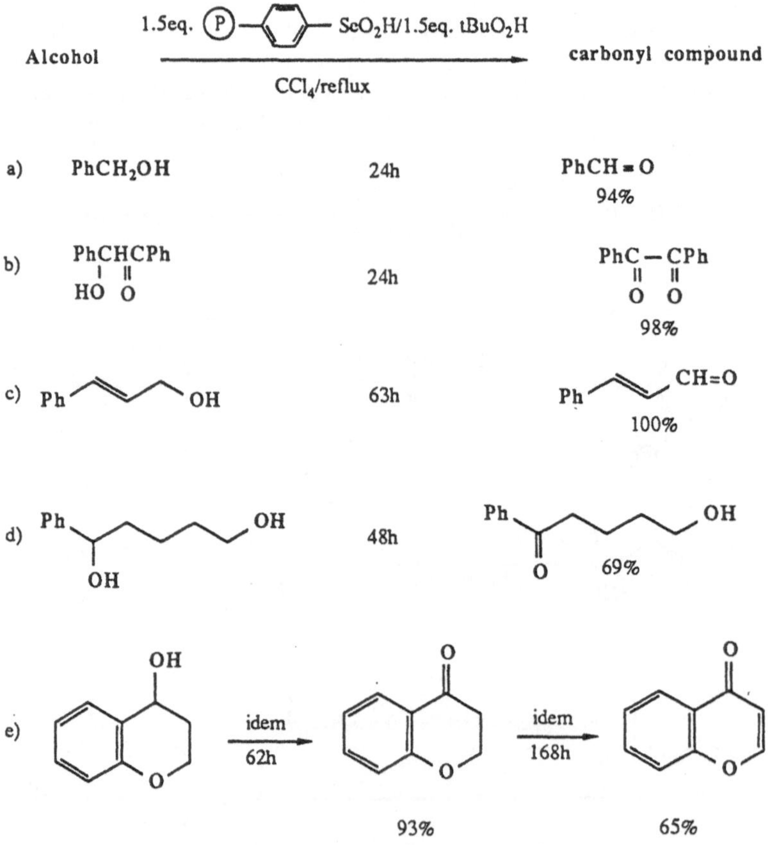

Alcohol $\xrightarrow[\text{CCl}_4/\text{reflux}]{\text{1.5eq. (P)}-\langle\rangle- \text{SeO}_2\text{H/1.5eq. tBuO}_2\text{H}}$ carbonyl compound

a)   PhCH₂OH   24h   PhCH=O
                                                                94%

b)   PhCHCPh
       | || 24h   PhC—CPh
       HO O || ||
                                                               O   O
                                                               98%

c) Ph ⌒⌒ OH   63h   Ph ⌒⌒ CH=O
                                                              100%

d) Ph ⌒⌒⌒ OH   48h   Ph ⌒⌒⌒ OH
       | ||
       OH   O   69%

e)

        idem   idem
        62h   168h

        93%   65%

Scheme 120 [308]

88

Under these conditions allylic alcohols are more readily oxidized than the corresponding saturated alcohols [305–307]. This method [305, 307] is particularly suitable for the selective oxidation of β-hydroxy alkyl sulfides, β-hydroxy alkyl selenides and γ-(phenylseleno)-allyl alcohols to the corresponding hetero-substituted carbonyl compounds without reaction at the sulfur or the selenium atom [@269, 307, 309] (Scheme 119e, f).

Polymer bound phenylseleninic acid/tBuO$_2$H reagent has a similar reactivity although the reaction usually requires much longer time [308] (compare Scheme 120 to Scheme 119). Benzyl alcohols are selectively oxidized in the presence of primary alkyl ones (Scheme 120d) and 4-chromanol is selectively converted to 4-chromanone [308] after 62hrs but further dehydrogenation leading to 4-chromone [308] occurs if the reaction is continued for an additional 168h (Scheme 120e). However cis-1,4-butenediol and cyclopentanol are inert to this reagent [308]. Polymer bound benzeneseleninic acid [308] reactivity differs from that of benzeneseleninic acid or of BSA. Phenol oxidation did not occur with this reagent [308] while dihydroxy aromatic compounds are cleanly oxidized to quinones (Scheme 121a, b) and 1,5-dihydroxynaphthalene is slowly oxidized (235h, 80 °C) to juglone [305] (70% yield) readily isolated from the polymer which can be recycled. Other methods afford substantially lower yields or require the use of quite expensive and toxic reagents [@308].

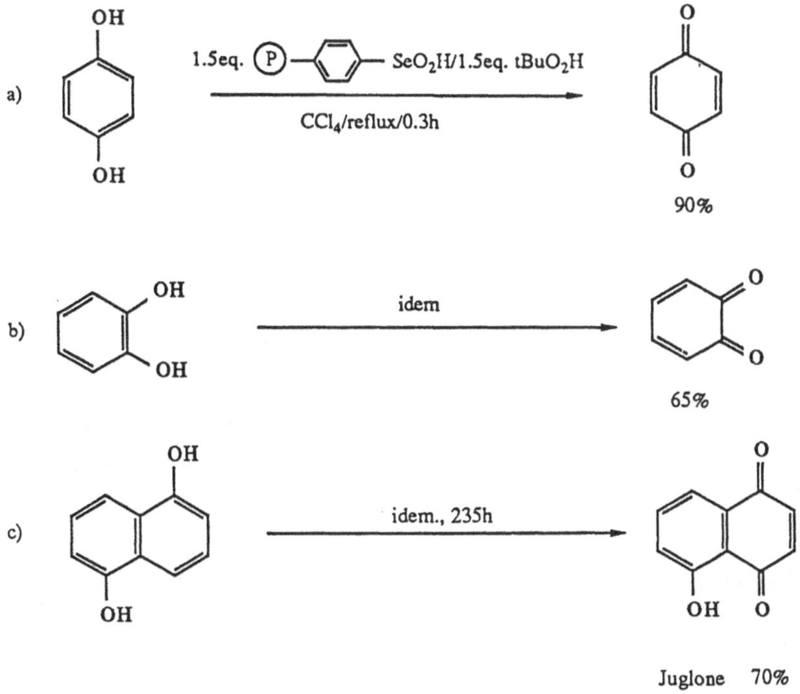

Scheme 121 [308]

6.5 Oxidative Conversion of Thiols to Disulfides, Sulfides to Sulfoxides and of Phosphines to Phosphinoxides

Benzeneseleninic acid reacts smoothly with thiols [310, 311] and depending on the conditions used produces a mixture of disulfides [311] and diselenide, or a mixture of disulfides and selenenosulfides [310] (Scheme 122). The reaction rate depends upon the structure of the thiol; it is much slower (10^{-4}) with t-butyl thiol than with n-butyl thiol [310]. The reaction is thought to proceed through thiolseleninates to give selenosulfides as the first isolable products (Scheme 122a), which on standing lead to disulfides and diselenides [310] (Scheme 122b).

a) $\text{PhSeO}_2\text{H} + 5\text{RSH} \xrightarrow[\text{HClO}_4/20°\text{C}]{\text{Aq. Dioxane}} \left[\begin{array}{c} \text{O} \\ \| \\ \text{PhSeSR} \end{array} \right] \longrightarrow \text{PhSeSR} + (\text{RSSR} + \text{H}_2\text{O})$

R = nBu	0.2h	93%	[310]
R = tBu	1h	87%	[310]

b) $6\ \text{RSH} + 2\ \text{ArSeO}_2\text{H} \xrightarrow[24h]{C_6H_6} 3\ \text{RSSR} + \text{ArSeSeAr} + 4\text{H}_2\text{O}$

R	Ar			
4-MePh	2-NO$_2$Ph	93.5%	94%	[311]
4-MePh	4-Cl-2-NO$_2$Ph	100%	100%	[311]

Scheme 122

p-Chloro benzeneseleninic acid in the presence of p-toluene sulfonic acid oxidizes di-n-butylsulfide and dibenzylsulfide to the corresponding sulfoxides [312] quantitatively. The presence of a strong acid was found to be essential for the success of the reaction [312]. This is not the case with triphenyl phosphine which is oxidized [312] by the same reagent without the requirement of a strong acid, although the reaction is more rapid in its presence [312].

6.6 Oxidation of Amines

Benzeneseleninic acid and its anhydride efficiently oxidize amines and lead to dehydrogenated products [313]. Certain primary amines, not susceptible to form enamines [314] (such as 2-phenylethyl amine) are converted to ketones in high yields (Scheme 123). Benzylamine however produces benzonitrile in low yield

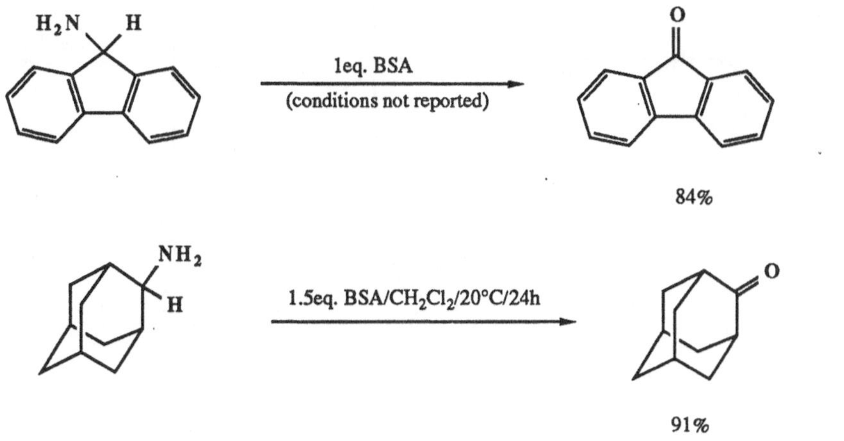

Scheme 123 [314]

[314] (26%) when 1 mol- equiv. of BSA is used and in much higher yields [314] (96%) with 2 mol. equiv. of BSA.

4-Phenylbutylamine reacts [313] with BSA and produces a mixture of 4-phenylbutyronitrile, 4-phenyl-2,2-bis(phenylseleno)butanal and diphenyl diselenide (Scheme 124a, b). Variation of the conditions enables the reactions to be directed [313] to a certain extent either towards nitrile (slow addition of the amine to a solution of BSA in pyridine at 70 °C) or towards the doubly selenylated aldehyde (inverse addition of the anhydride on the amine). On the other hand the formation of a doubly selenylated nitrile from phenylalanine has been explained [313] by a decarboxylative elimination (Scheme 124c).

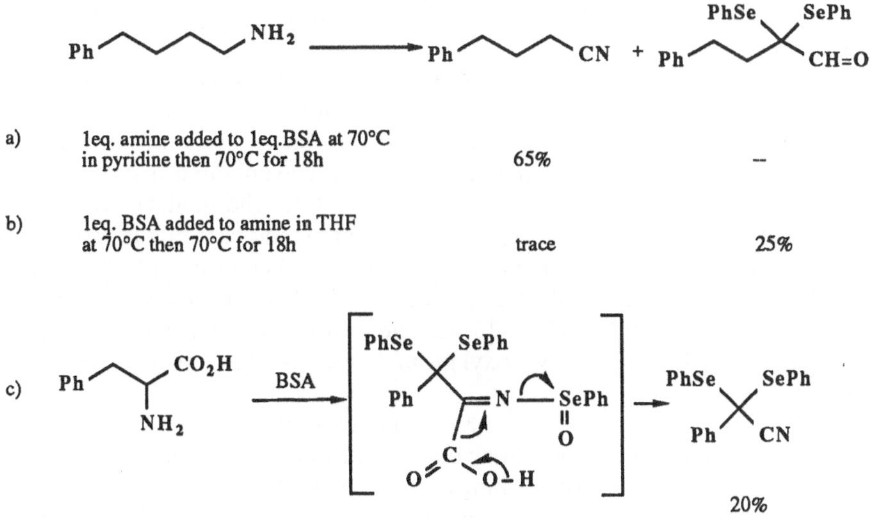

a) 1eq. amine added to 1eq.BSA at 70°C
    in pyridine then 70°C for 18h   65%   —

b) 1eq. BSA added to amine in THF
    at 70°C then 70°C for 18h   trace   25%

c)

20%

Scheme 124 [313]

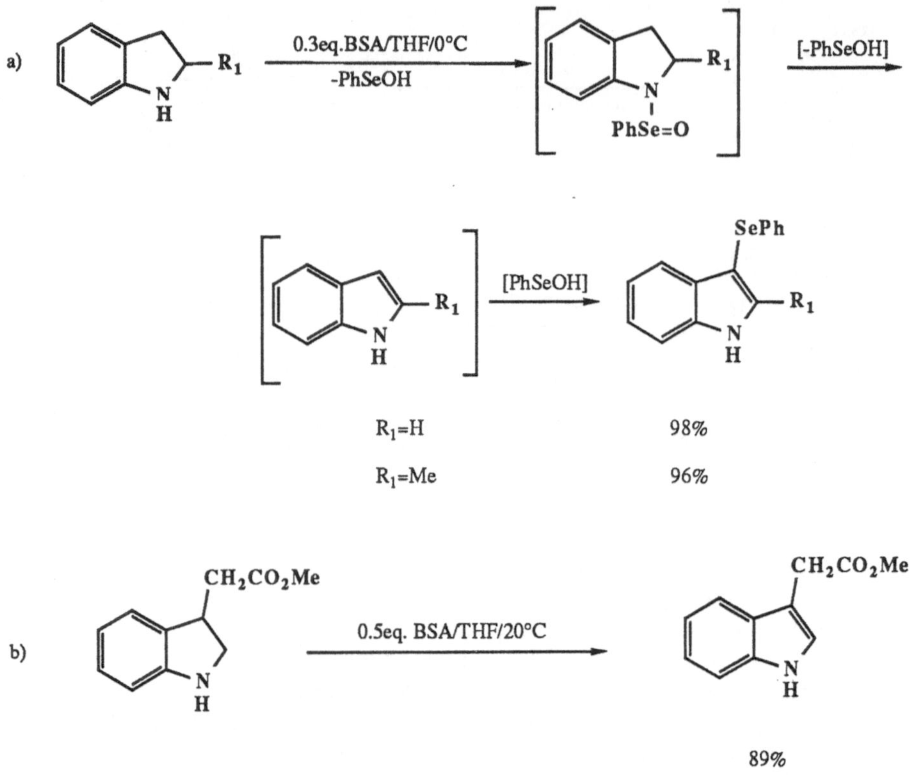

6. Reactions Involving Benzeneseleninic Anhydride (BSA) and Related Reagents

a)

b)

CH$_2$CO$_2$Me

0.5eq. BSA/THF/20°C

89%

Scheme 125 [315]

Indolines (Scheme 125) give the corresponding indoles [313, 315] when the beta position is substituted and phenylseleno indoles when this position is unsubstituted [313, 315]. Part of the indoles produced in the process reacts with the electrophilic Se(II) species (PhSeOH) concomitantly formed and lead to the 3-phenylselenylated derivative. Although the latter can be reduced back to indoles with nickel boride [315, 316] the yields are far from quantitative.

It was later suggested [317] that addition of sacrificial enamine as scavanger of the Se(II) species should allow a better yield of indole. Indole itself was found [317] to react fast enough to protect more complicated indoles from functionalisation by Se(II) (Scheme 126). The novel procedure has been successfully applied to the synthesis of various indole derivatives [316, 317] and used as a key step for the total synthesis of ergot alkaloids [316]. The transformation leaves untouched an olefinic linkage [316] or an hydroxyl group [316] linked to a benzylic position [317] (Scheme 126a).

Other secondary amines are also dehydrogenated [@318] with benzeneseleninic anhydride or acid under mild conditions. When the reactions are performed in the presence of sodium cyanide or of trimethylsilyl cyanide α-cyanoamines [318] (Scheme 127), which can be regarded as protected imines or as a source of α-amino acids [318], are produced in good yields.

92

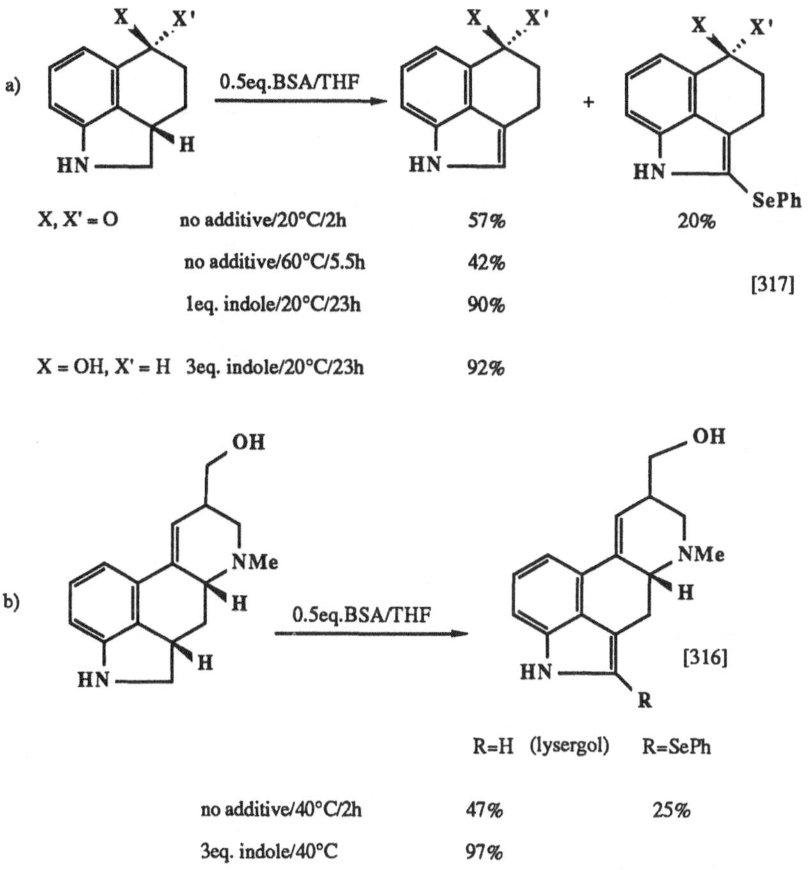

a)			
X, X' = O	no additive/20°C/2h	57%	20%
	no additive/60°C/5.5h	42%	
	1eq. indole/20°C/23h	90%	[317]
X = OH, X' = H	3eq. indole/20°C/23h	92%	

	R=H (lysergol)	R=SePh
no additive/40°C/2h	47%	25%
3eq. indole/40°C	97%	

Scheme 126

6.7 Oxidation of Hydrazines

BSA reacts vigorously with hydrazines at room temperature to form products determined by the position and nature of the substituents in the substrate [298, 319]. Thus hydrazine hydrate generates diimide which has been used [298, 320] for the in situ reduction of azobenzene to hydrazo benzene (100% yield in DMF) and of cinnamic acid to hydrocinnamic acid (90% yield in pyridine). 1,2-Dialkyl or 1,2-diaryl hydrazines cleanly gave the corresponding azo compound [298, 319, 320] (Scheme 128a) whereas monosubstituted aryl and acyl hydrazines lead mainly [298, 320] to the corresponding selenides or selenolesters (Scheme 128b, c).

6. Reactions Involving Benzeneseleninic Anhydride (BSA) and Related Reagents

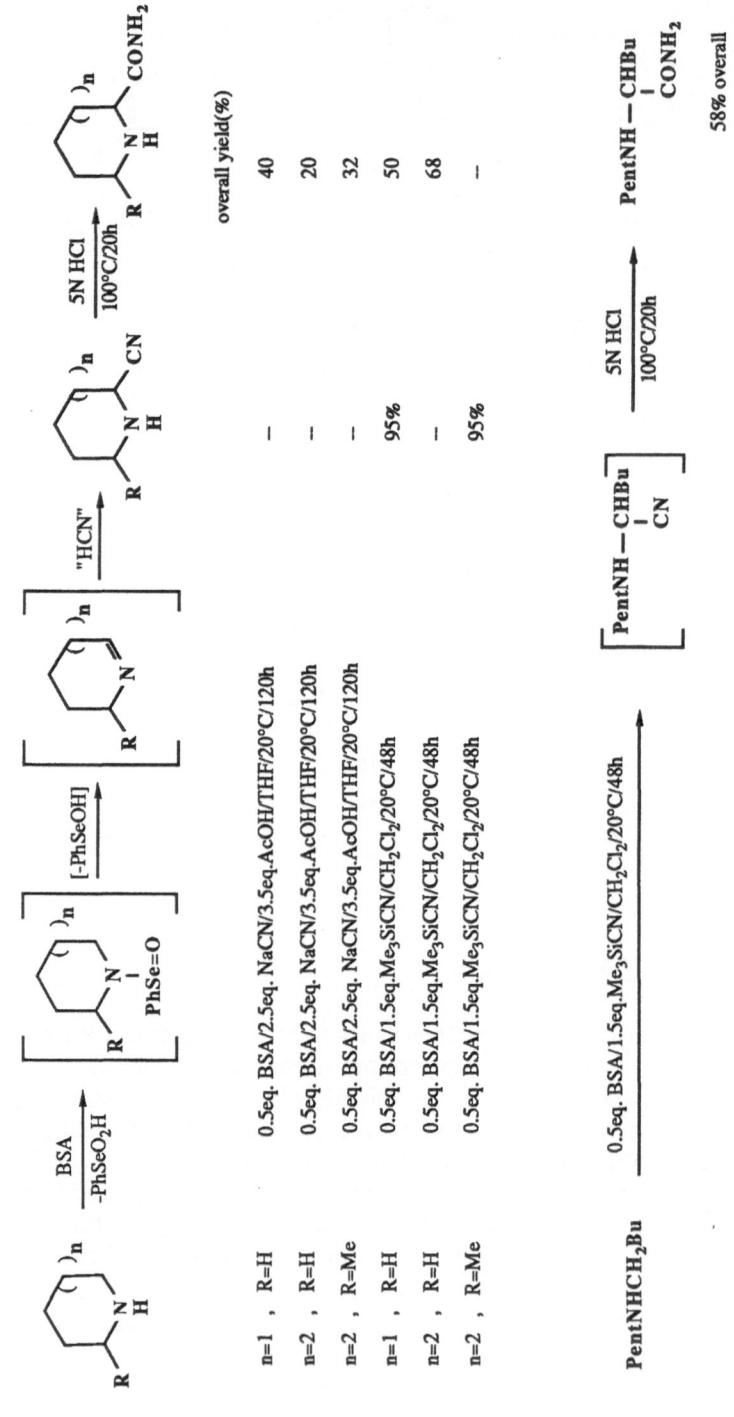

Scheme 127 [318]

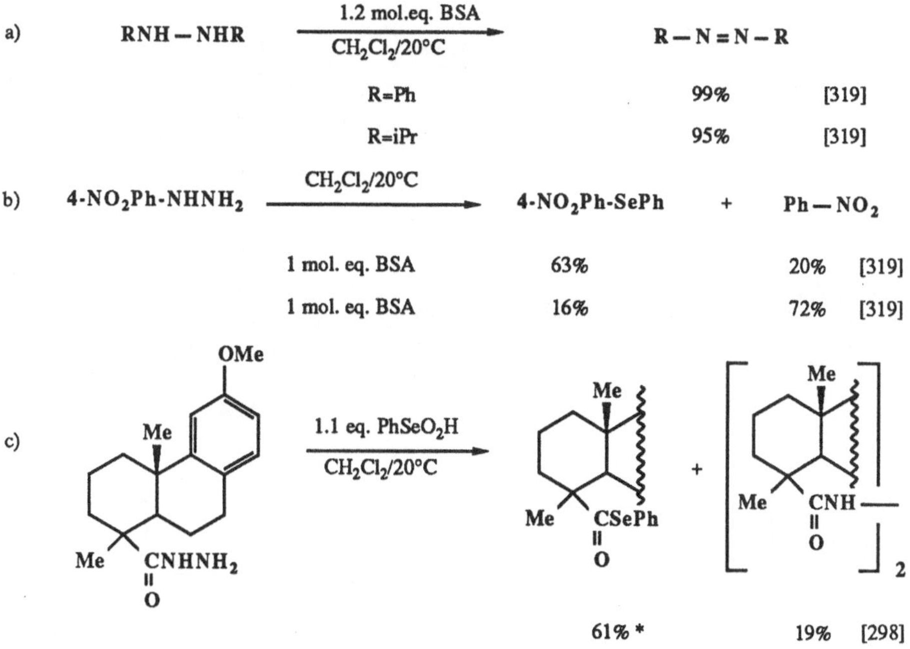

a)   RNH — NHR $\xrightarrow[\text{CH}_2\text{Cl}_2/20°\text{C}]{\text{1.2 mol.eq. BSA}}$ R — N = N — R

R=Ph	99%	[319]
R=iPr	95%	[319]

b)   4-NO$_2$Ph-NHNH$_2$ $\xrightarrow{\text{CH}_2\text{Cl}_2/20°\text{C}}$ 4-NO$_2$Ph-SePh   +   Ph — NO$_2$

1 mol. eq. BSA	63%	20%	[319]
1 mol. eq. BSA	16%	72%	[319]

c)

$\xrightarrow[\text{CH}_2\text{Cl}_2/20°\text{C}]{\text{1.1 eq. PhSeO}_2\text{H}}$

61% *   19%   [298]

* Yield increased to 71% with added PhSeSePh in reaction

Scheme 128

6.8 Oxidation of Hydroxylamines

Hydroxylamines are also oxidized with BSA immediatly to nitroso derivatives [319] (Scheme 129). However the corresponding *O*-methyl hydroxylamines are particularly unreactive, even under vigorous conditions [319].

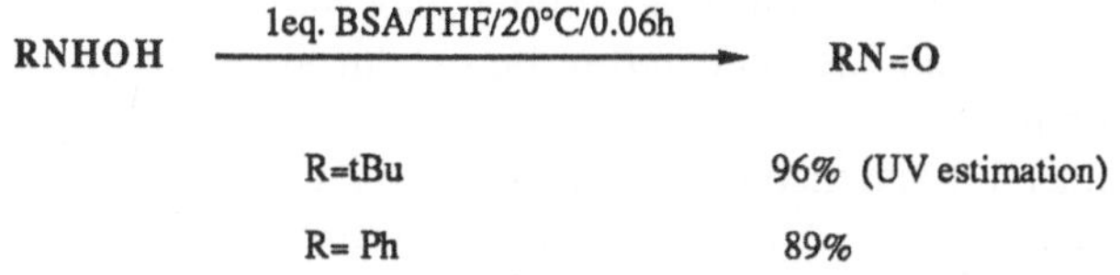

RNHOH $\xrightarrow{\text{1eq. BSA/THF/20°C/0.06h}}$ RN=O

R=tBu	96% (UV estimation)
R= Ph	89%

Scheme 129 [319]

6.9 Oxidation of the Alkyl Chain of Aromatic and Heteroaromatic Compounds

Benzylic oxidation of a variety of aromatic and heteroaromatic compounds [296, 321] (Scheme 130) can be achieved with BSA and lead to the corresponding carbonyl compounds. For simple substrates the oxidation normally proceeds,

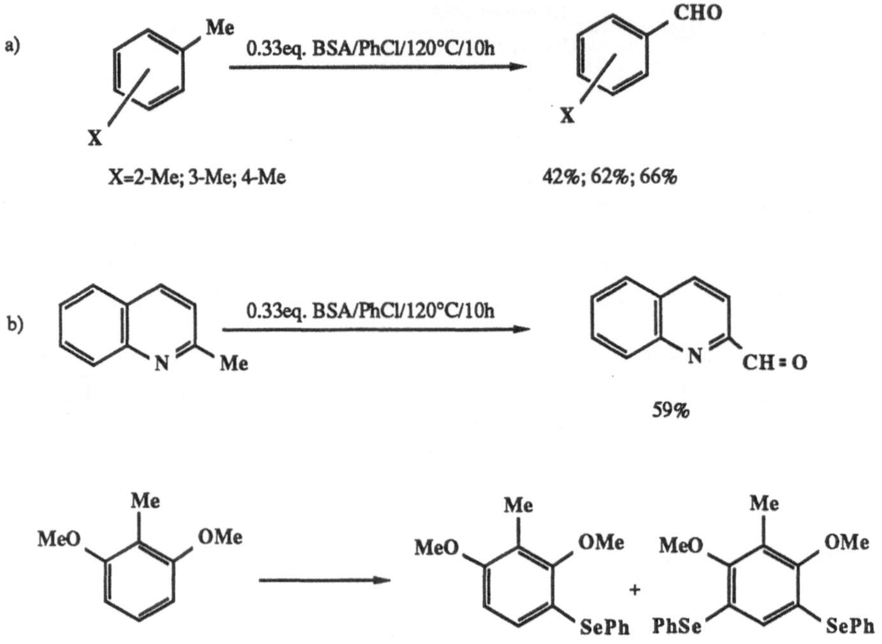

6. Reactions Involving Benzeneseleninic Anhydride (BSA) and Related Reagents

a)

X=2-Me; 3-Me; 4-Me

42%; 62%; 66%

b)

59%

c)	0.33eq. BSA/PhCl/110°C/17h	49%	...
d)	1eq. BSA/PhCl/110°C/17h	85%	5%
e)	0.33eq. BSA/0.67eq.(PhSe)$_2$/120°C/17h	91%	0.8%

Scheme 130 [296,321]

with 0.33 mol. equiv. of the reagent, either neat or in chlorobenzene at 100–130 °C. Greater quantities of BSA must be avoided especially when methyl groups are involved since the aldehyde expected to be produced can be further oxidized to the corresponding acid.

Toluene (120 °C, 1 week), ethyl benzene (120 °C, 4 days) and p-nitro toluene are not readily oxidized [269, 321]. 3-Methyl-pyridine failed to react [321] but xylenes (Scheme 130a), 1-methyl naphthalene, 9,10-dimethyl anthracene, 2-methyl quinoline and 3-methyl isoquinoline are efficiently oxidized [296, 321] (Scheme 130b). Aromatic rings susceptible to electrophilic attack, such as p-methoxy toluene, and 1,3-dimethoxy 2-methyl benzene [321], lead to selenylated products (Scheme 130c–e). These have been expected to be formed [321] on reaction of an electrophilic "reduced form of the anhydride". Consequently the yields of these selenides have been dramatically improved by perfoming the reactions with a 1:2 mixture of benzeneseleninic anhydride and diphenyl diselenide which is thought to produce benzeneselenenic anhydride (PhSeOSePh) [321] (Scheme 130e).

96

6.10 Regeneration of Carbonyl Compounds

6.10.1 From Thioketones, Hydrazones and Oximes

BSA is an effective reagent for the mild regeneration of the carbonyl group of aldehydes and ketones from thioketones [322, 323], semicarbazones, oximes and hydrazones [319, 324].

The first reaction is particularly efficient with thioketones which do not bear an alpha methylene group [322, 323] (Scheme 131a). In the absence of substitutents in the α position however the yield in ketone is lowered due to the formation of products arising from reaction of the ketone with the selenenyl species concomitantly produced [322, 323] (Scheme 131b).

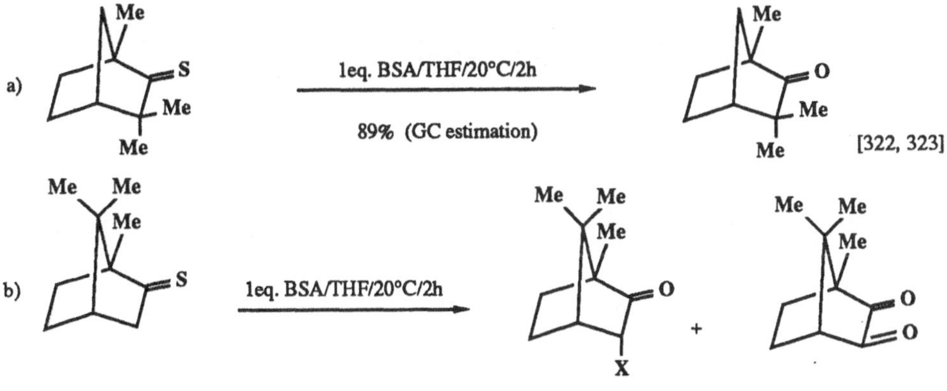

a) 1eq. BSA/THF/20°C/2h

89% (GC estimation)

[322, 323]

b) 1eq. BSA/THF/20°C/2h

+

X

Scheme 131   X = H 9% , X = Se(O)Ph 36%   54%   [322]

BSA has also been used successfully for the regeneration of carbonyl compounds from their semicarbazones and oximes [319, 324] (Scheme 132a). Methoxyoximes, however do not react even under more vigorous conditions [319] (Scheme 132a). Tosylhydrazones have been efficiently transformed to the parent aldehydes and ketones with BSA [319, 321] (Scheme 132a–d). The reaction occurs under particularly mild conditions with tosylhydrazones derived from aldehydes (Scheme 132c, d). Similar reactions take place with arylhydrazones such as phenylhydrazones, 4-nitrophenylhydrazones and 2,4-dinitrophenylhydrazones) derived from ketones (Scheme 132b) but they do not occur with N,N-dimethylhydrazones [319, 321]. Particularly interesting is the observation that both the p-nitrophenylhydrazone of cholesta-1,4-dien-3-one and cholesta-1,4,6-trien-3-one (directly prepared by a dehydrogenation procedure of cholestanone p-nitrophenyl hydrazone) are cleanly transformed to the parent unsaturated ketones in 73 and 74% yield respectively on reaction with BSA [@319]. This transformation is remarkable since many other methods had failed [319]. It does not occur for example with benzeneseleninic acid or with SeO$_2$ (Section 9.6) [319].

The reaction takes a different course with aryl hydrazones derived from aldehydes since acylazo derivatives are instead produced (Scheme 132 e compare scheme 133 d. The use of BSA offers advantages over other reagents such as N-bromosuccimide which allows [319] the same transformation but often in lower yields [@319].

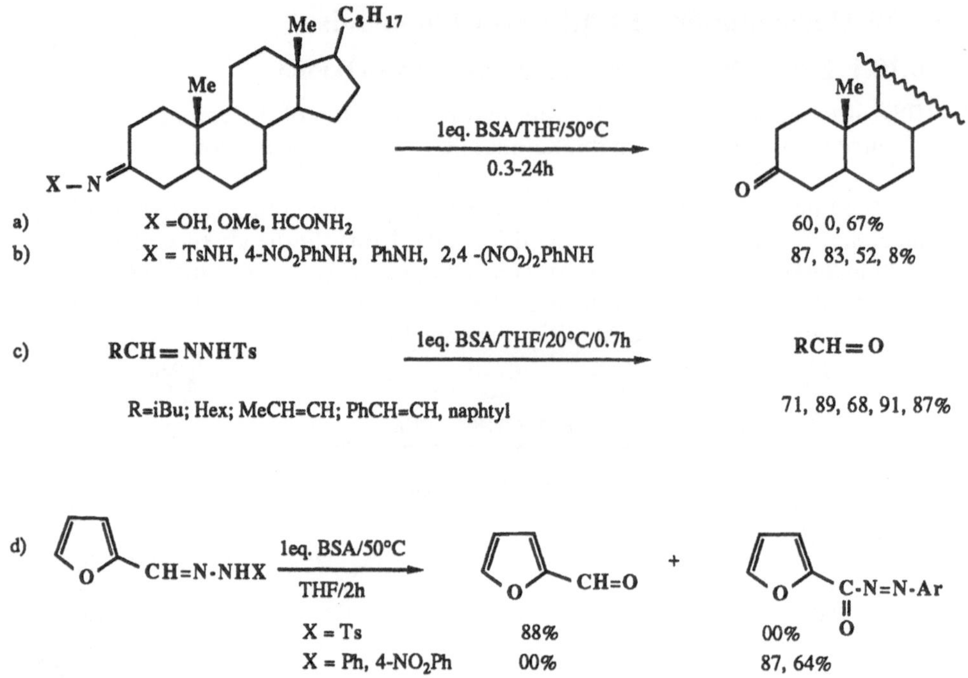

6. Reactions Involving Benzeneseleninic Anhydride (BSA) and Related Reagents

a)   X = OH, OMe, HCONH$_2$ 60, 0, 67%
b)   X = TsNH, 4-NO$_2$PhNH, PhNH, 2,4-(NO$_2$)$_2$PhNH   87, 83, 52, 8%

c)   RCH=NNHTs $\xrightarrow{\text{1eq. BSA/THF/20°C/0.7h}}$ RCH=O

            R=iBu; Hex; MeCH=CH; PhCH=CH, naphtyl   71, 89, 68, 91, 87%

d)

Scheme 132 [319, 324]

All these results have been rationalized as described in the Scheme 133. The initial reaction ov benzeneseleninic anhydride takes place on nitrogen to form a species which subsequently undergoes a [2, 3] sigmatropic rearrangement followed by a fragmentation reaction leading to nitrogen and the expected carbonyl compoud or by the elimination of benzeneselenol leading to the acylazo derivative. With aldehydes the leaving group ability of X in the initial compound directs the process. Good leaving groups give back aldehydes poor leaving groups fragment to acylazo compounds.

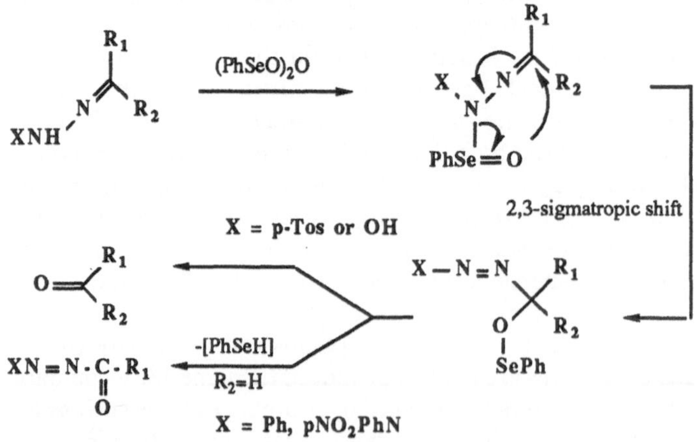

Scheme 133 [319]

98

6.10.2 From Thioacetals and Selenoacetals

BSA is particularly efficient for the synthesis of aldehydes and ketones from 1,3-dithiolan derivatives [325, 326, @327] (Scheme 134) and 1,3-dithians [325, 327]

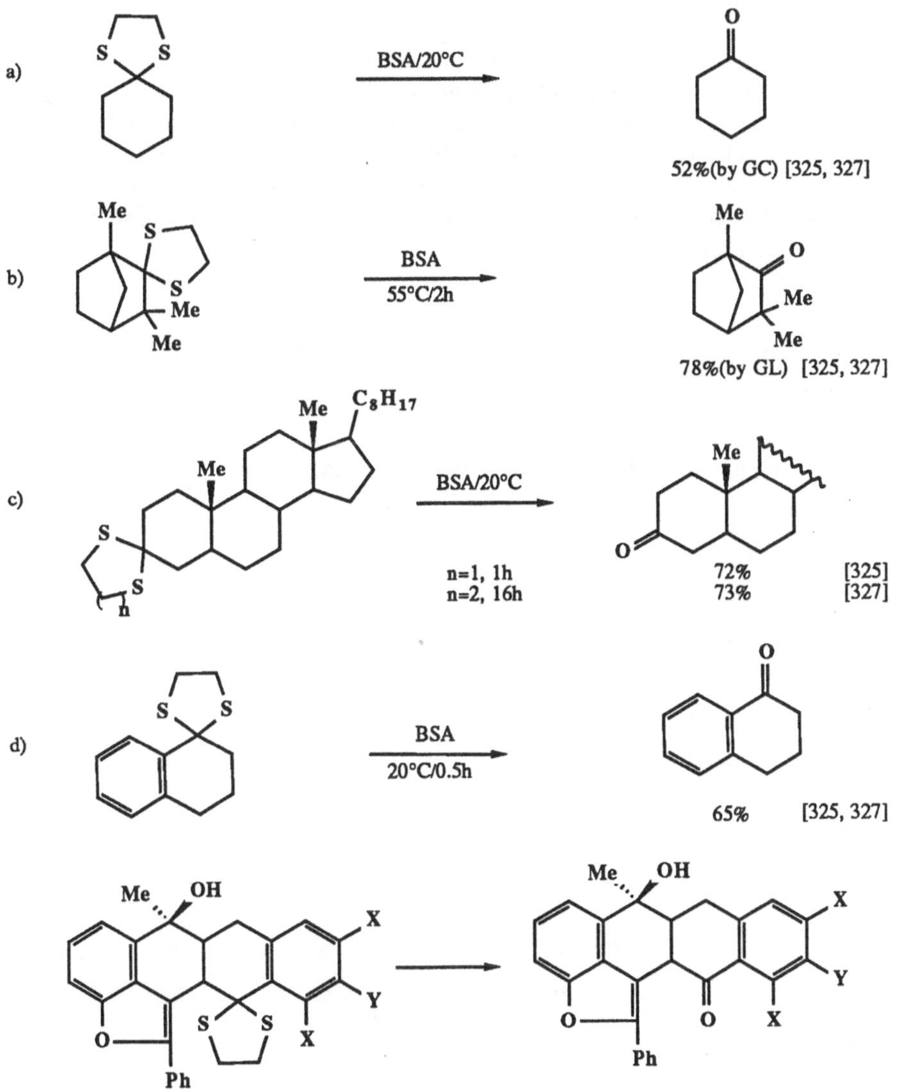

a)

BSA/20°C

52%(by GC) [325, 327]

b)

BSA
55°C/2h

78%(by GL) [325, 327]

c)

BSA/20°C

n=1, 1h
n=2, 16h

72%   [325]
73%   [327]

d)

BSA
20°C/0.5h

65%   [325, 327]

e)	X=Y=H	1eq. BSA/THF/Pyridine/40°C/50h	78%	[325, 326]
f)	X=OH, Y=CO₂Me	1eq. BSA/THF/20°C/16h	64%	[326]
g)	X=OH, Y= CONH₂	1eq. BSA/THF/20°C/16h	00%	[326]

Scheme 134

(Scheme 134c). This reaction has been successfully used for the deprotection of potential precusors of tetracyclines [326, 327] where other common procedures failed [@325]. In one exceptional case (Scheme 134f) the reaction proceeded in the presence of a free diphenolic ester [326] (Scheme 134f to g).

Methylseleno- and phenylselenoacetals [@328] are also efficiently deprotected with 1 mol. equiv. of BSA in THF (Scheme 135). The corresponding carbonyl compounds are formed in high yields and under mild conditions in the case of phenyl and methylselenoacetals derived from aldehydes and of methylselenoacetals derived from acyclic ketones (Scheme 135a, b). Methylselenoacetals derived from cyclic ketones and phenylselenoacetals derived from ketones also lead to ketones [328] but appreciable amounts of vinylselenides are also recovered (Scheme 135c–f). This side reaction can be minimized if the reaction is performed, for short reaction times [328], at 60 °C instead of 20 °C (Scheme 135 compare e to f and c to d).

a) nDec—CH(SeR)$_2$ $\xrightarrow{\text{1eq. BSA/THF/60°C/6h}}$ nDec–CH:O

R=Me,Ph   81, 84%

Nonyl—C(SeR)$_2$ (with Me substituent) $\xrightarrow{\text{1eq. BSA/THF}}$ Nonyl–C(=O)–Me   +   Vinyl selenides

			Ketone	Vinyl selenides
b)	R=Me	20°C/3h	81%	0%
c)	R=Ph	20°C/2.5h	77%	23%
d)	R=Ph	60°C/0.5h	84%	15%

MeSe   SeMe (cyclohexane with tBu) $\xrightarrow{\text{1eq. BSA/THF}}$ cyclohexanone with tBu + MeSe cyclohexene with tBu

	conditions	ketone	vinyl selenide
e)	20°C/4h	35%	24%
f)	60°C/1h	60%	15%

Scheme 135 [328]

6.10.3 From Xanthates, Thioesters, Thiocarbonates, Thioamides and Their Telluro Analogues

Xanthates [322, 323], thioesters [322, 323], thiocarbonates [322, 323], thioamides [322, 323] and tellurocarbonyloxy compounds [329] also react with BSA. The reaction takes place at room temperature and gives the parent oxo carbonyl

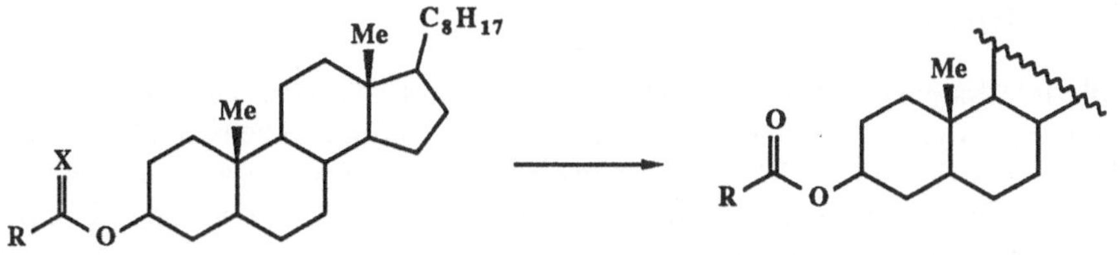

	R	X	Conditions		Yield(%)	
a)	R=SMe	X=S	1eq. BSA/THF/20°C/12h		67	[322, 323]
b)	R=Ph	X=S	"	3.5h	69	[322, 323]
c)	R=Ph	X=Se	"	0.6h	83	[322, 323]
d)	R=tBu	X=Te	"	1h	96	[329]
e)	R=Ph	X=Se	2.1eq. SeO₂/20°C/72h		97	[322]

f)

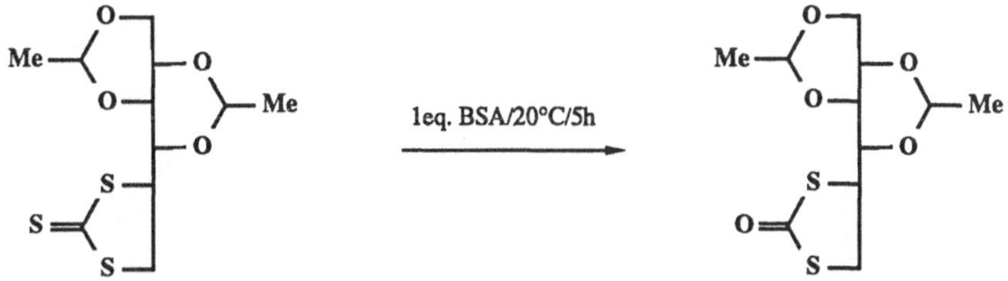

1eq. BSA/20°C/5h

64%   [322, 323]

Scheme 136

derivative [322] in all cases in high yields and with greater efficiency than SeO$_2$ (Scheme 136a–d, f compare also e to c).

6.11 Synthesis of α-Selenocarbonyl Compounds from BSA and:

6.11.1 Vinylsulfides and Vinylselenides

Although the lack of reactivity of BSA towards isolated double bonds has been noticed, it was found that vinyl sulfides and vinyl selenides [330] (Scheme 137a, b) react under mild conditions with BSA to produce α-selenocarbonyl compounds in good yields.

6.11.2 Allylalcohols

The olefinic system present in O-silylated allyl alcohols also reacts with BSA [331, 332] but diphenyl diselenide must be present in the medium. The reactive species in this case is thought to be phenylselenenic anhydride which adds across the carbon-carbon double bond and leads regiospecifically to the corresponding β-silyloxy-α-phenylselenocarbonyl compounds [331, 332] (Scheme 137c, d).

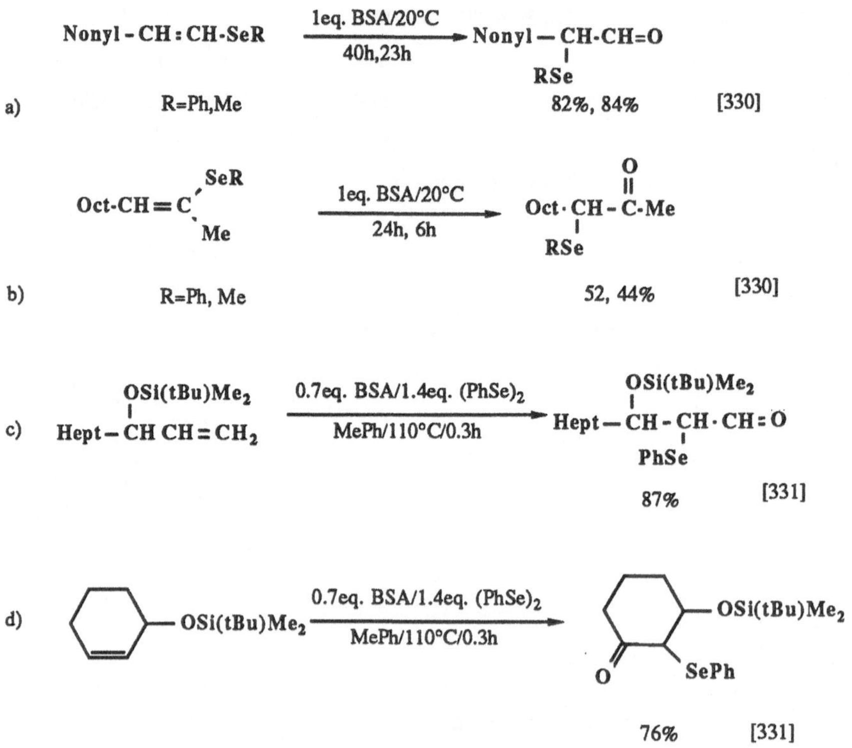

a)

$$\text{Nonyl-CH:CH-SeR} \xrightarrow[\text{40h,23h}]{\text{1eq. BSA/20°C}} \text{Nonyl}-\underset{\underset{\text{RSe}}{|}}{\text{CH}}\text{-CH=O}$$

R=Ph,Me   82%, 84%   [330]

b)

$$\text{Oct-CH=C}\overset{\text{SeR}}{\underset{\text{Me}}{\diagup}} \xrightarrow[\text{24h, 6h}]{\text{1eq. BSA/20°C}} \text{Oct}\cdot\underset{\underset{\text{RSe}}{|}}{\text{CH}}\text{-}\overset{\overset{\text{O}}{\|}}{\text{C}}\text{-Me}$$

R=Ph, Me   52, 44%   [330]

c)

$$\text{Hept}-\underset{\underset{\text{OSi(tBu)Me}_2}{|}}{\text{CH}}\text{ CH}=\text{CH}_2 \xrightarrow[\text{MePh/110°C/0.3h}]{\text{0.7eq. BSA/1.4eq. (PhSe)}_2} \text{Hept}-\underset{\underset{\text{OSi(tBu)Me}_2}{|}}{\text{CH}}\text{-}\underset{\underset{\text{PhSe}}{|}}{\text{CH}}\cdot\text{CH=O}$$

87%   [331]

d)

$$\xrightarrow[\text{MePh/110°C/0.3h}]{\text{0.7eq. BSA/1.4eq. (PhSe)}_2}$$

76%   [331]

Scheme 137

6.12 Synthesis of Allylalcohols from Allylsilanes

It was also found [333] that the allyl silane **36** reacts smoothly at 20 °C with BSA in CH_2Cl_2 containing a catalytic amount of $BF_3.OEt_2$ and gives the α-acetoxy allylic alcohol via a [2,3] sigmatropic rearrangement of the intermediate **37** (Scheme 138). This proved an efficient route to **38** a regioisomer of **35.** Both compounds are available from 7-trimethylsilylsubstituted bicyclo [4.1.0] carbinols **34** (Scheme 138).

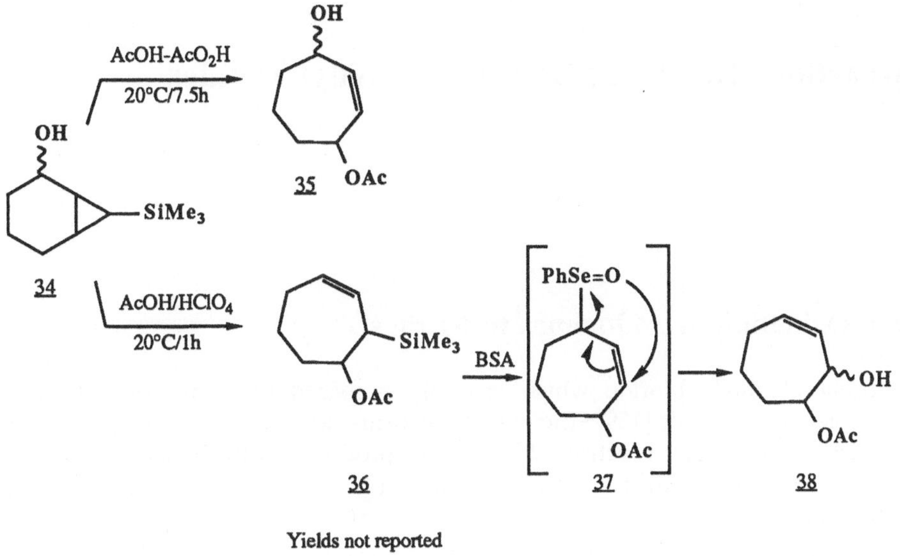

Yields not reported

Scheme 138 [333]

Chapter 7

Reactions Involving Benzeneseleninyl Halides

7.1 Oxidation of Aldoximes to Nitriles

Benzeneseleninyl chloride which is easily prepared by ozonolysis of benzeneselenenyl chloride [159] efficiently transforms aldoximes to nitriles [251, 334]
(Scheme 139b). The reaction which also proceeds with benzeneselenenyl
chloride (Scheme 139a) is faster with aromatic (25 °C, 0,25 h) than with aliphatic
compounds [334].

$$RCH = NOH \longrightarrow RC \equiv N$$

a) R=Ph, 4-MeOPh, Hex   PhSeCl/CHCl$_3$/25°C   86, 93, 83%

b) R=Ph, 4-MeOPh, Hex   PhSeOCl/CHCl$_3$/25°C   85, 88, 81%

Scheme 139 [334]

7.2 Oxidation of Amines

Benzeneseleninyl chloride reacts with amines but whereas it produces a complex
mixture of products with those amines capable of forming enamines [335], 1,1-

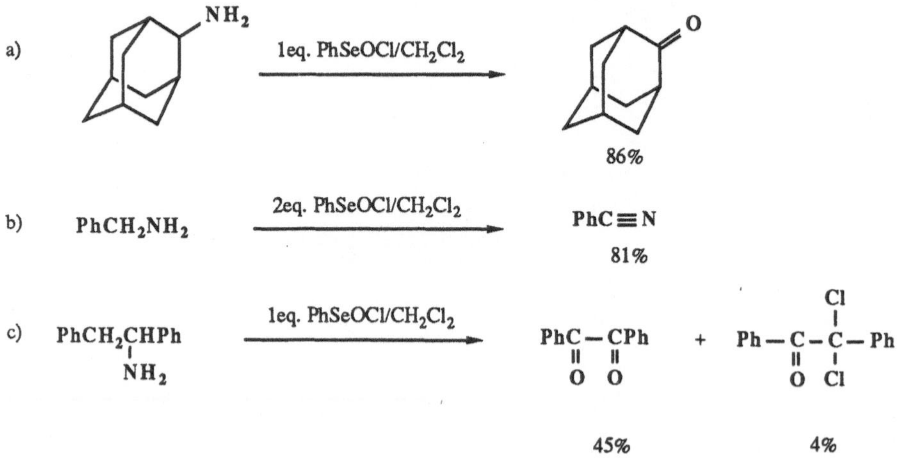

a)   1eq. PhSeOCl/CH$_2$Cl$_2$ 86%

b)   PhCH$_2$NH$_2$ 2eq. PhSeOCl/CH$_2$Cl$_2$ PhC≡N   81%

c)   PhCH$_2$CHPh
            |
           NH$_2$ 1eq. PhSeOCl/CH$_2$Cl$_2$ PhC—CPh   +   Ph—C—C—Ph
                                                            ‖ ‖   ‖ |
                                                            O O   O Cl
                                                            45%   4%

Scheme 140 [335]

104

diphenyl methyl amine, 9-aminofluorene and 2-aminoadamantane lead almost quantitatively to benzophenone, fluorenone and adamantanone [335] respectively (Scheme 140a). Closely related results have been reported when benzeneseleninic anhydride was used [314] instead of benzeneseleninyl chloride (see Sect. 6.6). However on reaction with benzeneseleninyl chloride benzylamine produces benzonitrile [335] (Scheme 140b) and 1,2-diphenyl ethyl amine does not lead to deoxybenzoin but gives [335] a mixture of bibenzil and of α,α-dichloro-benzoin, arising probably via a Pummerer type pathway [335] (Scheme 140c) in rather low yields.

7.3 Dehydrogenation of Carbonyl Compounds to Enones

Benzeneseleninyl chloride permits the C-seleninylation of enolates of ketones and of other carbonyl compounds [159]. The resulting compounds undergo selenenic acid elimination in situ and produce α,β-unsaturated carbonyl compounds in reasonable yields (Scheme 141) but sometimes lower than those obtained by the two step procedure (Procedure A). This involves the α-selenenylation of the ketone or its enolate, and the oxidation of the resulting α-selenoketone leading to the corresponding selenoxide which then eliminates [159]. (These processes will be discussed in Volume II). The use of benzeneseleninyl chloride is recommended when the selenide oxidation required in the Procedure A fails because of competing or preferential oxidation elsewhere in the molecule (for related results using BSA see Sect. 6.4.1).

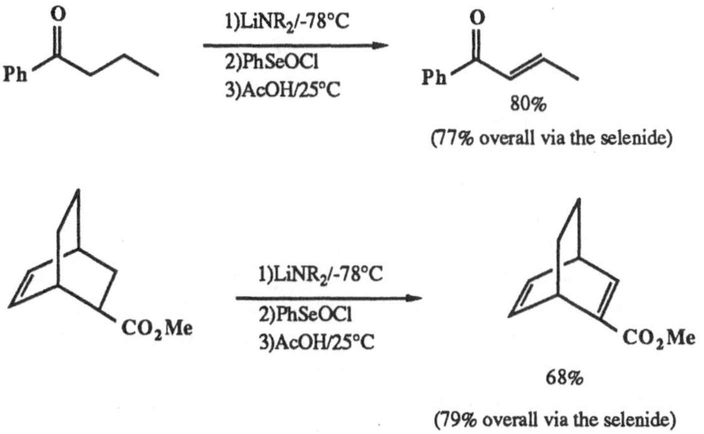

Scheme 141 [159]

Chapter 8

Reactions Involving Perseleninic Acids

Reaction of arseneseleninic acids with hydrogen peroxide produces new species which possess oxidative properties different from thoses expected from seleninic acid and hydrogen peroxide when used alone. The new species which have been described as perseleninic acids [but may also have a peroxo structure, Scheme 142] have been successfully used for the epoxidation of olefins as well as for the Baeyer-Villiger oxidation of cyclanones, possess a reactivity different from that of the corresponding benzeneselenonic acids [312].

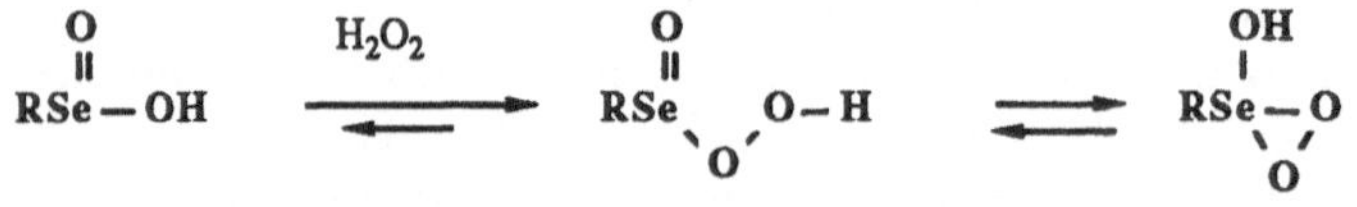

$$
\underset{\text{RSe}-\text{OH}}{\overset{\overset{\text{O}}{\parallel}}{}} \quad \xrightarrow{\text{H}_2\text{O}_2} \quad \underset{\text{RSe}}{\overset{\overset{\text{O}}{\parallel}}{}}\underset{\text{O}}{\overset{}{}}\text{O}-\text{H} \quad \rightleftharpoons \quad \underset{\text{RSe}-\text{O}}{\overset{\text{OH}}{\overset{|}{}}}\underset{\text{O}}{\overset{}{}}
$$

R=Ph; 4-NO$_2$Ph; 2,4-(NO$_2$)$_2$Ph; 2,4,6-(Me)$_3$Ph

Scheme 142 [338]

The first results in this field have been obtained simultaneously by five research groups [336–340] which used an excess (5 to 8 mol. equiv.) of hydrogen peroxide to perform the synthesis of α,β-unsaturated ketones [159, 336] from α-arylselenoketones and of allyl alcohols [337, 341–344] from β-hydroxyselenides (Volume II), being part of complex molecules bearing additional carbon-carbon double bonds. These were unexpectedly oxidized to the corresponding epoxides probably by a perseleninic acid species formed on oxidation, by the excess of hydrogen peroxide used, of the unstable selenenic acids generated as by-products [159, 338, 345, 346]. These side reactions can be suppressed if smaller quantities of hydrogen peroxide are used (i. e. 2 equiv. H$_2$O$_2$ are enough to permit both the selenoxide formation and the transformation of the selenenic acid concomitantly produced to the seleninic acid). The aptitude of the new reagent formed from seleninic acid and hydrogen peroxide to oxidize various functional groups was later tested.

8.1 Oxidation of Olefins to Epoxides

It was described by Grieco [336], Sharpless [338] and Reich [339] that areneseleninic acids/hydrogen peroxide mixtures are able to epoxidize carbon-

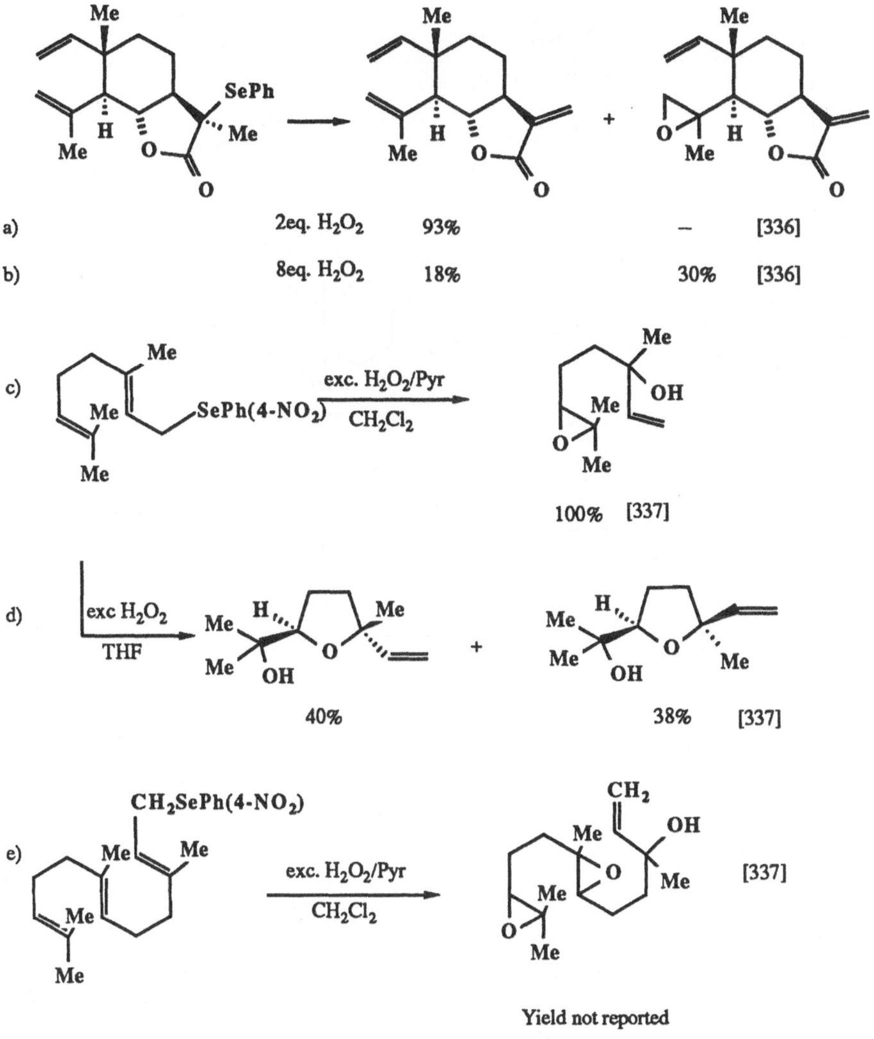

Scheme 143

carbon double bonds. Although several conditions have been tested which lead to closely related results, subtile differences have appeared from time to time (Schemes 143, 144, 145).

The following conditions have been successfully tested

(i) Benzeneseleninic acid (1.4 equiv.)/hydrogen peroxide (50%, 1.4 equiv.), the reaction being conducted in methanol in the presence of a phosphate buffer and silicagel which proved essential for the success of the reaction (Method A)

(ii) 2-nitrobenzeneseleninic acid [338] and 2,4-dinitrobenzeneseleninic acid

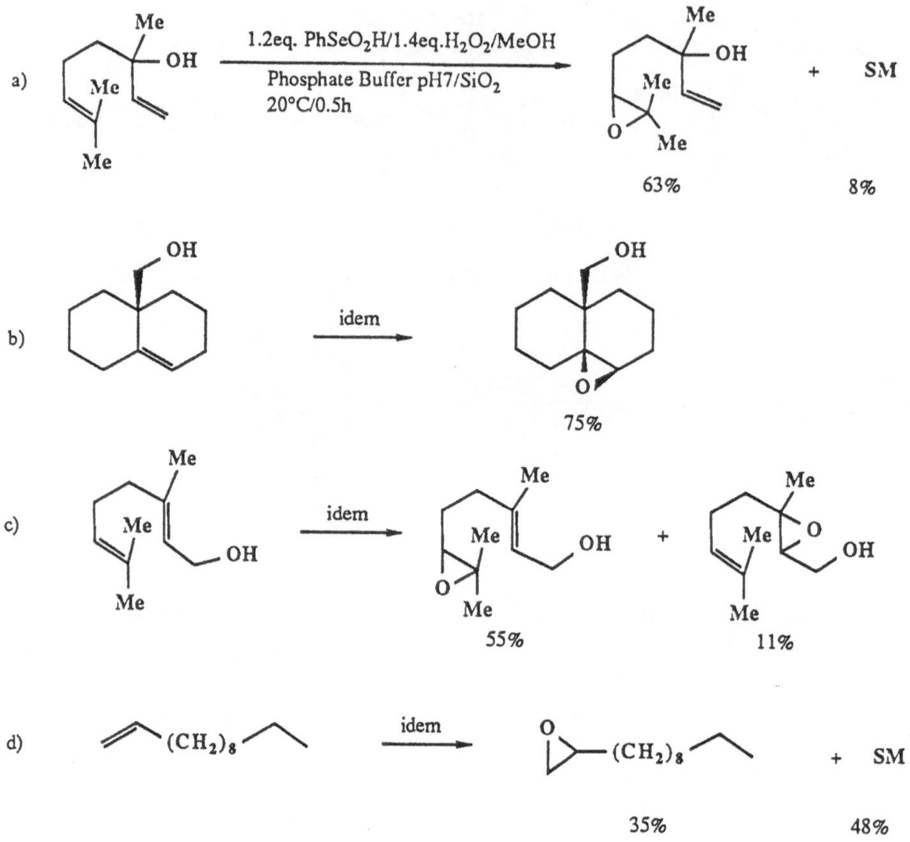

8. Reactions Involving Perseleninic Acids

Scheme 144 [336]

(0.1 equiv.)/hydrogen peroxide (30%, 2 equiv. and anhydrous MgSO$_4$) in methylene chloride (Method B)
(iii) 2-nitrobenzeneseleninic acid [338, 339] in catalytic amounts (0.01 equiv.)/ hydrogen peroxide (98%, 2 equiv.) in methylene chloride (Method C).

Several features are common to all these reagents and conditions. For example the order of reactivity of olefins parallels their nucleophilicity[1]. Thus using Condition A, disubstituted olefins are selectively oxidized in the presence of a monosubstituted one or an α,β-unsaturated carbonyl compound [336] (Scheme 143b, c) and although less reactive than alkylsubstituted olefins, allylalcohols have been oxidized to the corresponding epoxides under Condition A (Scheme 144), or Condition B [338] (Scheme 145d).

The simplicity of the procedures [308, 336, 338, 339] and the fact that H$_2$O$_2$ is less expensive than peracetic and perbenzoic acids (the most commonly used

---

[1] Reich however reported [339] a reversed order of reactivity.

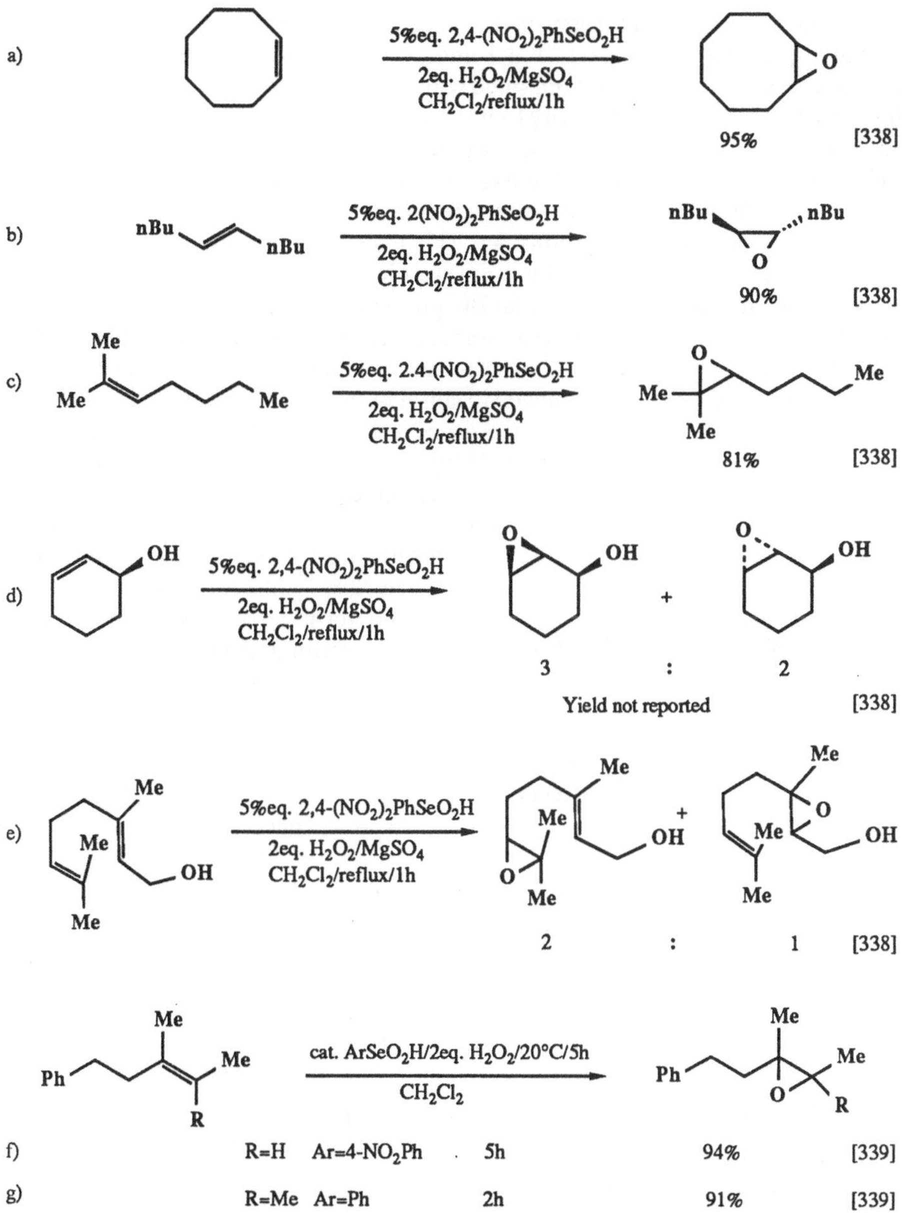

a)

5%eq. 2,4-(NO₂)₂PhSeO₂H

2eq. H₂O₂/MgSO₄
CH₂Cl₂/reflux/1h

95% [338]

b)

5%eq. 2(NO₂)₂PhSeO₂H

2eq. H₂O₂/MgSO₄
CH₂Cl₂/reflux/1h

90% [338]

c)

5%eq. 2.4-(NO₂)₂PhSeO₂H

2eq. H₂O₂/MgSO₄
CH₂Cl₂/reflux/1h

81% [338]

d)

5%eq. 2,4-(NO₂)₂PhSeO₂H

2eq. H₂O₂/MgSO₄
CH₂Cl₂/reflux/1h

3 : 2

Yield not reported [338]

e)

5%eq. 2,4-(NO₂)₂PhSeO₂H

2eq. H₂O₂/MgSO₄
CH₂Cl₂/reflux/1h

2 : 1 [338]

cat. ArSeO₂H/2eq. H₂O₂/20°C/5h

CH₂Cl₂

f) R=H Ar=4-NO₂Ph . 5h 94% [339]

g) R=Me Ar=Ph 2h 91% [339]

Scheme 145

reagents for large scale laboratory epoxidation) are the attractive aspects of these catalytic epoxidations [@338]. However, as shown in Scheme 145d, epoxidation of 2-cyclohexen-1-ol using Condition B occurs almost randomly giving a 3:2 mixture of the syn and anti epoxyalcohols. This contrasts with the highly stereoselective epoxidation of the same compound with peracids [349], or with transition metal/alkyl hydroperoxide reagents [347, 348]. Under Condition B geraniol (Scheme 145e) gave a 2:1 mixture of 6,7-oxido geraniol/2,3-oxido geraniol resembling the result obtained with carboxylic peracid [347] but contrasting with the high chemoselectivity found when alkyl hydroperoxide reagents [347] are used in the presence of vanadium or molybdenum catalysts. These two results differ from those reported in related cases by Grieco (compare Scheme 144b and c to Scheme 145d and e) who used stoichiometric quantities of benzeneseleninic acid in protic solvents (conditions A). This may be due to the difference in the solvents and in the seleninic acids employed in these two series of experiments.

It has been anticipated [338] that the putative active epoxidizing species (the areneperseleninic acid or its peroxo isomer) was chiral. Since this chirality resides at the selenium center this would place the assymetry one atom closer to the site of oxygen transfer than can be achieved in chiral peracids. However the epoxidation of prochiral allylic alcohols with chiral areneseleninic acid/H_2O_2 mixture unfortunately yields racemic epoxyalcohols [338].

Finally polystyrene bound benzeneseleninic acid/H_2O_2 reagent promotes epoxidation of olefins [308]. Under these conditions the reactivity of the olefins follows their nucleophilicity; i.e. tri- and tetrasubstituted olefins are the most reactive ones and produce [308] epoxides in 80% yield after 36 to 60 hrs of reaction (Scheme 146c, d). Disubstituted and terminal olefins require much longer reaction times (72 to 120 h) and diols are formed instead [308] in several instances (Scheme 146a, b).

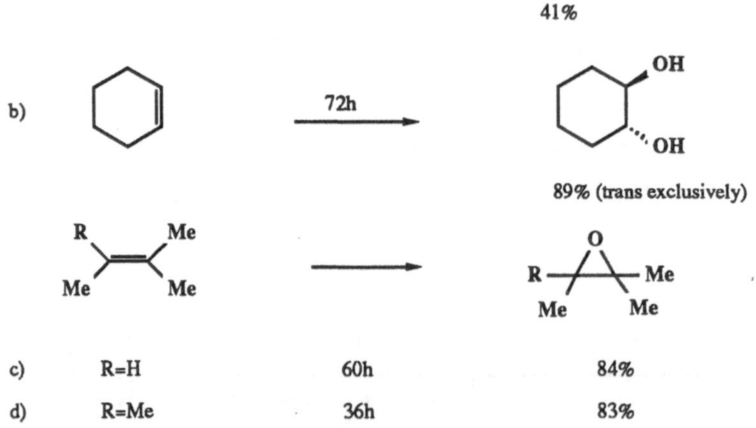

Scheme 146 [308]

8.2 Baeyer-Villiger Type Oxidation of Ketones to Esters and Lactones

The ability of benzeneseleninic acid/hydrogen peroxide mixture to promote Baeyer-Villiger reaction was demonstrated independently by Sharpless [88] and Williams [340] during a trial oxidation of α-phenylselenocyclopentanones to cyclopentenones (Scheme 147). They found that an α,β-unsaturated lactone is formed instead of the expected cyclopentenone. Williams [340] also observed that hydrogen peroxide alone was unable to perform the Baeyer-Villiger oxidation of a related cyclopentanone which did not bear the phenylseleno group. Furthermore, α,β-unsaturated lactones were not formed when ozone or periodic

a)

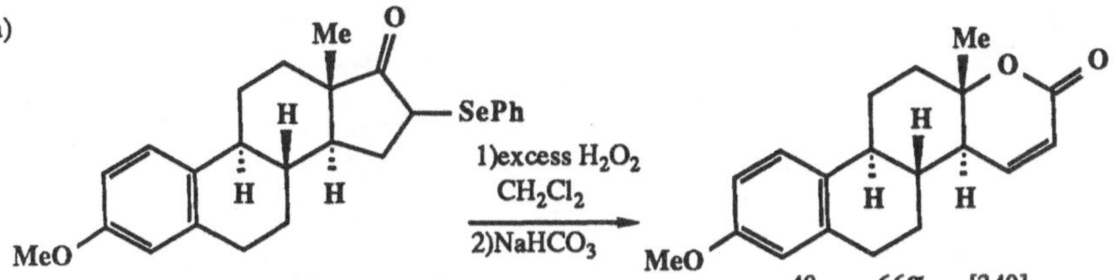

39 40 66% [340]

b)

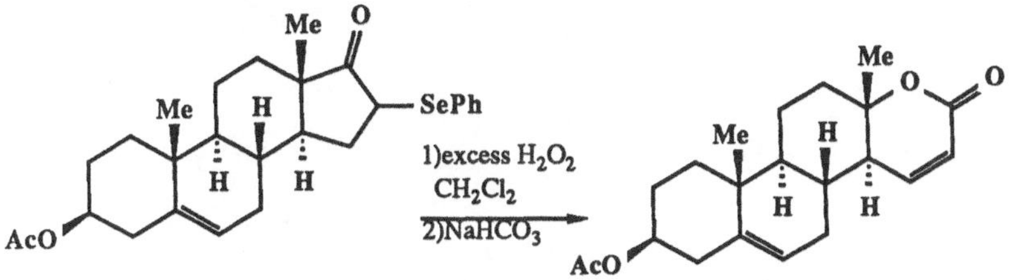

60% [340]

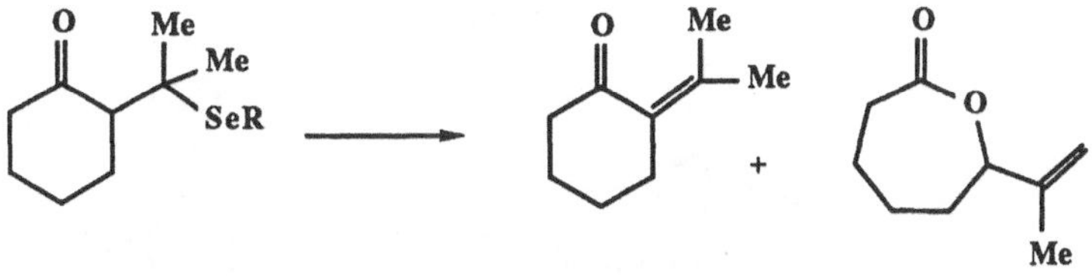

41 42 [350]

c)	R=Ph	10eq. H_2O_2/THF/0°C/0.2h	00%	62%
d)	Me	" "	14%	58%
e)	Me	1.2eq. mCPBA/-78°C/O.2h	72%	00%
f)	Me	4eq. $tBuO_2H/Al_2O_3$/THF/2h	80%	00%

Scheme 147

8. Reactions Involving Perseleninic Acids

acid were used, instead of H_2O_2 with the α-phenylselenocyclopentanone **39** (Scheme 147a). The formation of the lactone **42** (Scheme 147c to f) from the β-selenoketone **41** and excess of H_2O_2 (10 equiv.) may well be a related transformation [350].

At the same time the perseleninic acid promoted Baeyer-Villiger reaction was disclosed by Grieco [351] on non-selenylated ketones. A series of cyclanones (5 or 6 membered) were successfully transformed to the corresponding lactones on reaction with benzeneseleninic acid/H_2O_2 mixtures [351] (Scheme 148a–c) and later with polymer bound perseleninic acid [308] (Scheme 148d). The reaction is

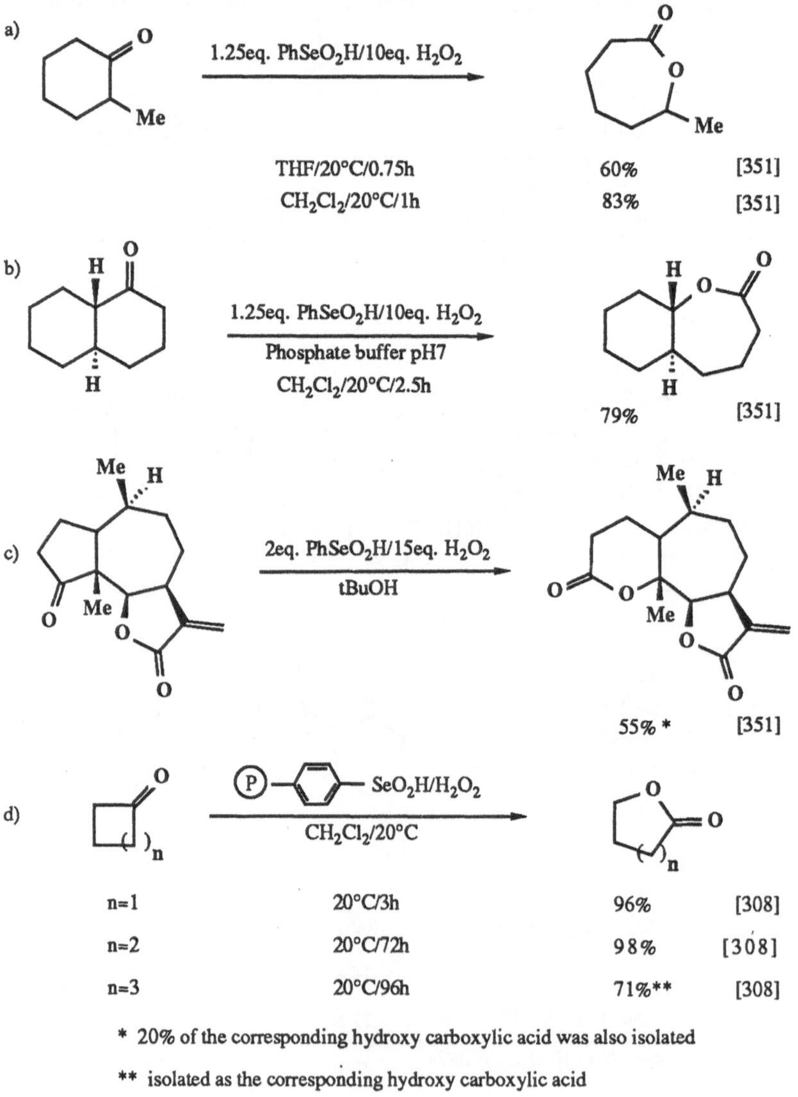

a)
1.25eq. PhSeO$_2$H/10eq. H$_2$O$_2$

THF/20°C/0.75h	60%	[351]
CH$_2$Cl$_2$/20°C/1h	83%	[351]

b)
1.25eq. PhSeO$_2$H/10eq. H$_2$O$_2$
Phosphate buffer pH7
CH$_2$Cl$_2$/20°C/2.5h

79%   [351]

c)
2eq. PhSeO$_2$H/15eq. H$_2$O$_2$
tBuOH

55% *   [351]

d)
(P)—⟨⟩—SeO$_2$H/H$_2$O$_2$
CH$_2$Cl$_2$/20°C

n=1	20°C/3h	96%	[308]
n=2	20°C/72h	98%	[308]
n=3	20°C/96h	71%**	[308]

* 20% of the corresponding hydroxy carboxylic acid was also isolated

** isolated as the corresponding hydroxy carboxylic acid

Scheme 148

112

usually performed in methylene chloride, t-butanol or THF and occurs rapidly (0,75–3 h) under mild conditions (20 °C). It seems to possess all the characteristics of the Baeyer-Villiger reaction: the more substituted carbon atom migrates and the reaction occurs with retention of the configuration at the migrating carbon atom [340, 351]. Finally, lactone formation is selectively observed [340, 351] (Schemes 147b, 148c) even from compounds also bearing a carbon-carbon double bond, susceptible to be oxidized by the reagent.

8.3 Oxidation of Selenides to Selenoxides and Selenones and of Sulfides to Sulfoxides and Sulfones

Hydrogen peroxide is able to perform, in the presence of stoichiometric or catalytic amounts of areneseleninic acids (or diaryl diselenides), the oxidation of selenides to selenones [352] (Scheme 149), of sulfides to sulfones (Schemes 150a, b, 151) [339, 353–357] and of triphenylphosphine to its oxide [312]. The reactions usually proceed at 20 °C in chlorinated solvents. Reich described that o-nitrobenzeneseleninic acid is the most valuable among the seleninic acids tested for the oxidation of sulfides but we observed that benzeneseleninic acid works equally well [352].

$$\text{RSePh} \xrightarrow[20°C/CH_2Cl_2]{n_1 \text{ eq.PhSeO}_2\text{H}/n_2 \text{ eq.H}_2\text{O}_2(30\%)} \text{RSe(O}_2)\text{Ph}$$

R	n_1	n_2	t(h)	Yield (%)
Bu	2.1	10	5	80
Dec	2.1	10	8	97

Scheme 149 [352]

a) $\text{Ph-S-iPr} \xrightarrow[CH_2Cl_2/20°C/3h]{0.01\text{eq.2-NO}_2\text{PhSeO}_2\text{H}/3\text{eq. H}_2\text{O}_2} \text{Ph-S(O}_2)\text{-iPr}$

98%

b) $\text{Bu-S-Bu} \xrightarrow[CH_2Cl_2/20°C/3h]{0.01\text{eq.2-NO}_2\text{PhSeO}_2\text{H}/3\text{eq. H}_2\text{O}_2} \text{Bu-S(O}_2)\text{-Bu}$

100%

c) $\text{Ph-S-Me} \xrightarrow{\text{controlled conditions}} \text{Ph-S(O)-Me}$

95%

Scheme 150 [339]

8. Reactions Involving Perseleninic Acids

Most of the work has been performed on sulfides [339, 353–357]. These can be oxidized [339], under carefully controlled conditions, to the corresponding sulfoxides (Scheme 150c).

The oxidation of sulfides to sulfones occurs chemoselectively [312] leaving untouched a carbonyl [357], an olefinic [353–357] or an hydroxyl [353] group when present in the molecule (Scheme 151). It is interesting to notice that when seleninic acids are omitted, selenides and sulfides are exclusively oxidized to selenoxides and sulfoxides (reactions performed at 20 °C). In the latter case the over oxidation to sulfones requires heating for several hours in acetic acid.

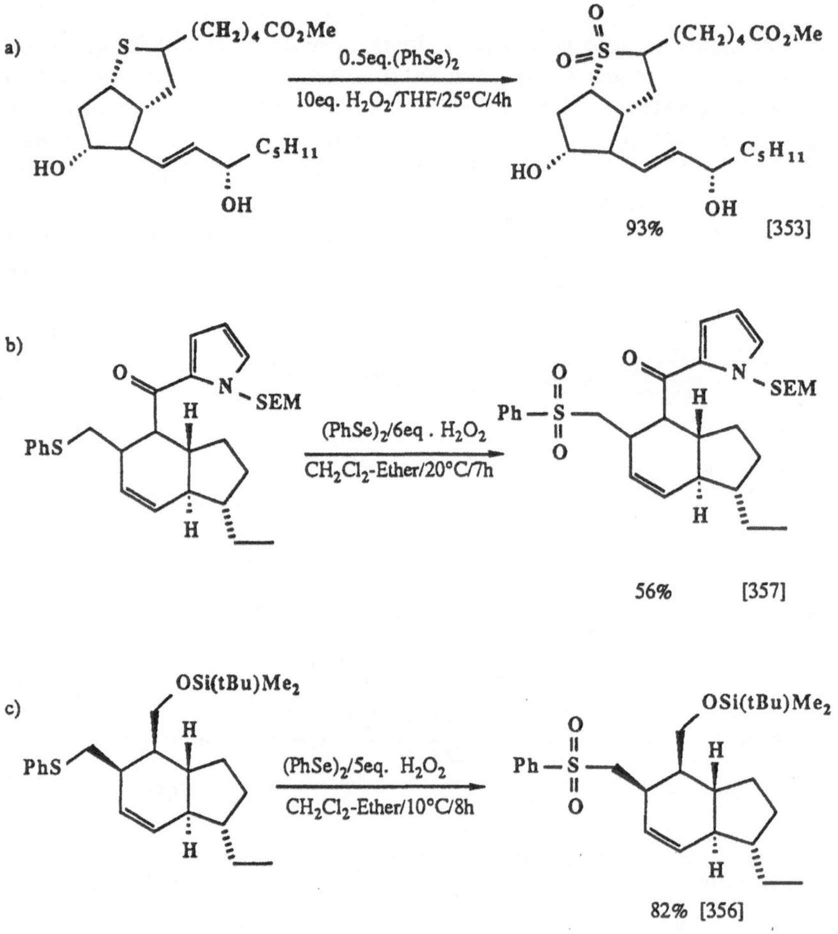

Scheme 151

Chapter 9

Selenium Dioxide Oxidations

Selenium dioxide is one of the limited number of chemicals which permits the oxidation of functionalized hydrocarbons at the carbon atom linked to olefinic (Scheme 152), aromatic and acetylenic groups or when it is attached to the carbonyl group of aldehydes, ketones, esters or amides (see below).

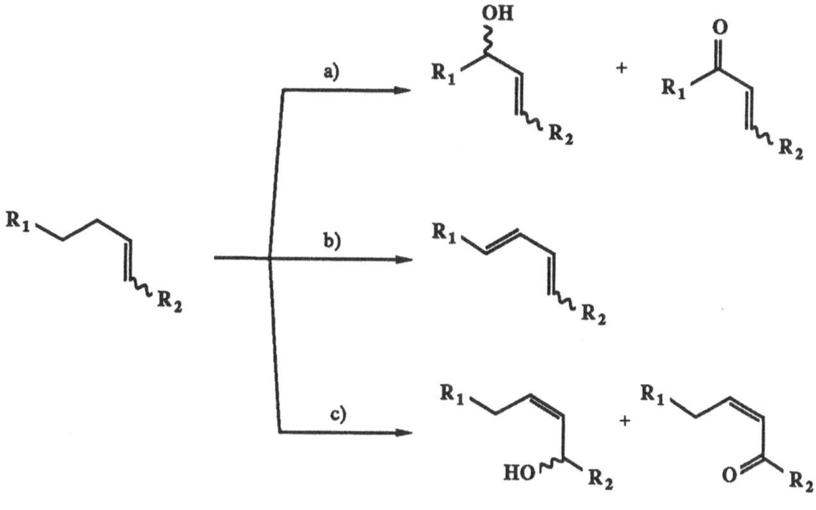

Scheme 152

Depending upon the conditions used and the structure of the substrate, selenium dioxide reacts alpha to the functional groups, such as hydroxyl, carbonyl or olefins. Several reviews have extensively covered the subject since the discovery of the reaction in 1932 [*63, *80–*82, *84, *85, *89, *358–*360].

These reactions have been achieved with stoichiometric or higher amounts of SeO_2 in alcoholic solvents (usually t-butanol), in dioxane or in acetic acid. More recently selenium dioxide/t-butyl hydroperoxide in methylene chloride has met with substantial success for the allylic oxidation of olefins and acetylenes especially using catalytic amounts of selenium dioxide [*96]. Other catalytic systems able to perform oxidation of organic molecules such as SeO_2/bis(p-methoxyphenyl)selenoxide [361] and SeO_2/iodoxybenzene [295] have been mentioned although they have not been extensively used in synthesis.

9. Selenium Dioxide Oxidations

The course of the reactions is dramatically changed if performed in the presence of mineral acids [362–366] (Schemes 153c) or hydrogen peroxide (Schemes 153d, 154f) [367–369]. In the latter case perselenious acid is the reactive species. Typical examples involving olefins under different experimental conditions are shown in Schemes 153, 154 [362–371].

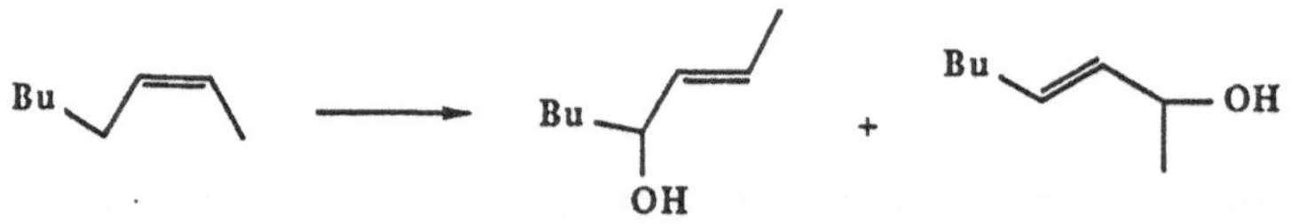

a)	0.5eq. SeO$_2$/tBuOH/reflux	18%	94	:	6	[370]
b)	0.5eq. SeO$_2$/2eq. tBuO$_2$H CH$_2$Cl$_2$/20°C	64%	82	:	18	[370]

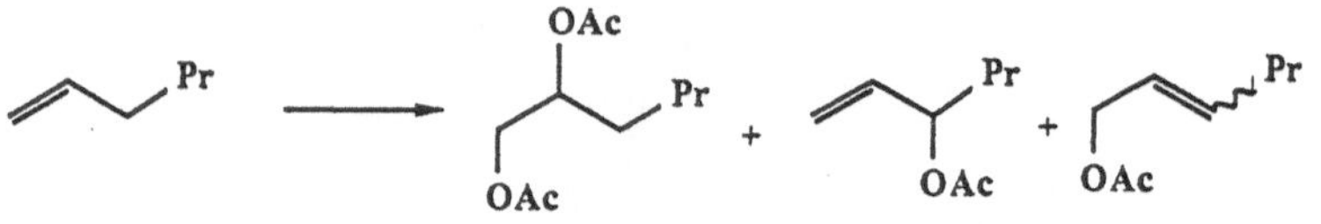

c)	SeO$_2$/AcOH/Ac$_2$O cat. H$_2$SO$_4$/105°C/10h	35%	12%	3% [362]

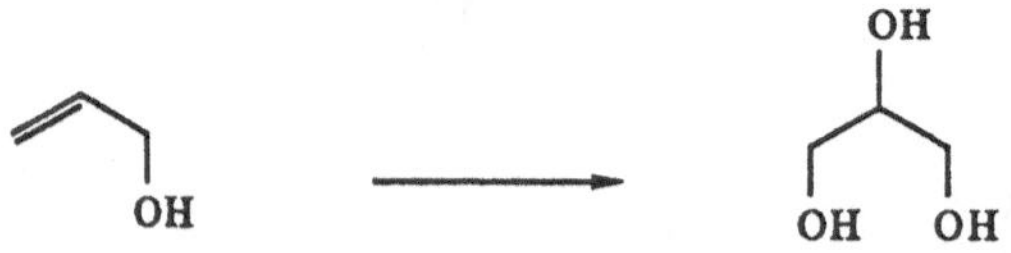

d)	cat SeO$_2$/H$_2$O$_2$/70°C/18h then 100°C/2h	36% (67% of S.M. consumed)	[367]

Scheme 153

9.1 Reactivity of SeO$_2$ with Alkenes

9.1.1 Oxidation of Alkenes with SeO$_2$ or SeO$_2$/tBuO$_2$H

9.1.1.1 Scope and Limitation

Olefins react with selenium dioxide at reflux in acetic acid, ethanol, t-butanol or dioxane used as the solvent (conventional procedure: C. P.) [*63, *80, *82, *84, *89, *358, *372, *373] or at room temperature with stoichiometric and even with

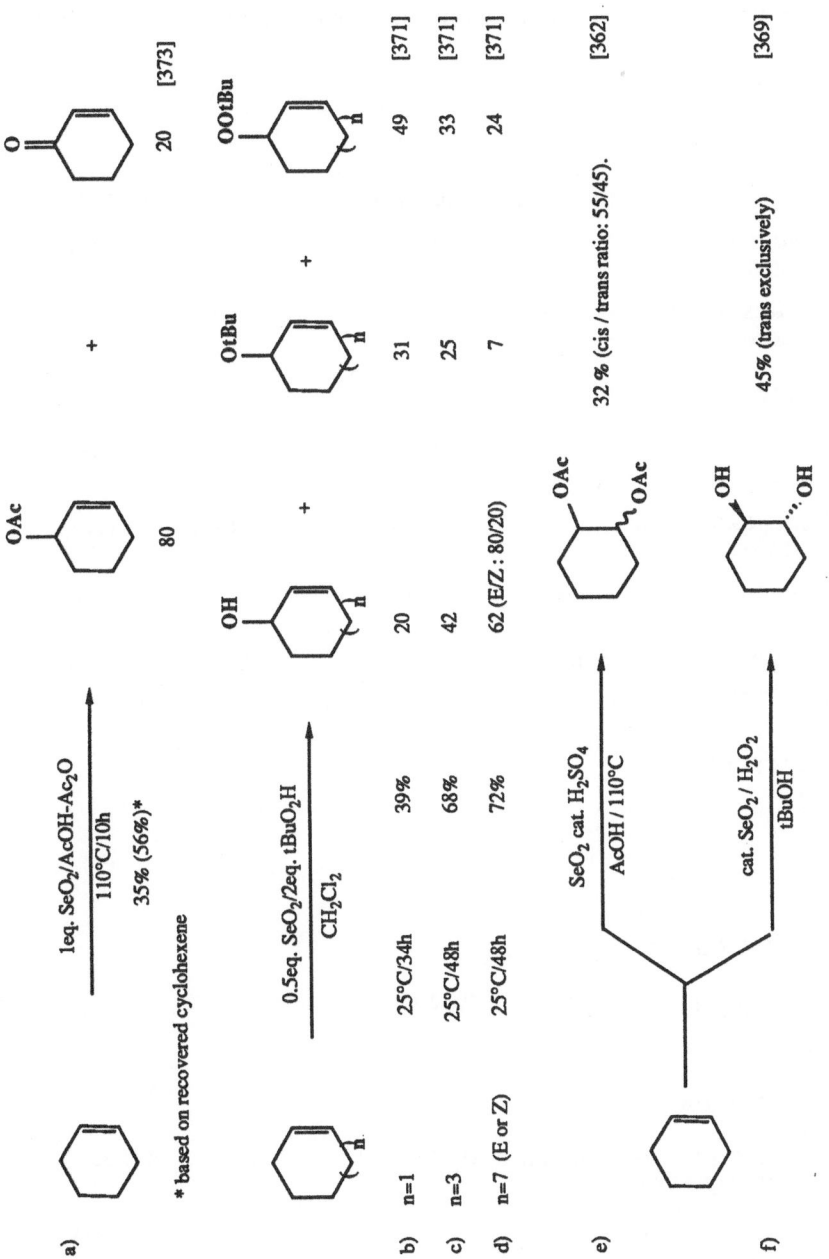

Scheme 154

117

9. Selenium Dioxide Oxidations

catalytic amounts of selenium dioxide and t-butyl hydroperoxide in non polar solvents such as methylene chloride (the Sharpless procedure: S. P.) [*96, 370, 371]. Compounds resulting from their allylic oxidation such as allyl alcohols and/ or α,β-unsaturated aldehydes or ketones (Scheme 152a) are formed. In some

R= Et

		Yield % (E/Z ratio)	Yield % (E/Z ratio)	
a)	1eq. SeO$_2$/EtOH-H$_2$O reflux/1.5h	70(98/2)	8(98/2)	[387]
b)	2eq. SeO$_2$/EtOH-H$_2$O reflux/10h	---	65(99/1)	[387]

R= CH$_3$C(OCH$_2$)$_2$

		Yield % (E/Z ratio)	Yield % (E/Z ratio)	
c)	1.4eq. SeO$_2$/EtOH-H$_2$O reflux	---	33(100/0)	[387]
d)	1.4eq. SeO$_2$/Dioxane reflux/5h	---	51(100/0)	[388]
e)	1eq. SeO$_2$/Pyridine/EtOH-H$_2$O reflux/3h	42(100/0)	...	[378]
f)	1eq. SeO$_2$/Pyridine/EtOH-H$_2$O reflux/9h	---	39(100/0)	[378]

R= CH$_3$CHCH=CH$_2$

		Yield % (E/Z ratio)	Yield % (E/Z ratio)	
g)	0.05eq. SeO$_2$/4eq. tBuO$_2$H CH$_2$Cl$_2$/25°C/7h	---	50*	[376]

R= CH$_2$=C-CH=CH$_2$

		Yield % (E/Z ratio)	Yield % (E/Z ratio)	
h)	0.5eq. SeO$_2$/EtOH-H$_2$O reflux/1h	19(100/0)	...	[377]

R= CH$_3$C=CHCH$_2$OA

		Yield % (E/Z ratio)	Yield % (E/Z ratio)	
i) A=Ac	0.03eq. SeO$_2$/4eq. tBuO$_2$H/H$^+$ CH$_2$Cl$_2$/25°C/7h	48	8	[377, 381]
j) A=THP	0.5eq. SeO$_2$/ Pyridine EtOH-H$_2$O/reflux/4h	35	trace	[378]

* After further oxidation with MnO$_2$

Scheme 155

118

special cases conjugated dienes (Scheme 152b) or rearranged allyl alcohols (Scheme 152c) can be produced instead.

The reactivity of olefins follows their nucleophilicity (Scheme 155 compare entries g,h to a,b). Trisubstituted olefins are more reactive than α,α-dialkyl substituted systems [374, 375] which in turn are more reactive than α,β-disubsti-

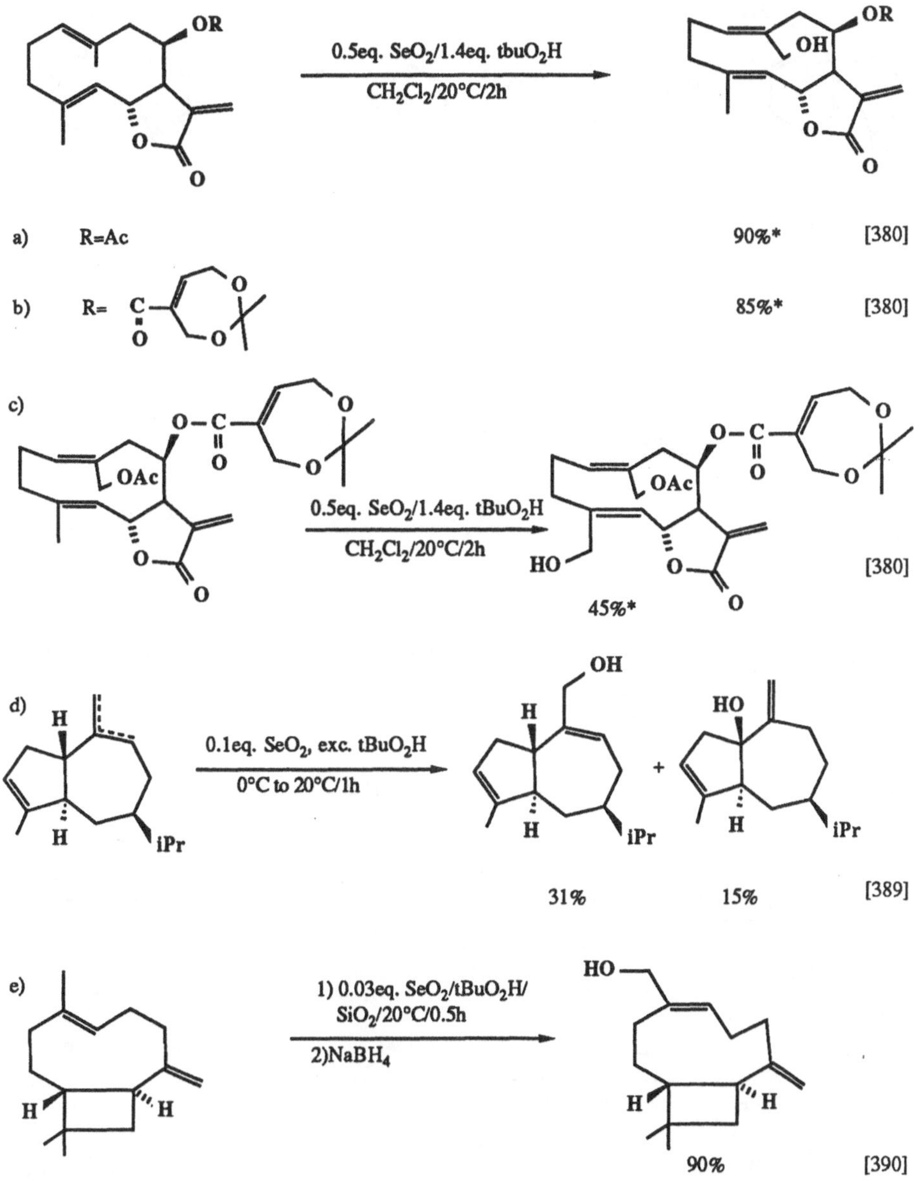

0.5eq. SeO$_2$/1.4eq. tbuO$_2$H

CH$_2$Cl$_2$/20°C/2h

a) R=Ac 90%* [380]

b) R= 85%* [380]

c)

0.5eq. SeO$_2$/1.4eq. tBuO$_2$H

CH$_2$Cl$_2$/20°C/2h

45%* [380]

d)

0.1eq. SeO$_2$, exc. tBuO$_2$H

0°C to 20°C/1h

31% 15% [389]

e)

1) 0.03eq. SeO$_2$/tBuO$_2$H/ SiO$_2$/20°C/0.5h

2)NaBH$_4$

90% [390]

* The corresponding aldehyde is also formed in 5% yield

Scheme 156

119

tuted, mono-substituted olefins being the least reactive of the series [376] (Scheme 155g,h). Conjugated dienes [377], allylic ethers [378–380] and acetates [377, 380] as well as α,β-unsaturated carbonyl compounds [381–386] are less reactive than the corresponding alkenes (Schemes 155h–j, 156–158). Some

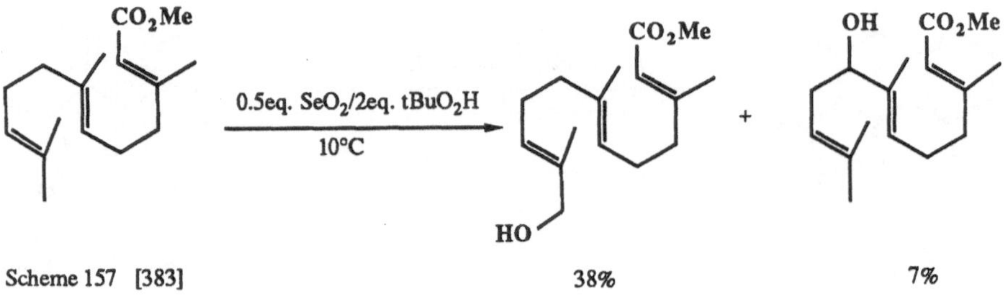

Scheme 157 [383]   38%   7%

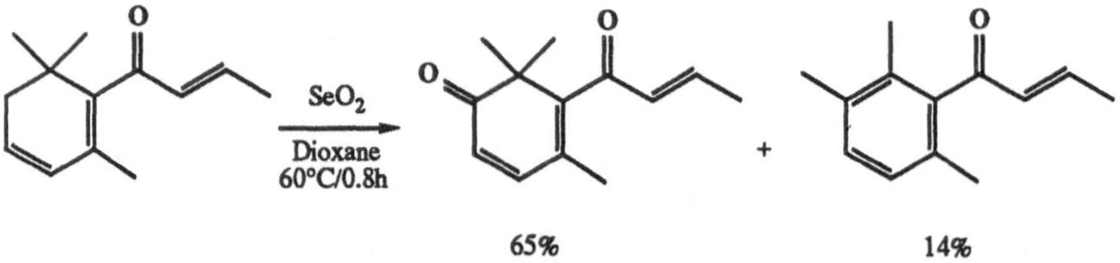

Scheme 158 [385]

a)

15%   50%

b)

47%   11%   10%

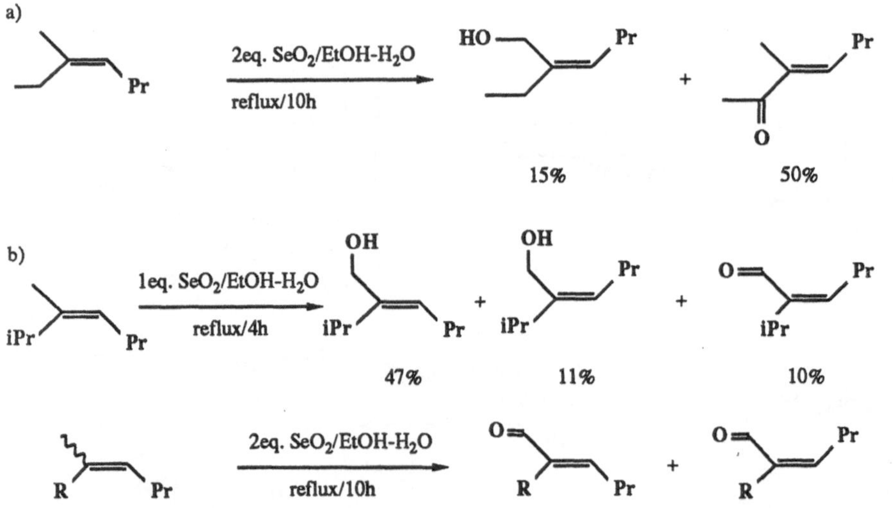

c) R=iPr (Z or E)	58% overall	80	20
d) R=cHex	65% overall	80	20

Scheme 159 [387]

120

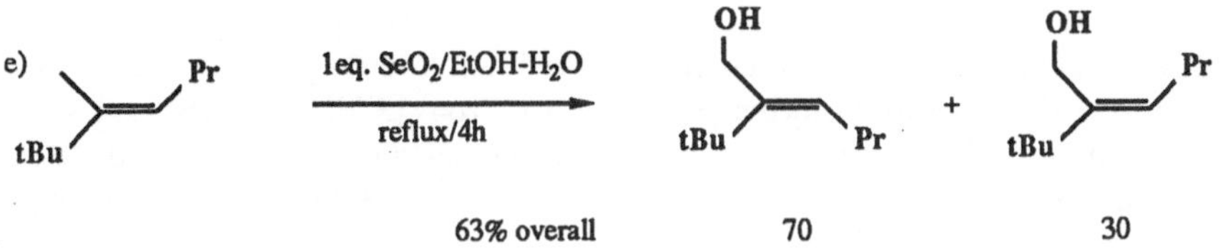
e)
Pr
tBu
1eq. SeO2/EtOH-H2O
reflux/4h
OH
tBu
Pr
+
OH
Pr
tBu
63% overall
70
30

Scheme 159 [387] (contd)

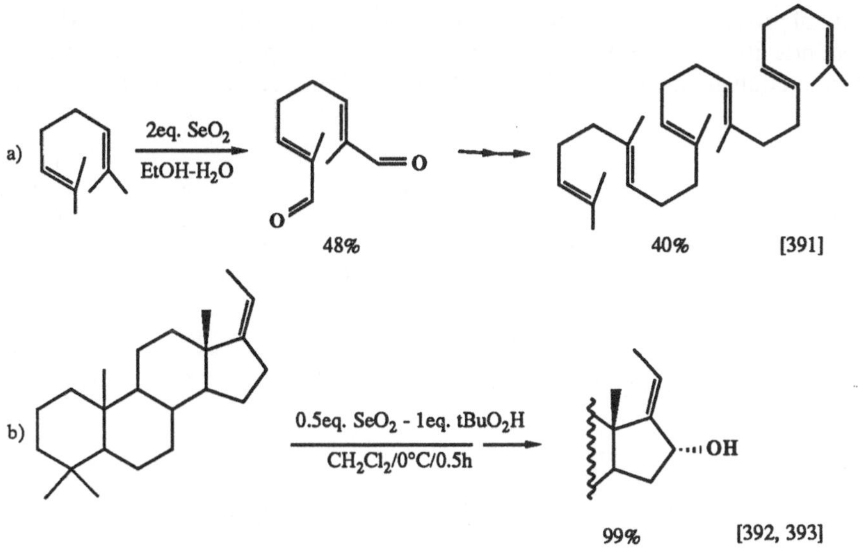
a)
2eq. SeO2
EtOH-H2O
O
O
48%
40%
[391]
b)
0.5eq. SeO2 - 1eq. tBuO2H
CH2Cl2/0°C/0.5h
OH
99%
[392, 393]

Scheme 160

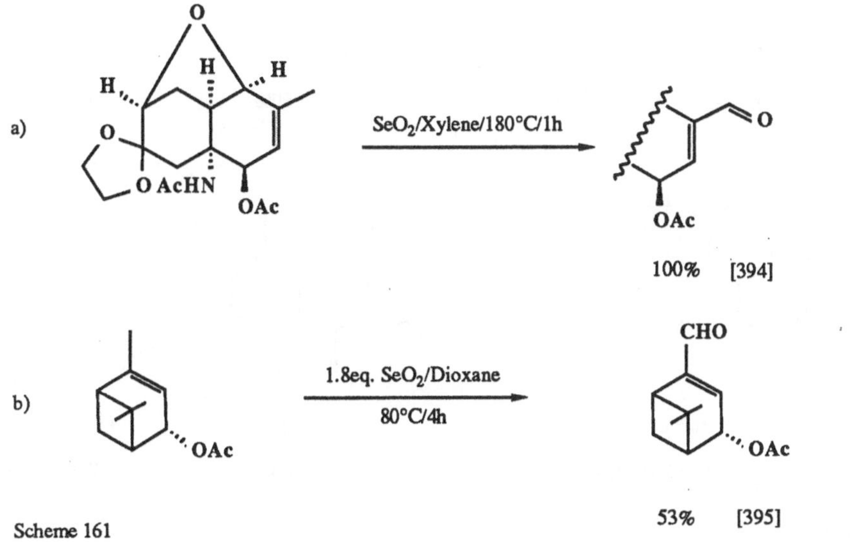
a)
O
H
H
H
O
O AcHN
OAc
SeO2/Xylene/180°C/1h
O
OAc
100%
[394]
b)
OAc
1.8eq. SeO2/Dioxane
80°C/4h
CHO
OAc
53%
[395]

Scheme 161

9. Selenium Dioxide Oxidations

representative examples of oxidation using SeO_2 are presented in the following schemes: trisubstituted olefins (Schemes 155–157, 159–161) [376–379, 381, 383, 387–397], α,α-disubstituted olefins (Schemes 162–164) [370, 374, 396–399], α,β-disubstituted olefins (Schemes 153a, b, 165) in the acyclic series [362, 370, 402], cycloalkenes (Schemes 154a–d, 166a, c) [371, 373, 400, 403] and monosubstituted olefins (Schemes 153c, 167) [362, 370, 404].

In most cases allylic oxidation takes place giving a mixture of allyl alcohols (or derivatives) and of α,β-unsaturated carbonyl compounds (Schemes 154–156, 159, 160b, 162, 163, 166c) [370, 373, 376–379, 381, 387, 392, 393, 403]. The relative amount of products depends upon the conditions used. α,β-Unsaturated carbonyl compounds are mainly formed using the conventional procedure when at least one molar equivalent of SeO_2 is used for long reaction times (8–10 h) whereas allyl

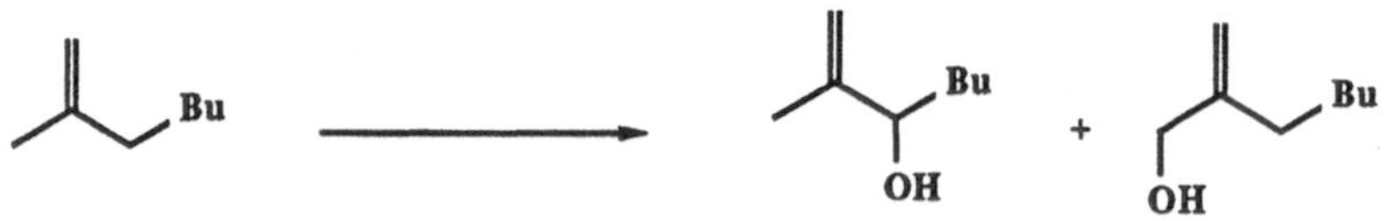

a)	0.5eq. SeO₂/tBuOH/reflux/	22% overall	85	15	[370]
b)	0.5eq. SeO₂-tBuO₂H/CH₂Cl₂/20°C	56% overall	56	15	[370]

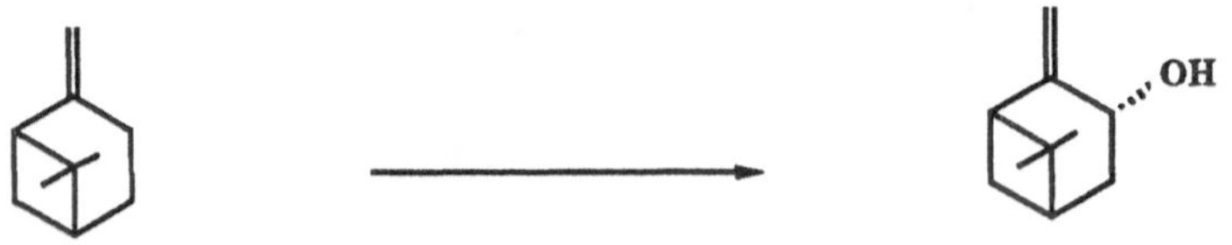

c)	0.4eq. SeO₂/EtOH/reflux/4h	58%	[370, 396]
d)	0.07eq. SeO₂-5eq.tBuO₂H/Hexane-0.5eq. AcOH then NaBH₄	86%	[370]
e)	0.03eq. SeO₂-1.3eq. H₂O₂/tBuOH/50°C	50%	[397, 398]

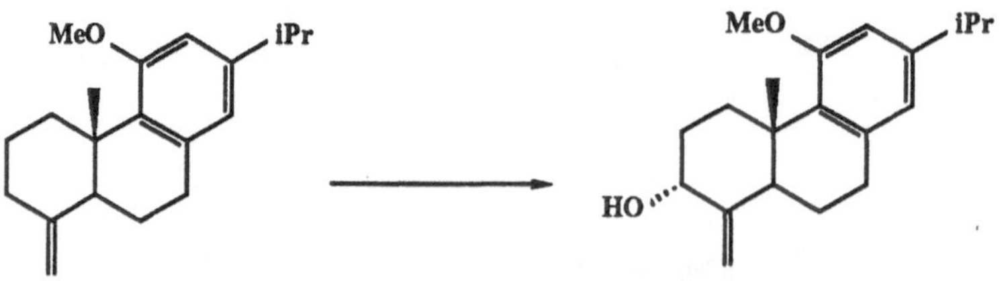

f)	SeO₂/EtOH-H₂O/reflux/80h	55%	[399]
g)	0.03eq. SeO₂/5eq. tBuO₂H/H⁺cat/20°C/28h	63%	[399]

Scheme 162

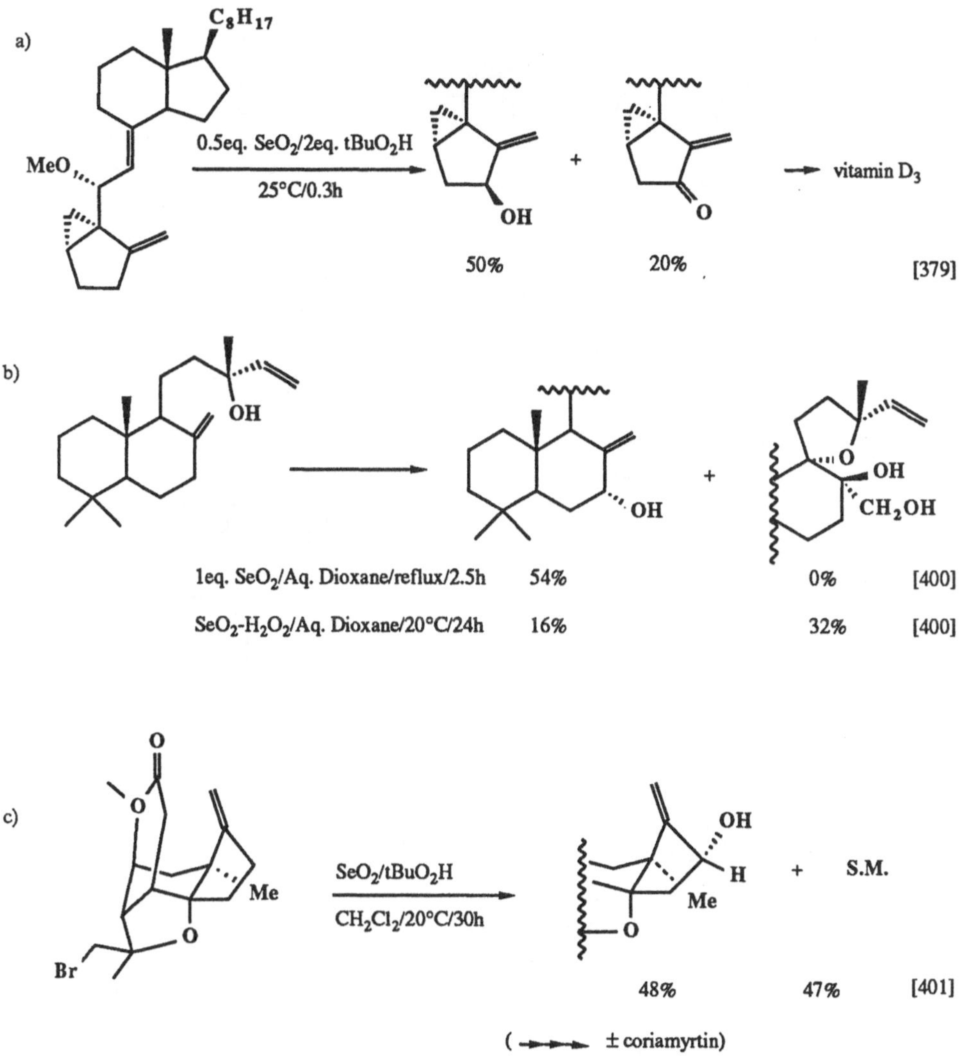

a)

C₈H₁₇

MeO

0.5eq. SeO₂/2eq. tBuO₂H

25°C/0.3h

OH

50%   20%   → vitamin D₃   [379]

b)

OH

1eq. SeO₂/Aq. Dioxane/reflux/2.5h   54%   0%   [400]

SeO₂-H₂O₂/Aq. Dioxane/20°C/24h   16%   32%   [400]

OH

CH₂OH

c)

Me

SeO₂/tBuO₂H

CH₂Cl₂/20°C/30h

OH

H   +   S.M.

Me

O

48%   47%   [401]

(⟶ ± coriamyrtin)

Scheme 163

alcohols are obtained when smaller amounts of SeO₂ are reacted for short periods (1–4 h) (compare Scheme 155a to b and e to f) [387] or when the Sharpless procedure is used (Schemes 153b, 155g, i, 156a–d, 157, 162, 163a) [370, 376, 377, 380, 381, 383, 389, 399]. On the basis of these results we suggest to search for the best conditions which permit the allylic oxidation of the olefin regardless of the compound(s) formed and then to oxidize (MnO₂) or to reduce (NaBH₄) the mixture to the desired enone or allylalcohol.

The case of *t*-butyl-trans-chrysanthemate is quite unusual since upon reaction with equivalent amounts of SeO₂ in dioxane the expected (E) unsaturated aldehyde and a product resulting from the oxidation of both methyl groups of the trisubstituted olefin are concomitantly formed (Scheme 168) [405, 406].

123

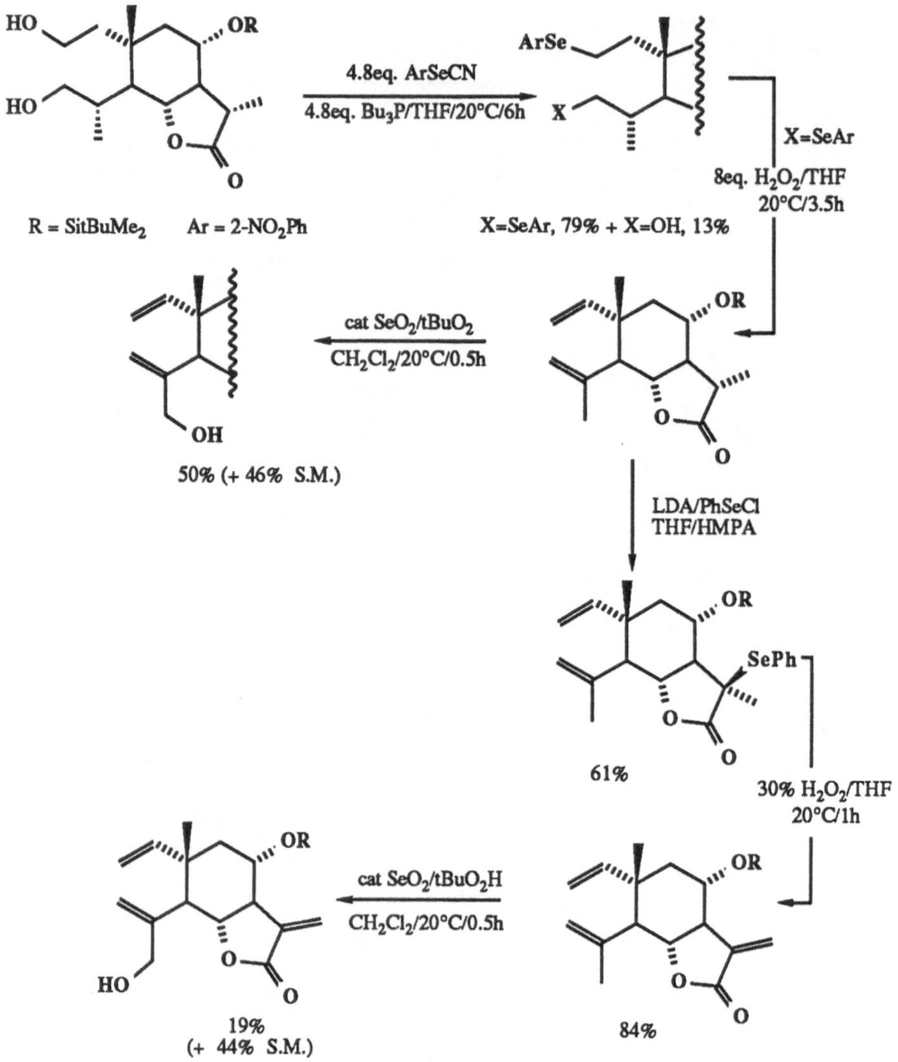

9. Selenium Dioxide Oxidations

Scheme 164 [374]

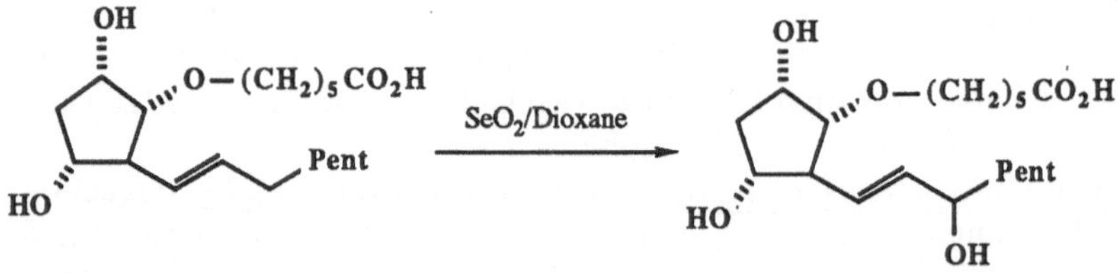

Scheme 165 [402]

124

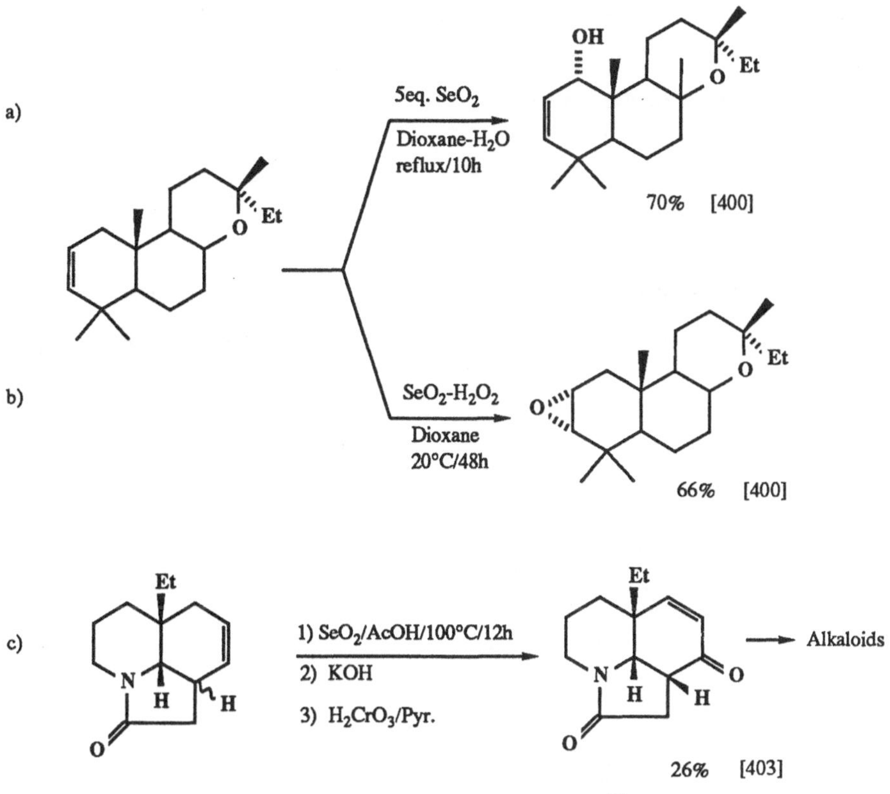

Scheme 166

The reaction proceeds without regiochemical isomerisation of the carbon-carbon double bond in the case of trisubstituted and α,α-disubstituted olefins (Schemes 155–157, 159, 160) [380, 387, 390–393] (Schemes 162–164) [370, 374, 379, 396–401] whereas some isomerisation occurs with α,β-disubstituted or monosubstituted olefins (Schemes 153a, b) [370, 400, 402] and (Scheme 167c) [370]. In a few cases belonging to the last two series of olefins (Schemes 166c and 167b), allylic rearrangement can however take place [370, 403] (see also Section 9.1.1.2). This has been advantageously used [403] for the synthesis of 43 (Scheme 166c) a useful intermediate for alkaloid synthesis. These rearrangements if undesired can be minimized or avoided by using the Sharpless procedure [370] (Scheme 167 compare c to b).

The Sharpless procedure [*96, 370, 383, 393, 402] proved [370, 376, 379–381, 383, 390, 392, 393, 399, 402] highly superior to the conventional procedure when applied to acyclic olefins (Schemes 153b, 155, 156, 157, 160, 162, 163, 165, 167) especially for α,β-disubstituted and for monosubstituted alkenes (Schemes 153b,

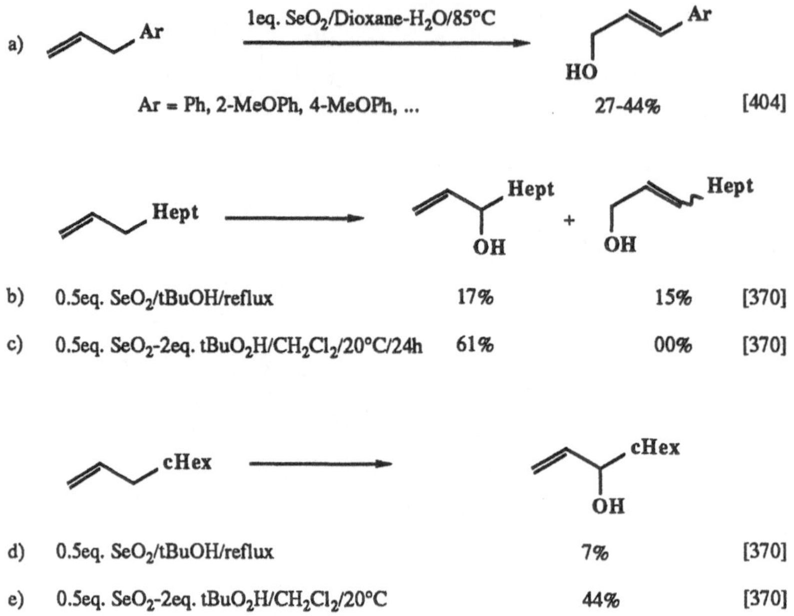

9. Selenium Dioxide Oxidations

a)

1eq. SeO$_2$/Dioxane-H$_2$O/85°C

Ar = Ph, 2-MeOPh, 4-MeOPh, ...

HO

27-44%   [404]

			Hept	Hept		
		OH		OH		
b)	0.5eq. SeO$_2$/tBuOH/reflux	17%		15%	[370]
c)	0.5eq. SeO$_2$-2eq. tBuO$_2$H/CH$_2$Cl$_2$/20°C/24h	61%		00%	[370]

cHex

OH

| d) | 0.5eq. SeO$_2$/tBuOH/reflux | 7% | [370] |
| e) | 0.5eq. SeO$_2$-2eq. tBuO$_2$H/CH$_2$Cl$_2$/20°C | 44% | [370] |

Scheme 167

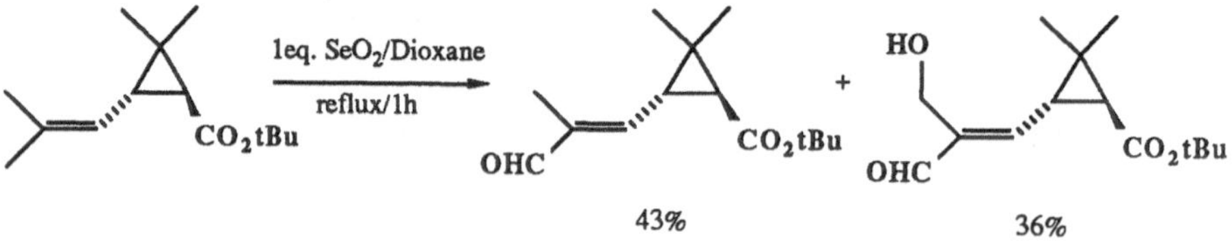

1eq. SeO$_2$/Dioxane

reflux/1h

CO$_2$tBu   OHC   CO$_2$tBu   HO   OHC   CO$_2$tBu

43%   36%

Scheme 168   [405, 406]

165, 167) [370, 402] as well as for exocyclic olefins (Schemes 160b, 162, 163a, c) [370, 379, 392, 393, 399, 401] and for large ring cycloalkenes (>8 atoms) [380, 390, 407] (Scheme 156). It is however not interesting to use it with small and medium ring cycloalkenes (<8 atoms, see below). The Sharpless procedure offers also the following advantages:

(i) it often requires only catalytic amounts of SeO$_2$ (although in some cases, 0,5 molar equiv. give better results);
(ii) it takes place at room temperature (20 °C instead of 80–120 °C required in the conventional procedure) (Schemes 153, 155, 162, 163, 167);
(iii) it avoids the formation of colored and malodoriferous selenium (II) species, always formed in the conventional procedure but which are oxidized back to SeO$_2$ by the t-butyl hydroperoxide used;
(iv) the yields in allyl alcohols are increased and the amount of α,β-unsaturated carbonyl compounds arising from over oxidation is lowered;

126

a)   R = H ;   SeO₂-tBuO₂H/CH₂Cl₂/20°C/336h   50%

b)   R = H ;   SeO₂/CH₂Cl₂/25°C/528h   0%

c)   R = CH₂tBu ;   DDQ/THF-H₂O   5%

SeO₂/Dioxane
reflux/0.5h

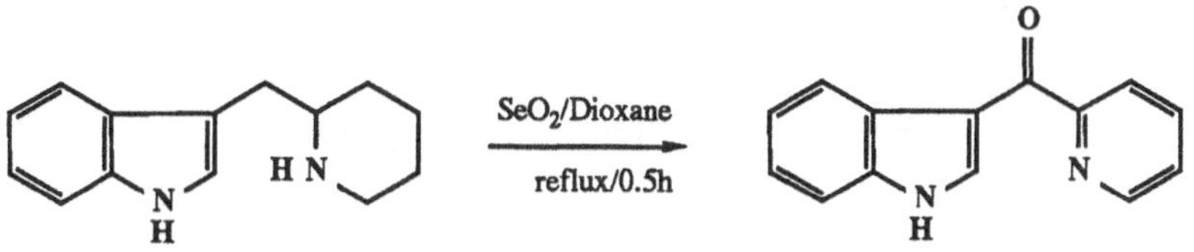

90%

e)   R = CH₂Ph   SeO₂-tbuO₂H/CH₂Cl₂/20°C/192h   0%

f)   R = CH₂Ph   SeO₂/Dioxane/reflux/240h   47%

Scheme 169 [408, 409]

(v) it prevents the formation of allyl alcohols resulting from the migration of the original carbon-carbon double bond;

(vi) it proved particularly suitable for the allylic oxidation of alkyl substituted indoles (Scheme 169) [408, 409] which are not oxidized further.

The allylic oxidation of small and medium sized cycloalkenes with SeO₂ is usually more difficult than that of acyclic and large ring derivatives [380, 390, 407]. Unsubstituted cycloalkenes are sometimes oxidized at their allylic position using the conventional procedure [373, 400, 403, 431] (Schemes 154a, 166a) but allylic rearrangement often takes place [403] especially when acidic conditions are used (Scheme 166c). Alkylsubstituted cycloalkenes are usually oxidized at the exocyclic allylic position and give allylic alcohols or aldehydes [380, 394, 395] (Scheme 161). Dienes [410, 434] (Scheme 170a, b, c, d) or aromatic compounds [411] (Scheme 170e) are often formed during the oxidation of polycyclic compounds when one of the olefinic carbon atoms belongs to two rings.

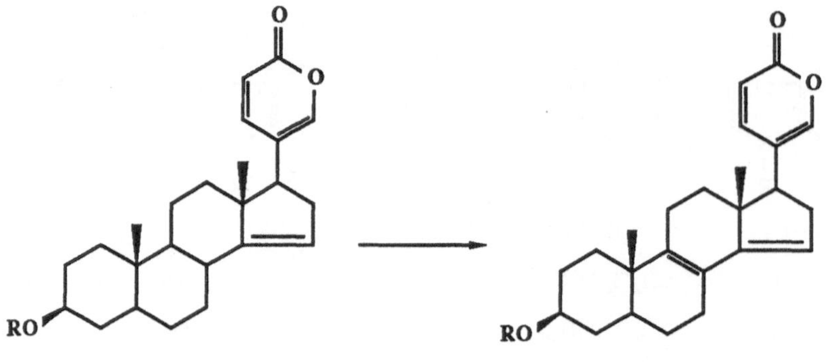

9. Selenium Dioxide Oxidations

a)	R = H	excessSeO$_2$/Dioxane/reflux/12h	75%	[434]
b)	R = H	SeO$_2$/Dioxane - AcOH - H$_2$O/80°C/5h	78%	[434]
c)	R = Ac	SeO$_2$/Dioxane/reflux/12h	66%	[434]

d) 3eq. SeO$_2$/EtOH - H$_2$O 37°C/16h 60% [410]

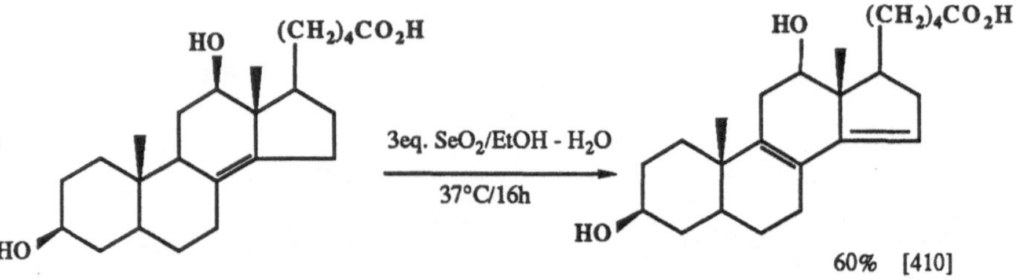

ß-dihydroxy choladienic acid

e) SeO$_2$

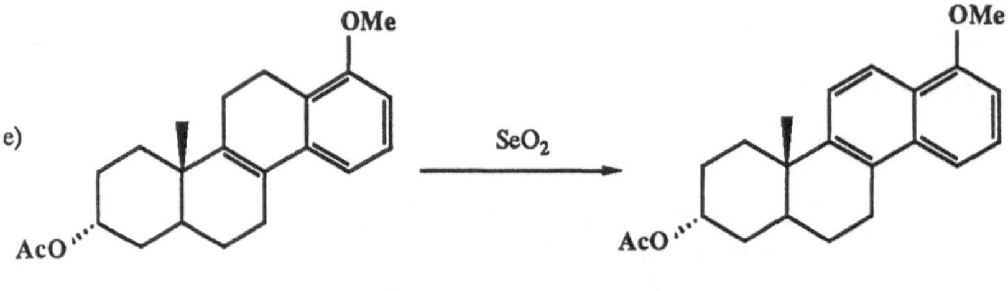

Toluene - Pyridine/reflux/12h	79%	[411]
Ethanol/reflux/24h	00%	[411]

Scheme 170

The Sharpless procedure, which proved particularly useful for the allylic oxidation of acyclic and exocyclic olefins, is not satisfactory when the olefinic linkage is part of six or seven membered cyclic systems [*96, 371, 412] (Scheme 154b, c, d). In these cases t-butyl allyl ethers and peroxides are also formed in substantial quantities besides the expected allyl alcohols [*96, 371]. Their percentage decreases when the size of the ring increases (Scheme 154d). Thus germacrolides are successfully oxidized to allyl alcohols using the Sharpless procedure (Scheme 156a–c) [*96, 380]. In the case of humulenes and caryophyllene [390] the same procedure leads to lower yields of allyl alcohols since epoxides and t-butyl allyl ethers are usually formed. It was anticipated that the more hindered is the starting material or the higher is the strain involved during the process the lower are the yields of allyl alcohols [389, 390] (Scheme 156d). The formation of the side products mentionned above can be avoided by a slight modification of the Sharpless procedure which involves the use silica gel supported SeO$_2$ [390] (Scheme 156e).

9.1.1.2 Regiochemistry

As already pointed out, the reaction usually proceeds without apparent migration of the double bond (although it occurs by an ene reaction followed by a [2,3] sigmatropic rearrangement of the intermediate allylic seleninate) and its regiochemistry can, in most cases be predicted on the basis of the rules edicted by Guillemonat [373].

As expected monoalkylsubstituted olefins lead, on reaction with SeO$_2$, to allyl alcohols possessing a terminal double bond [370, 374] (Scheme 167). These alcohols are often accompanied, under the conventional procedure, by isomeric allyl alcohols arising from the migration of the double bond [370]. This is particularly the case when the allylic position is linked to an aromatic ring (Scheme 167a) [404]. The best selectivity is found when the Sharpless procedure is used [370] (Scheme 167c, e) since no rearrangement is observed.

α,α-Dialkylsubstituted olefins also produce allyl alcohols possessing a terminal double bond (Schemes 162, 163) [*89, 370, 383, 397, 399] but now the

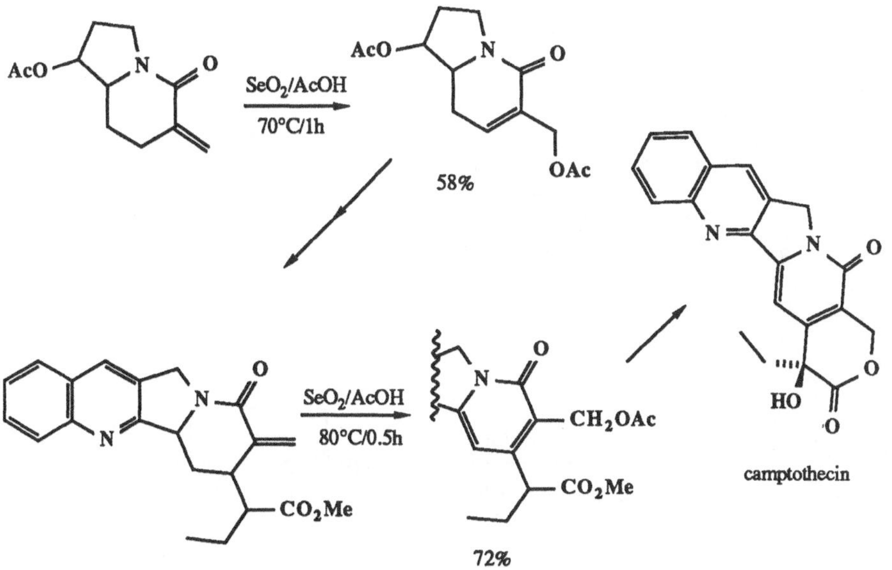

Scheme 171 [413]

129

oxidation takes place selectively on the more substituted carbon atom in the case of acyclic olefins (Scheme 162a, b) [370] and usually on the less substituted one when exocyclic olefins are reacted (Schemes 162c–g, 163) [370, 379, 396, 397, 399, 400, 401] (for an exception see Scheme 171) [413].

In the case of α,β-disubstituted olefins, a methylene group is more easily oxidized than a methyl group (Scheme 153a, b) [370] or a methine group

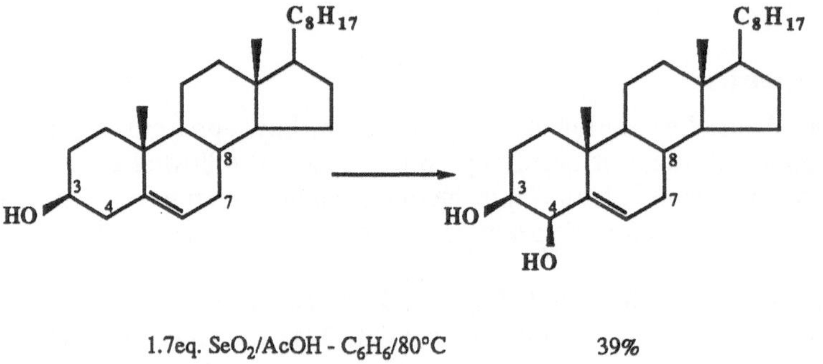
C8H17
HO
3
4
7
8
HO
HO
1.7eq. SeO2/AcOH - C6H6/80°C
39%

Scheme 172 [435]

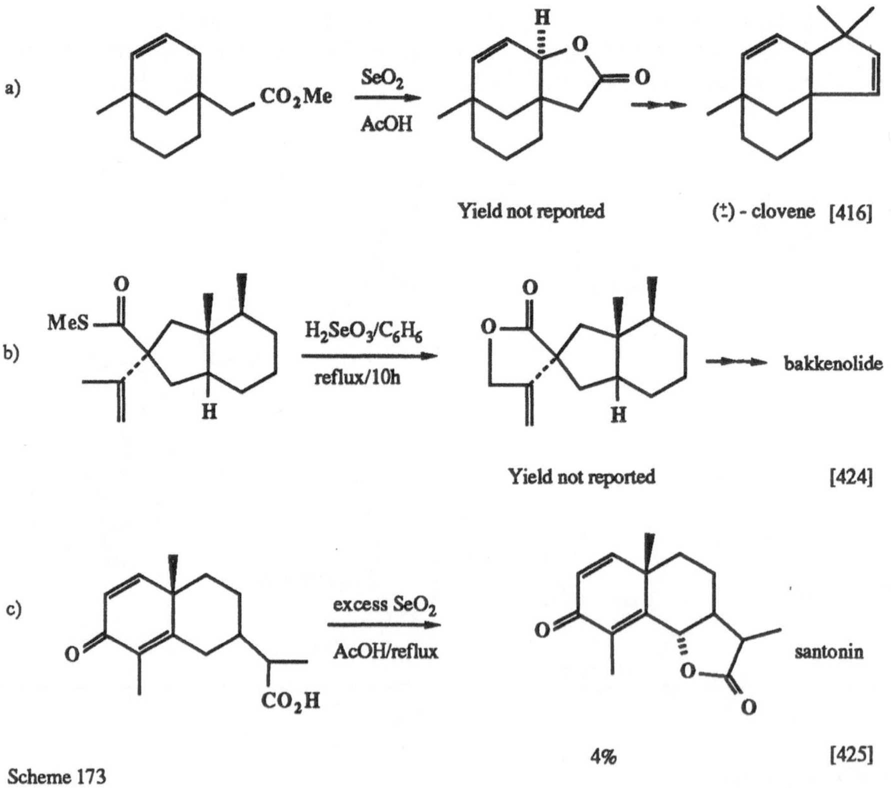
a)
CO2Me
SeO2
AcOH
H
O
O
Yield not reported
(±) - clovene [416]
b)
MeS
O
H
H2SeO3/C6H6
reflux/10h
O
O
H
bakkenolide
Yield not reported
[424]
c)
O
CO2H
excess SeO2
AcOH/reflux
O
O
O
santonin
4%
[425]

Scheme 173

(Scheme 165) [402]. This feature has been used for the regioselective synthesis of a 7-oxaprostaglandin [402] (Scheme 165).

An even more remarkable regioselectivity is observed with trisubstituted olefins since the reaction not only takes place on the more substituted side of the double bond (Schemes 155–157, 159) [*81, *96, 370, 376, 378, 383, 387] but also a

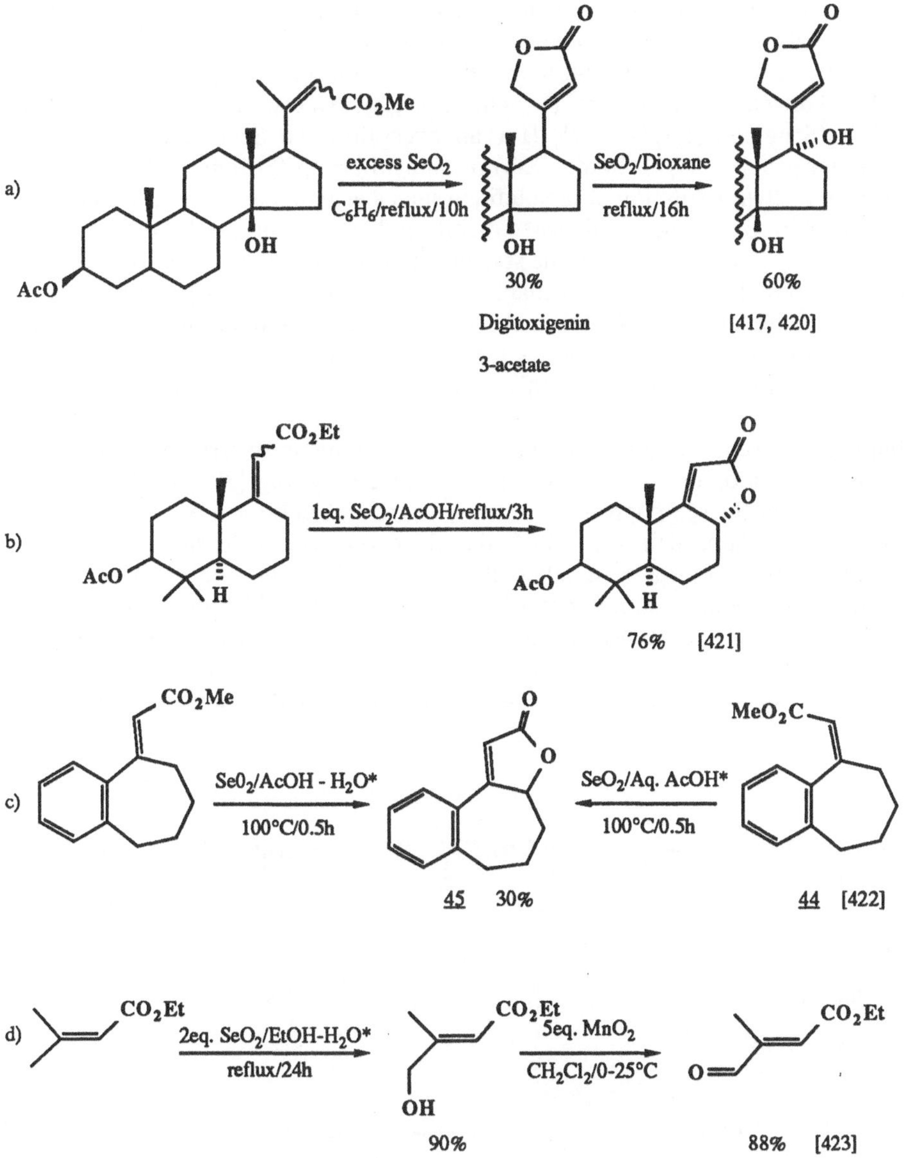

a)
CO2Me
excess SeO2
C6H6/reflux/10h
SeO2/Dioxane
reflux/16h
AcO
OH
OH
OH
30%
60%
Digitoxigenin
3-acetate
[417, 420]
b)
CO2Et
1eq. SeO2/AcOH/reflux/3h
AcO
H
76% [421]
c)
CO2Me
MeO2C
SeO2/AcOH - H2O*
100°C/0.5h
SeO2/Aq. AcOH*
100°C/0.5h
45 30%
44 [422]
d)
CO2Et
2eq. SeO2/EtOH-H2O*
reflux/24h
5eq. MnO2
CH2Cl2/0-25°C
OH
90%
88% [423]

* Similar results have been obtained in dioxane

Scheme 174

methyl group is selectively oxidized rather than a methylene [380, 381, 383, 387, 389, 414, 390, 391] (Schemes 156a–c, e, 157, 159a, 160a) or methine group [387, 415] (Scheme 159b, c, d). Finally a remarkable chemio-selectivity was observed with farnesyl compounds [381, 383] (Scheme 157) in which the terminal trialkyl substituted double bond is quite selectively oxidized [381, 383].

In some special cases a functionality present in the neighbourhood of the carbon-carbon double bond can dramatically influence the regioselectivity of the SeO$_2$ oxidation. This is particularly the case for cholesterol [435] in which the hydroxyl group present in the homoallylic position favors the formation of the 3,4-diol instead of the 3,7-diol (Scheme 172a, b). An ester group (Scheme 173a) [416], (Scheme 174) [417–422] (for an exception see Scheme 174d) [423], a thioester group [424] (Scheme 173b) or an acid group (Scheme 173c) [425], can direct the allylic oxidation of the olefin in such a way that a medium sized lactone with an exo- or an endocyclic carbon-carbon double bond can be formed [424], depending upon the nature of the starting material. It is worthwile to notice that lactone **45** (Scheme 174c) is available [422] from both cis and trans **44**.

Esters and ethers of allylic alcohols are less reactive than the corresponding alkyl substituted alkenes (Schemes 155i, j, 156a–c) [377, 378, 380, 381]. They are however oxidized by SeO$_2$ and this oxidation occurs at the allylic position away from the one already bearing the oxygen moiety [394, 395] (Scheme 161). This feature was used by Kishi [394] for the synthesis (Scheme 161a) of an intermediate in his synthesis of tetrodamine and also used by Takayanagi [395] for the synthesis of a mimic of the sex pheromone of the American Cockroach. If the only allylic hydrogen available is the one attached to the carbon on which the oxygen moiety is linked, the oxidation now takes place there [426]. The intermediate decomposes and produces simultaneously α,β-unsaturated aldehydes and alcohols (Scheme 175). This reaction is valuable for deprotecting hydroxyl groups previously masked as allyl ethers [426]. When applied to α-methyl allyl alcohols [427] and cinnamic alcohols [361] it leads to the corresponding aldehydes (Scheme 176a–c).

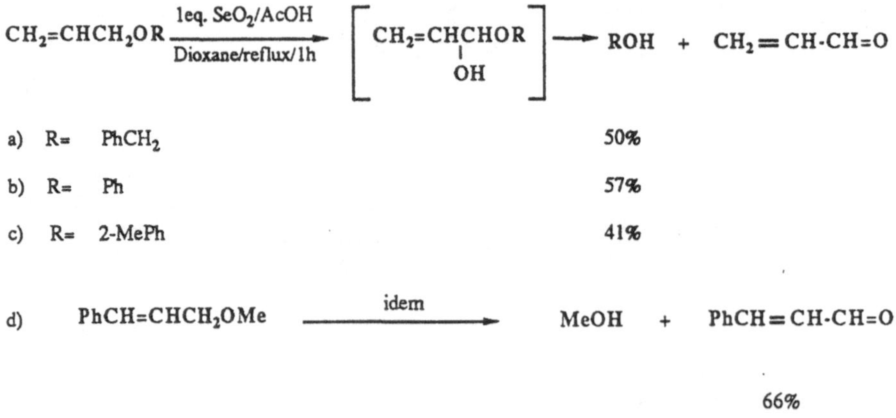

$$CH_2=CHCH_2OR \xrightarrow[\text{Dioxane/reflux/1h}]{\text{1eq. SeO}_2/\text{AcOH}} \left[\begin{array}{c} CH_2=CHCHOR \\ | \\ OH \end{array} \right] \longrightarrow ROH + CH_2=CH\text{-}CH=O$$

a)   R=   PhCH$_2$ 50%

b)   R=   Ph   57%

c)   R=   2-MePh   41%

d)   PhCH=CHCH$_2$OMe $\xrightarrow{\text{idem}}$ MeOH   +   PhCH=CH-CH=O

66%

Scheme 175 [426]

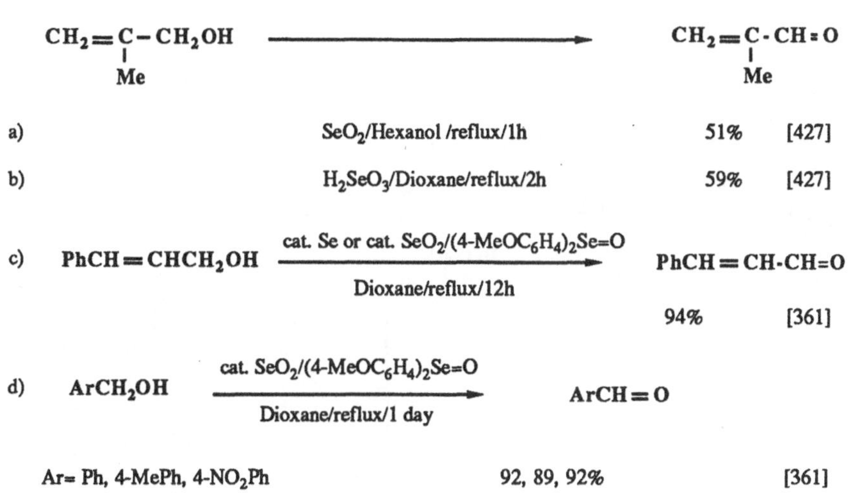

CH$_2$=C–CH$_2$OH   ⟶   CH$_2$=C-CH=O

    | |

   Me   Me

a)   SeO$_2$/Hexanol /reflux/1h   51%   [427]

b)   H$_2$SeO$_3$/Dioxane/reflux/2h   59%   [427]

c)   PhCH=CHCH$_2$OH   cat. Se or cat. SeO$_2$/(4-MeOC$_6$H$_4$)$_2$Se=O ⟶ PhCH=CH-CH=O

                                  Dioxane/reflux/12h

                                                  94%   [361]

d)   ArCH$_2$OH   cat. SeO$_2$/(4-MeOC$_6$H$_4$)$_2$Se=O ⟶ ArCH=O

                          Dioxane/reflux/1 day

    Ar= Ph, 4-MePh, 4-NO$_2$Ph   92, 89, 92%   [361]

Scheme 176

9.1.1.3 Stereochemical Considerations

The allylic oxidation of straight chain olefins with SeO$_2$ leads to allyl alcohols and to α,β-unsaturated carbonyl compounds possessing mainly the (E) stereochemistry (Schemes 153a, 155–157, 159, 160a, 165) [96, 370, 376, 378]. The stereoselectivity in these reactions is high particularly in the case of allyl alcohols resulting from the oxidation of trisubstituted olefins bearing an isopropylidene moiety (Schemes 155–157, 159, 160, 168) [*96, 370, 376, 378, 380, 381, 383, 388, 391, 402, 406, 428].

Careful investigation of the latter transformation, using labelled molecules [414, 428, 429] leads however to the observation [429] that both the methyl groups cis and trans to the alkyl group are oxidized. The oxidation of the latter being favored over the former. These results [429] are contrary to previous ones [414] in which these two methyl groups were found to be equally oxidizable.

9.1.1.4 Mechanistic Considerations

Since its discovery several different mechanisms have been proposed to explain the selenium dioxide mediated allylic oxidation of olefins. Wiberg and Nielsen [431] first proposed that the reaction involves the attack of the electrophilic selenium atom of SeO$_2$ by the carbon-carbon double bond with the implication of an intermediary organoselenium derivative. However the real breakthrough in the determination of the mechanism was later made by Sharpless and Lauer [344]. This mechanism shown in Scheme 177 involves two separate key steps:

(i) an ene reaction between the olefin and SeO$_2$ which should produce an allyl seleninic acid **47** (or ester).

(ii) a [2,3] sigmatropic rearrangement of **47** to the selenium ester **48** which is further hydrolysed to the allyl alcohol **49** or oxidized to the carbonyl compound **51**.

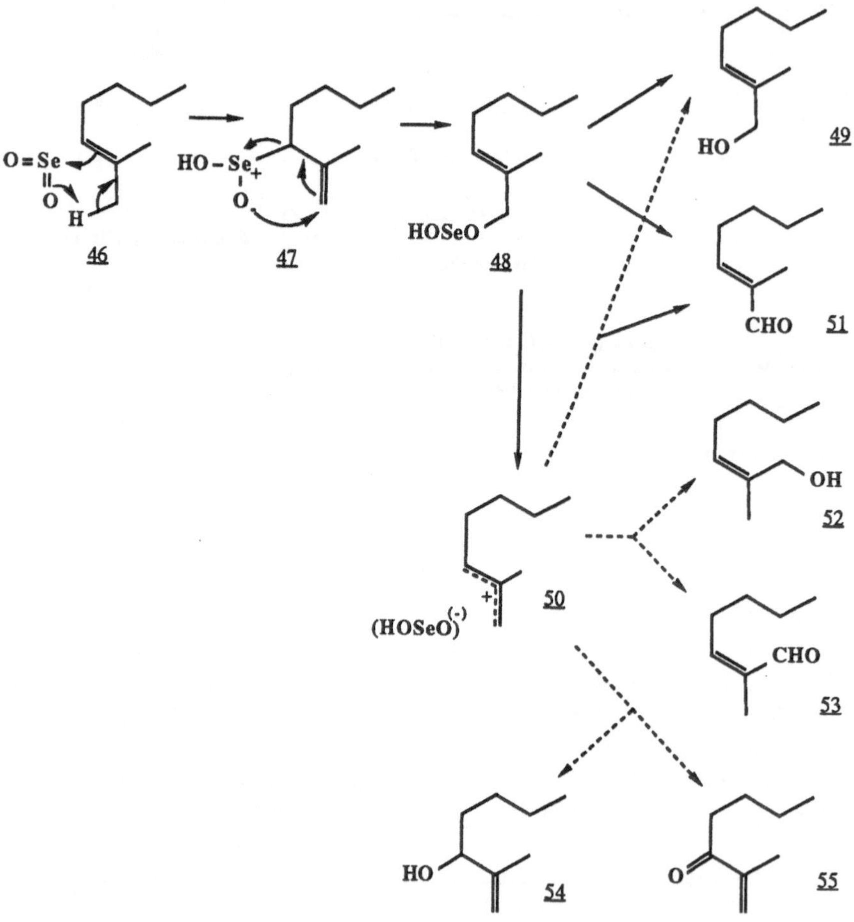

9. Selenium Dioxide Oxidations

Scheme 177

The ene reaction determines the site of oxidation and usually proceeds with a stereoselectivity determined by the steric interactions involved in the transition state.

The [2,3] sigmatropic shift assures the (E) configuration of the products 49 and 51. These guidelines apply particularly well for the oxidation of trisubstituted olefins [414, 429].

Sharpless and Lauer have not discarded in their original paper [344] a dissociation-recombination pathway that can lead to the other stereo- (52 and 53 Scheme 177, see for specific examples Scheme 159) [387] and regioisomers (54 and 55) which can arise from 50 and which are observed as by-products in some cases [370, 404] (Schemes 153a, b, 167b) but become the major ones in the case of medium sized (5–6 carbons) cyclic olefins (Schemes 166c) [403], some conjugated carbonyl compounds having an exocyclic double bond (Scheme 171) [413] as well as in the case of allylbenzenes [404] (Scheme 167a).

134

The mechanism presented in Scheme 177 was supported experimentally [432] by trapping of the intermediate allyl seleninic acid arising from the ene reaction between SeO$_2$ and 1-isopulegol as the seleninolactone **56** (Scheme 178). It was also found [344] that allylic seleninic acids prepared by an independent route [344] immediatly rearranged to allylic alcohols (or esters).

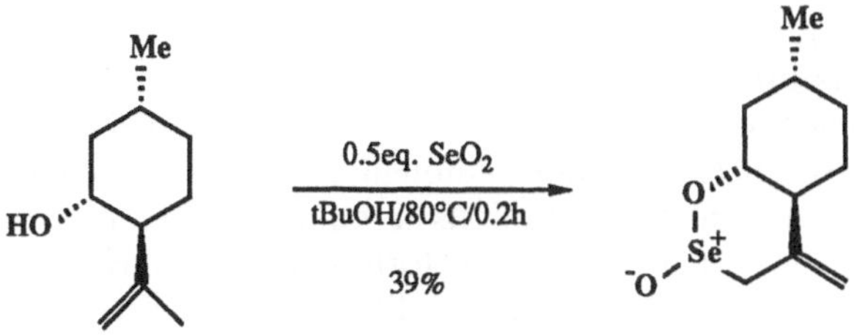

Scheme 178 [432]

9.1.1.5 Use of the Reaction for the Syntheses of Complex Molecules and as a Key Step in the Total Synthesis of Natural Products

Prior to 1960 selenium dioxide oxidation of olefins has been mainly used for the structure elucidation of natural products especially in the steroid area [*81, *89, *433] (Scheme 160b) [392, 393] (Scheme 174a) [417–420]. The propensity of selenium dioxide to effect dehydrogenation [*89, 434] of molecules possessing the cyclohexene moiety has often be used [96, 410, 411, 434, 435] for dehydrogenation of sterols and for the synthesis of bufadienolides.

The SeO$_2$ oxidation of olefins was used by Buchi [436] and Rapoport [384, 387] in a series of important syntheses, involving trisubstituted olefins, especially those possessing an isopropylidene moiety, which are known to produce allyl alcohols in a highly regio and stereoselective manner (Scheme 155a–f, h, j). The improved Sharpless method proved even more efficient in most of these cases (Scheme 155g, i) [376, 377, 381].

The SeO$_2$ allylic oxidation of olefins has been used inter alias as the key step in the synthesis of ganoderic acid [437], α-sinesal [438], d,l-serinin [384], squalene [391] (Scheme 160a), oxygenated germacrane sesquiterpenes [380, 390] (Scheme 156a, b, c), oxygenated humulenes [380, 407] (Scheme 156a–c), caryophyllenes [390] (Scheme 156e), elemanolides such as melitensin [374] (Scheme 164), decipiene diterpenes [415], tricyclic diterpenes [399] (Scheme 162f, g) mokupalide [383], hydroxylated vitamine D$_3$ [379] (Scheme 163a), oxa-prosta-glandin [402] (Scheme 165), acylindoles [408, 409] (Scheme 169), alkaloids [403] (Scheme 166c), tetrodamine [394] (Scheme 161a), clovene [416] (Scheme 173a), bakkenolide [424] (Scheme 173b), santonin [425] (Scheme 173c), camptothecin [413] (Scheme 171a), digitoxigenin [417–420] (Scheme 174a), onocerine [421], bilobanone [386], lycopodium alkaloids annotinine [439–441], coriamyrtin [401] (Scheme 163c) and quadrone [412].

135

9.1.2 Diol Formation from Alkenes and SeO$_2$/H$_2$O$_2$ or SeO$_2$/H$_2$SO$_4$

The presence of hydrogen peroxide or sulfuric acid besides selenium dioxide dramatically changes the course of the reaction since the double bond itself is oxidized rather than the allylic position.

The reaction of SeO$_2$/H$_2$SO$_4$ in acetic acid has been described [362–364] on a small number of olefins. The reaction occurs at 110 °C and leads to modest yields of 1,2-diacetates (Schemes 153c, 154e). The reaction is not stereoselective and in the case of cyclohexene a mixture of both stereoisomers is formed [362, 363] (Scheme 154e). Selenium dioxide oxidation of olefins has also been carried out in the presence of hydrogen peroxide. The reaction is thought to involve perselenious acid and is usually performed in alcoholic solutions at room temperature with catalytic amounts of SeO$_2$ [@367]. Diols are often formed [@367, 369, 442] (Scheme 153d, 154f) in these reactions. A remarkable difference of reactivity is observed between endo and exocyclic olefins. Epoxides [368, 400, 443] (Scheme 166b) and trans diols [368, 369, 444], resulting from the selenious acid catalysed ring opening of the epoxides [368] are obtained in the former cases whereas allylic oxidation of exocyclic olefins leads to the allyl alcohol [396–398] similar to the one obtained from the reaction of SeO$_2$ alone (Scheme 162e, compare entry d to e).

9.1.3 Oxidation of Functionalized Olefins (Such as Conjuguated Dienes, Polyenes, Enynes, Allylic Alcohols and Ethers as Well as α,β-Unsaturated Carbonyl Compounds and Enamines) with SeO$_2$

SeO$_2$ reacts with dienes but the nature of the products greatly depends upon the structure of the starting material. Open chain dienes cycloadd to SeO$_2$ and lead to products originally assigned [445] as selenones. The structure of the adduct from 2,3-dimethyl-1,3-butadiene and SeO$_2$ was later revised [446] and shown to be a se-

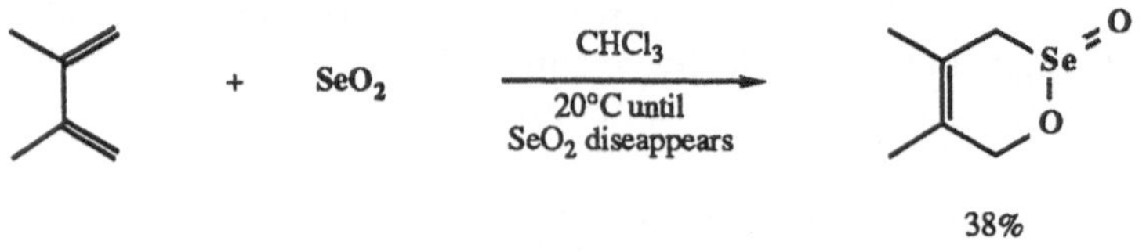

Scheme 179 [445, 446]

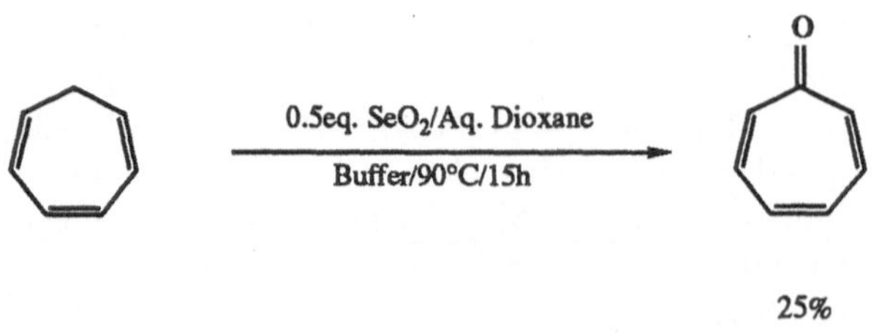

Scheme 180 [447]

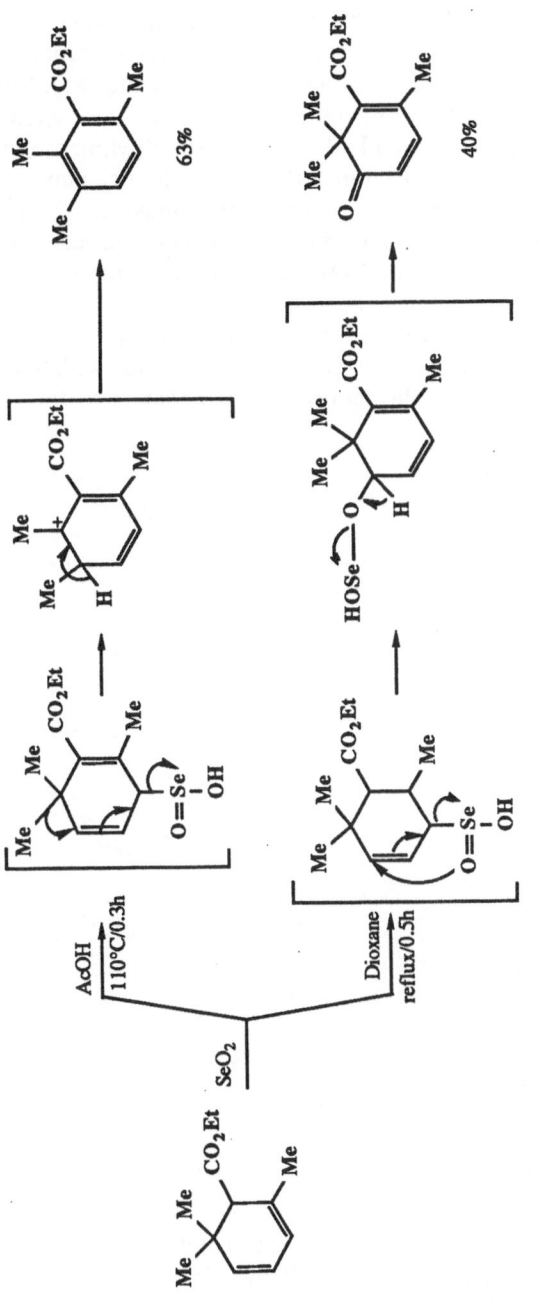

Scheme 181 [448]

leninolactone (Scheme 179). This was in fact the first example of trapping a SeO₂-olefin adduct [446] (for a related example see: Sect. 9.1.1, Scheme 178) [432].

Cyclic dienes and polyenes are usually oxidized to polyunsaturated ketones [385, 447, 448] (Schemes 158, 180, 181). These are often accompanied by aromatic compounds [385, 448], resulting from an alkyl migration. Their percentage is enhanced if acidic solvents are used [448] (Scheme 181). For example on reaction with SeO₂, β-damascone [385] produces a 65% yield of 2-keto-γ-damascone along with 14% yield of 1-(2,3,6-trimethylphenyl)but-2-en-1-one (Scheme 158). It is interesting to notice that the cyclohexadiene moiety is more easily oxidized than the enone system present in damascone [385]. On the other hand 1,3,5 cycloheptatriene is oxidized to tropone [447] on reaction with SeO₂ in buffered aqueous dioxan (Scheme 180). The buffer proved to be essential for the success of this reaction.

SeO₂, as expected, behaves differently with conjugated enynes than with dienes. Interestingly 2-methyl-1-hexene-3-yne on reaction with SeO₂/t-BuO₂H is selectively oxidized [449] on the methyl group attached to the carbon-carbon double bond (Scheme 182).

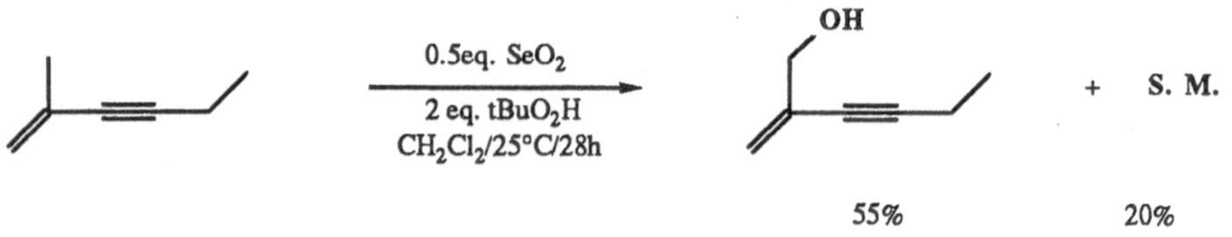

Scheme 182 [449]

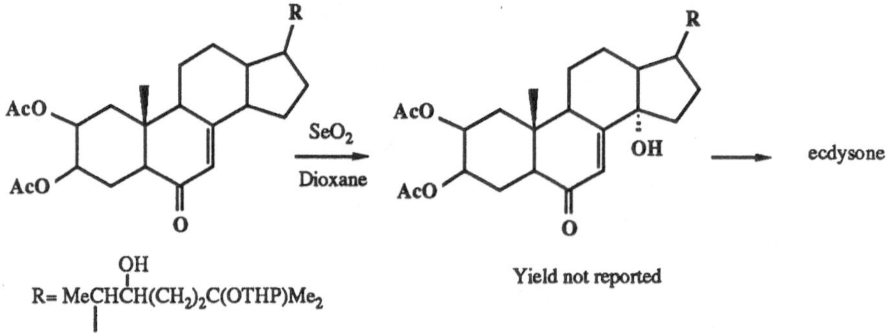

Scheme 183 [382]

Although less reactive than alkyl substituted olefins, allylic acetates [394, 395] (Scheme 161), allyl ethers [426] (Scheme 175), α,β-unsaturated ketones (Scheme 183) [382], α,β-unsaturated esters [417–423] as well as alkyl substituted lactones, pyrones (Scheme 184) [450–452] and indoles (Scheme 169) [408, 409] have been oxidized with SeO₂ or with SeO₂/t-BuO₂H mixtures. The reaction has

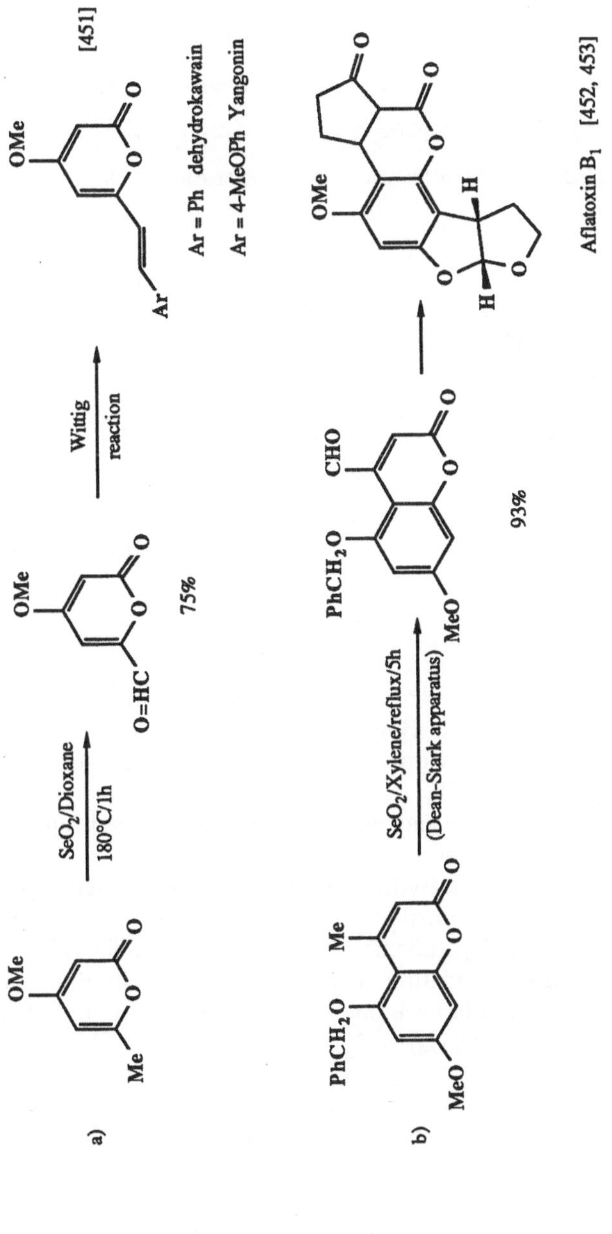

a)
OMe
Me
O
O
SeO2/Dioxane
180°C/1h
OMe
O=HC
O
O
75%
Wittig
reaction
OMe
Ar
O
O
[451]
Ar = Ph dehydrokawain
Ar = 4-MeOPh Yangonin
b)
PhCH2O
Me
MeO
O
O
SeO2/Xylene/reflux/5h
(Dean-Stark apparatus)
PhCH2O
CHO
MeO
O
O
93%
OMe
O
O
O
H
H
O
O
Aflatoxin B1
[452, 453]

Scheme 184

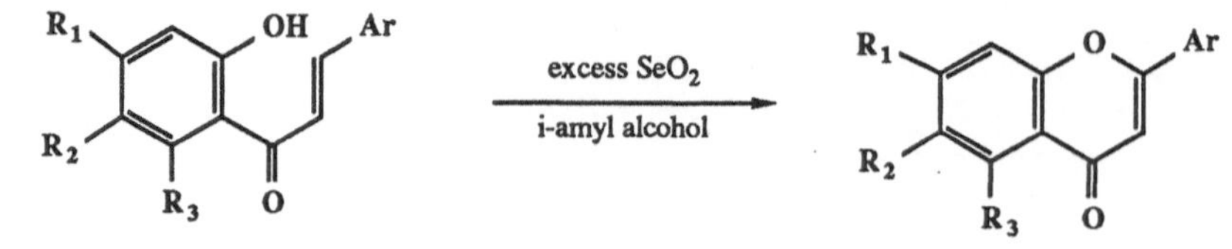

9. Selenium Dioxide Oxidations

	R_1	R_2	R_3	Ar	condition	Yield(%)	
a)	H	H	H	3-NO$_2$-4-BrPh	reflux/48h	80	[454]
b)	MeO	H	MeO	4-(PhCH$_2$O)Ph	" "	72	[455]
c)	MeO	MeO	MeO	3,4-(PhCH$_2$)$_2$Ph	reflux/6h	85	[456]

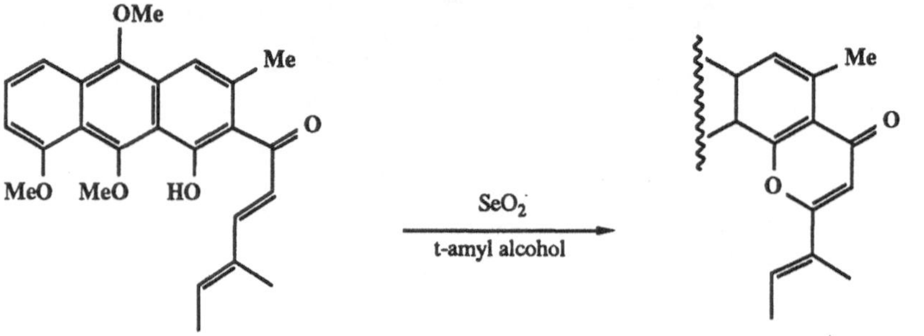

Yield not reported [457]

Scheme 185

been successfully used as the key step *inter alias* in the synthesis of several natural products such as tetrodamine from an allylic acetate [394] (Scheme 161); digitoxigenin [417–420], and α-onocerin [450] from an α,β-unsaturated ester (Scheme 174a); yangonin [451] (Scheme 184a) and aflatoxin B, [452, 453] (Scheme 184b) from a 2-pyrone and alpha coumarin respectively and α-ecdysone [382] (Scheme 183) from an enone.

The case of enones with an o-hydroxyaryl substituent attached to the carbonyl group (Scheme 185) deserves further comment since on reaction with SeO$_2$ a cyclization followed by a dehydrogenation leading to chromones [454–457] takes place rather than an allylic oxidation. This reaction was successfully used during the synthesis of O-methyl-kidamycinone [457] (Scheme 185d) and of genkwanin (Scheme 185b; Ar = p-HO-C$_6$H$_4$).

9.2 Reactivity SeO$_2$ Towards Acetylenic Hydrocarbons

Relatively few acetylenic compounds have been reacted with SeO$_2$. They usually produce propargylic alcohols if the propargylic position possesses an hydrogen [426, 449, 458, 459] (Schemes 186–188). Those which do not belong to this last subclass such as phenyl acetylene or diphenyl acetylene are inert under the usual

140

$$RCH_2C \equiv CH \longrightarrow \underset{\underset{OH}{|}}{RCHC \equiv CH}$$

a) R = Bu   0.5eq. SeO$_2$/EtOH/80°C   27%   [458]

b) R = C$_7$H$_{15}$ 0.5eq. SeO$_2$/3eq. tBuO$_2$H, CH$_2$Cl$_2$/20°C/25h   48%   [449]

                                                                          (+ 40% S.M.)

c)

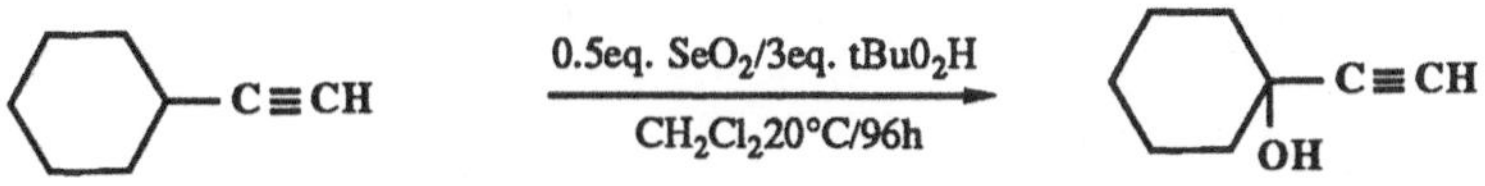

0.5eq. SeO$_2$/3eq. tBuO$_2$H
————————————————
CH$_2$Cl$_2$20°C/96h

                                                                          88%   [449]

                                                                          (+11% S.M.)

Scheme 186

$$PhC \equiv C - \underset{\underset{R_2}{\diagdown}}{\overset{\overset{R_1}{\diagup}}{CH}} \longrightarrow PhC \equiv C - \underset{\underset{R_2}{\diagdown}}{\overset{\overset{R_1}{\diagup}}{C}} - OH \ + \ PhC \equiv C - \underset{\underset{O}{\parallel}}{C} - R_2$$

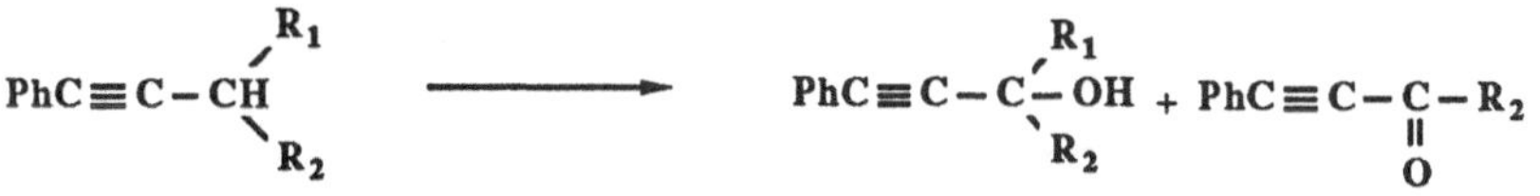

a) R$_1$ = R$_2$ = H   SeO$_2$/EtOH/reflux/10h*   0%*   0%   [459]

b) R$_1$ = H, R$_2$ = Me   0.5eq. SeO$_2$/2eq. tBuO$_2$H   54%   16%   [449]
                          CH$_2$Cl$_2$/20°C/48h

c) R$_1$ = H, R$_2$ = Me   SeO$_2$/EtOH/reflux/3h   25%   0%   [459]

* 10% propiophenone was isolated in addition to 50% S.M.

Scheme 187

conditions [365, 460] but can be oxidized [365, 460] to α-dioxo compounds when the reaction is performed at much higher temperature [460] (Scheme 189c) or in the presence of a mineral or an organic acid [365] (Scheme 189a, b, d, f). In fact under these conditions the oxidation reaction takes mainly place on the carbon-carbon triple bond even when propargylic hydrogens are present [365] (Scheme 189f, compare entry f to e).

Propargylic alcohols are usually obtained in modest yields when straight chain acetylenic compounds are reacted with SeO$_2$ under classical conditions [458, 459] (Schemes 186a, 187a, c) but the Sharpless method proved to be particularly suitable for the oxidation of terminal acetylenes [449] (Schemes 186b, 187b). Dialkyl-substituted alkynes however show a strong tendency to undergo α,α'-dioxygenation when the Sharpless method [449] is used (Scheme 188). The reactivity sequence for unsymmetrical compounds under the Sharpless method

141

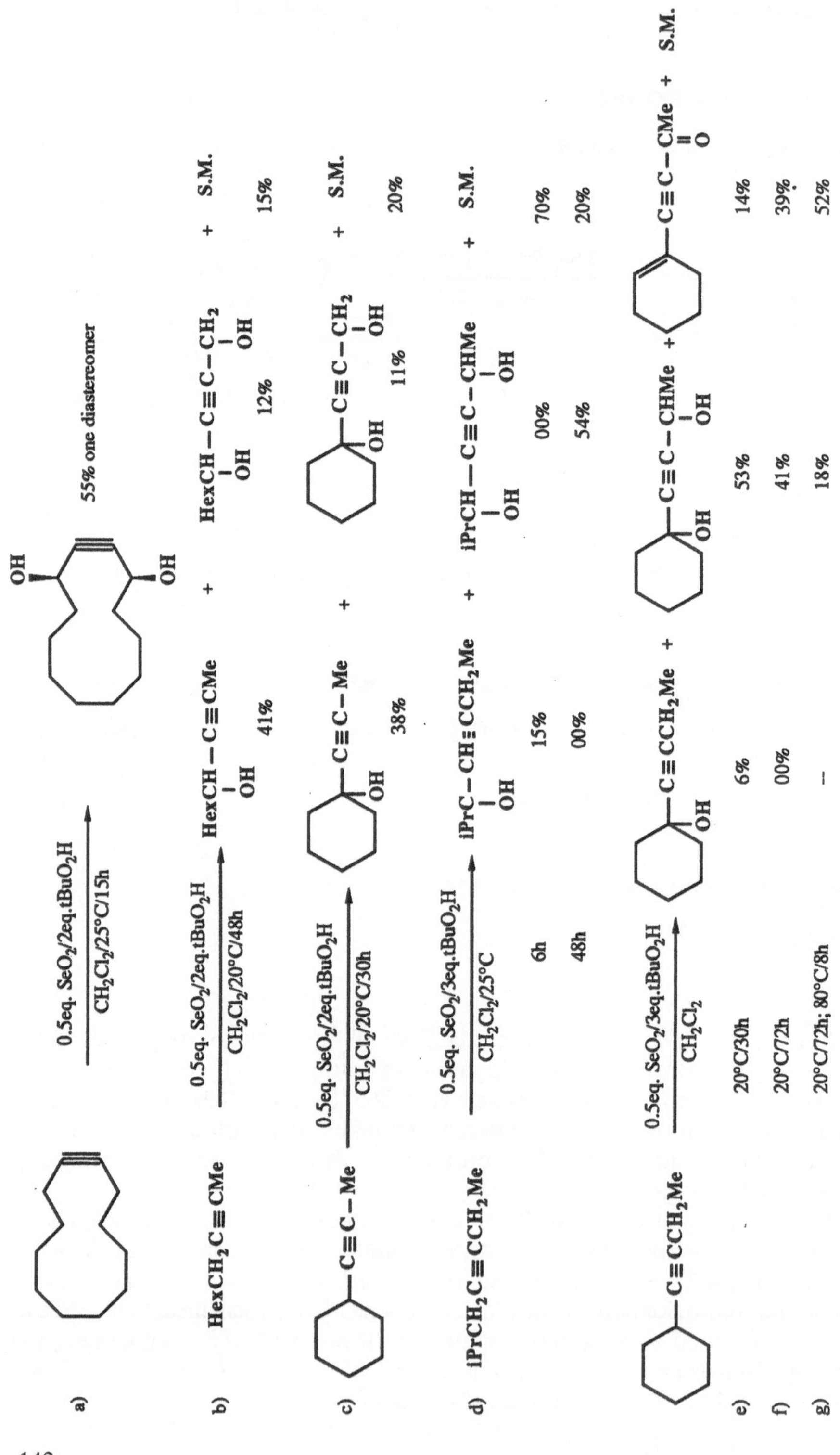

a) 0.5eq. SeO_2/2eq.$tBuO_2H$
CH_2Cl_2/25°C/15h

55% one diastereomer

b) 0.5eq. SeO_2/2eq.$tBuO_2H$
CH_2Cl_2/20°C/48h

$HexCH-C\equiv CMe$ + $HexCH-C\equiv C-CH_2$ + S.M.
OH 41%   OH   OH
12%   15%

c) 0.5eq. SeO_2/2eq.$tBuO_2H$
CH_2Cl_2/20°C/30h

$C\equiv C-Me$ + $C\equiv C-CH_2$ + S.M.
OH 38%   OH OH
11%   20%

d) 0.5eq. SeO_2/3eq.$tBuO_2H$
CH_2Cl_2/25°C

$iPrC-CH\equiv CCH_2Me$ + $iPrCH-C\equiv CCH_2Me$ + $iPrCH-C\equiv C-CHMe$ + S.M.
OH   OH   OH OH

6h   15%   00%   70%
48h   00%   54%   20%

e) 0.5eq. SeO_2/3eq.$tBuO_2H$
CH_2Cl_2

$C\equiv CCH_2Me$ + $C\equiv C-CHMe$ + $C\equiv C-CMe$ + S.M.
OH   OH   ‖
OH   O

e) 20°C/30h   6%   53%   14%
f) 20°C/72h   00%   41%   39%
g) 20°C/72h; 80°C/8h —   18%   52%

Scheme 188 [449]

[449] was found to be: CH$_2$ ~ CH > CH$_3$ (Scheme 188b, c, d). This order seems to be also valid when the classical procedure is used. Thus whereas 1-phenyl-1-butyne is rapidly oxidized to 1-phenyl-3-hydroxy-1-butyne [459] (25%) (Scheme 187c), 1-phenyl-1-propyne does not produce 3-hydroxy-1-phenyl-1-propyne but a small amount of propiophenone (Scheme 187a) [459] and 1-cyclohexenyl-1-butyne-3-one is formed as a by-product at 20 °C when 1-cyclohexyl-1-butyne is oxidized under Sharpless conditions. Its percentage can be increased at the expense of the propargylic alcohol when forcing conditions [449] are used (Scheme 188 compare entry g to e and f).

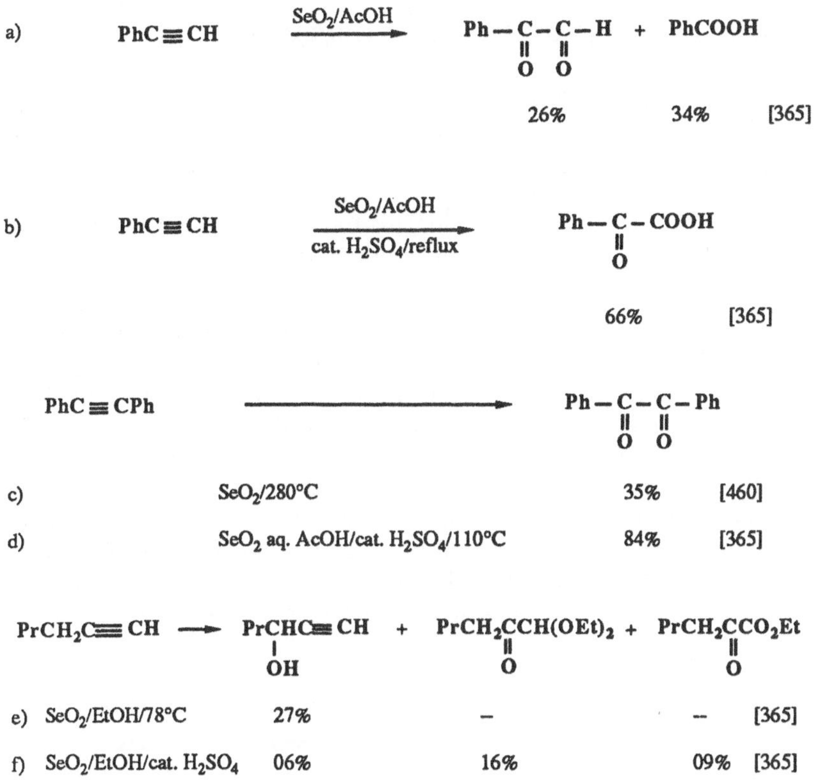

a) PhC≡CH $\xrightarrow{\text{SeO}_2/\text{AcOH}}$ Ph—C—C—H + PhCOOH
                                   ‖   ‖
                                   O   O

                                   26%   34%   [365]

b) PhC≡CH $\xrightarrow[\text{cat. H}_2\text{SO}_4/\text{reflux}]{\text{SeO}_2/\text{AcOH}}$ Ph—C—COOH
                                                        ‖
                                                        O

                                                 66%   [365]

PhC≡CPh ————————————→ Ph—C—C—Ph
                                       ‖   ‖
                                       O   O

c) SeO$_2$/280°C   35%   [460]

d) SeO$_2$ aq. AcOH/cat. H$_2$SO$_4$/110°C   84%   [365]

PrCH$_2$C≡CH —→ PrCHC≡CH + PrCH$_2$CCH(OEt)$_2$ + PrCH$_2$CCO$_2$Et
                     | ‖   ‖
                     OH   O   O

e) SeO$_2$/EtOH/78°C   27%   –   --   [365]

f) SeO$_2$/EtOH/cat. H$_2$SO$_4$ 06%   16%   09%   [365]

Scheme 189

Finally aryl propargylic ethers react rapidly with SeO$_2$ when the conventional procedure is used. The intermediary α-hydroxy ethers decompose and provide butynal and phenols in high yields [426] (Scheme 190). It should be pointed out that phenols are not further oxidized under these conditions (SeO$_2$, AcOH-dioxane, reflux, 1 h) [426].

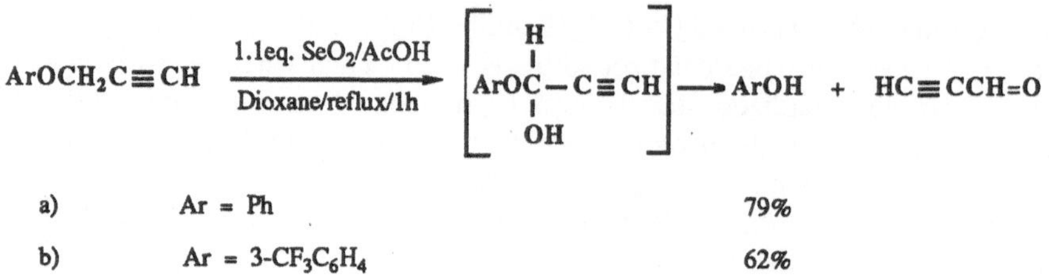

$$ArOCH_2C\equiv CH \xrightarrow[\text{Dioxane/reflux/1h}]{\text{1.1eq. SeO}_2\text{/AcOH}} \left[\begin{array}{c} H \\ | \\ ArOC-C\equiv CH \\ | \\ OH \end{array} \right] \longrightarrow ArOH \ + \ HC\equiv CCH=O$$

a)	Ar = Ph	79%
b)	Ar = 3-CF$_3$C$_6$H$_4$	62%

Scheme 190 [426]

9.3 Reactivity of SeO$_2$ with Carbonyl Compounds

9.3.1 Reactions Involving SeO$_2$

9.3.1.1 Scope and Limitation

Since its discovery by Riley in 1930 [461–463] the reaction of SeO$_2$ with aldehydes and ketones has been widely used in organic synthesis. Two kinds of selenium free compounds can result from this reaction:

A α-dicarbonyl compounds **58** (Scheme 191) resulting from the oxidation of a methyl or a methylene group attached to the carbonyl function;

B α,β-unsaturated compounds **59** arising from the dehydrogenation of their alkyl chain (Scheme 191). Recently ketoseleninic acids **57** (Scheme 191) were proposed [464] as key intermediates in these transformations. They are expected to be formed on addition of SeO$_2$ to the enol and to produce α-diketones [464] (such as **58** in Scheme 191) via a Pummerer type rearrangement and enones (such as **59** in Scheme 191) via a syn elimination between the seleninyl moiety and one of the β-hydrogens. The nature of the starting material seems to have a great influence on the course of the reaction but

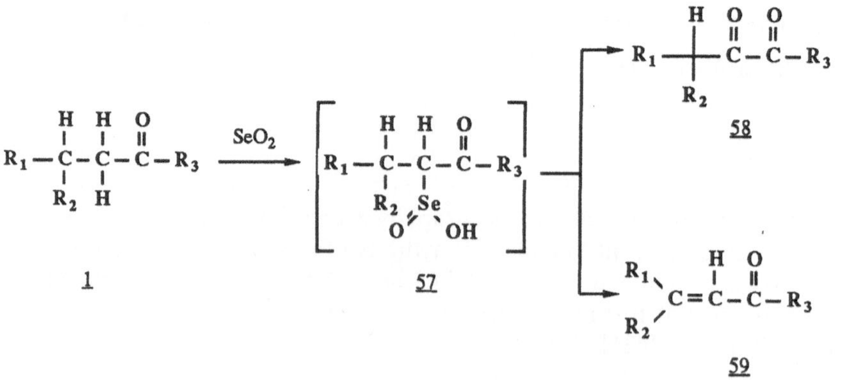

Scheme 191

usually one of the two types of compounds can be specifically obtained by using the right solvent. For example dioxane favors the formation of the dione whereas t-butanol favors the unsaturated carbonyl compound.

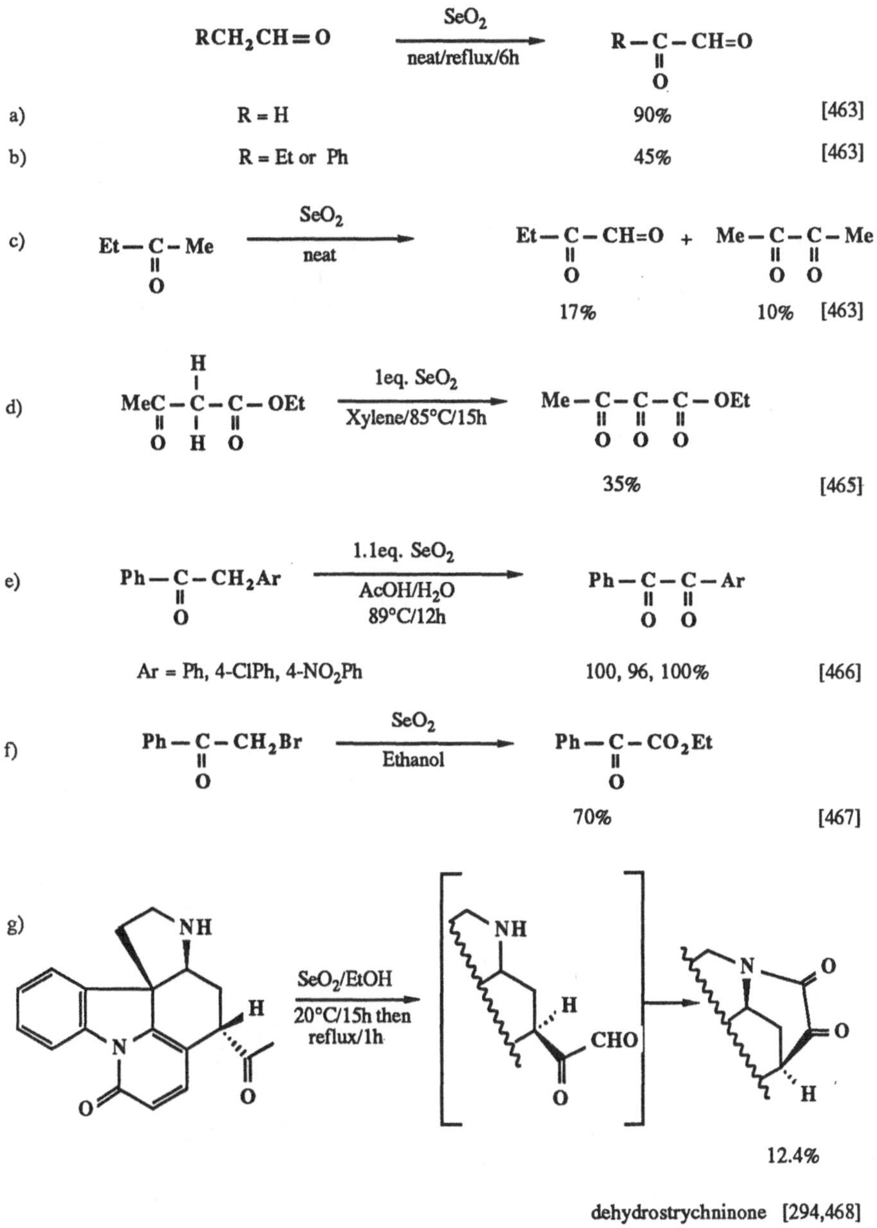

$$RCH_2CH=O \xrightarrow[\text{neat/reflux/6h}]{SeO_2} R-\underset{\underset{O}{\|}}{C}-CH=O$$

a)   R = H   90%   [463]

b)   R = Et or Ph   45%   [463]

c)

$$Et-\underset{\underset{O}{\|}}{C}-Me \xrightarrow[\text{neat}]{SeO_2} Et-\underset{\underset{O}{\|}}{C}-CH=O \; + \; Me-\underset{\underset{O}{\|}}{C}-\underset{\underset{O}{\|}}{C}-Me$$

                                             17%   10%   [463]

d)

$$MeC-\underset{\underset{H}{\overset{\overset{H}{|}}{|}}}{C}-\underset{\underset{O}{\|}}{C}-OEt \xrightarrow[\text{Xylene/85°C/15h}]{1eq.\ SeO_2} Me-\underset{\underset{O}{\|}}{C}-\underset{\underset{O}{\|}}{C}-\underset{\underset{O}{\|}}{C}-OEt$$

                                           35%   [465]

e)

$$Ph-\underset{\underset{O}{\|}}{C}-CH_2Ar \xrightarrow[\substack{AcOH/H_2O \\ 89°C/12h}]{1.1eq.\ SeO_2} Ph-\underset{\underset{O}{\|}}{C}-\underset{\underset{O}{\|}}{C}-Ar$$

        Ar = Ph, 4-ClPh, 4-NO$_2$Ph   100, 96, 100%   [466]

f)

$$Ph-\underset{\underset{O}{\|}}{C}-CH_2Br \xrightarrow[\text{Ethanol}]{SeO_2} Ph-\underset{\underset{O}{\|}}{C}-CO_2Et$$

                                           70%   [467]

g)

                                                        12.4%

dehydrostrychninone [294,468]

Scheme 192

9.3.1.2 Oxidation of Carbonyl Compounds to α-Dicarbonyl Compounds

Carbonyl compounds often lead [463–479] to α-dicarbonyl derivatives (Schemes 192, 193, 194). This process patented originally by Riley [461, 462] for the synthesis of glyoxals, was later used by Woodward [294, 468] in his synthesis of

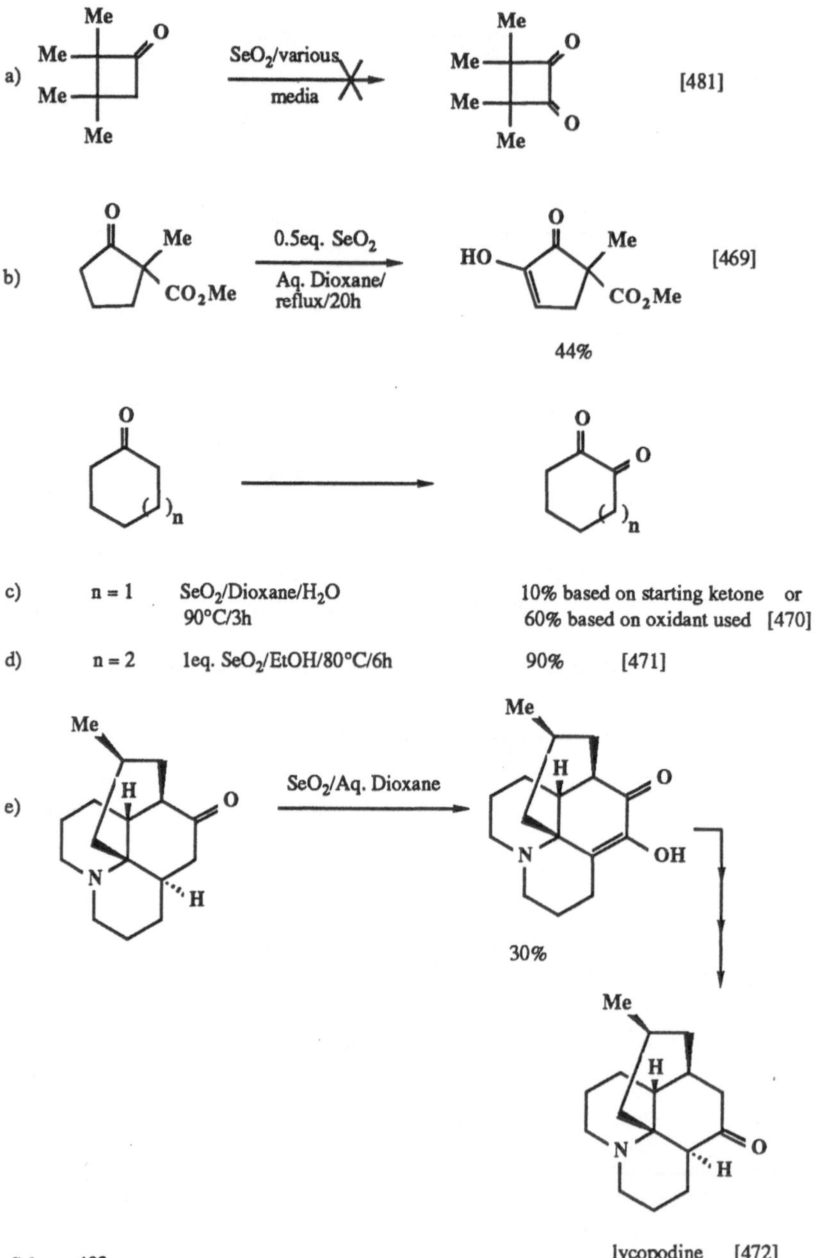

a) SeO₂/various media [481]

b) 0.5eq. SeO₂ Aq. Dioxane/ reflux/20h [469]

44%

c)   n = 1   SeO₂/Dioxane/H₂O   10% based on starting ketone   or
                    90°C/3h   60% based on oxidant used   [470]

d)   n = 2   1eq. SeO₂/EtOH/80°C/6h   90%   [471]

e) SeO₂/Aq. Dioxane

30%

lycopodine   [472]

Scheme 193

146

a) Y = O, X = H   excess SeO$_2$/EtOH/reflux/8h   27% + 73% SM   [473]

b)   "   "   excess SeO$_2$/Toluene/reflux   90% + 10% SM   [473]

c)   "   "   1eq. SeO$_2$/Ac$_2$O/reflux/6h   95%   [473, 474]

d) Y = 4-NO$_2$PhNH-N,
   X = H   SeO$_2$/AcOH/12h   95%   [475]

e) Y = O, X = OAc   1eq. SeO$_2$/AcOH/reflux/6h   00%   [475]

f)   "   "   3eq. SeO$_2$/Dioxane/140°C   41%   [475]

Scheme 194

strychnine (Scheme 192g). In this case the α-dicarbonyl compound was not isolated since it reacted intramolecularly with the amine present in the molecule. The resulting carbinolamine was further oxidized to the corresponding amide by SeO$_2$ (Scheme 192g). It is interesting to notice that epimerisation at the C-14 centre [294] concomitantly takes place during this transformation. Cyclopentanones [469, 473], cyclohexanones [470, 472–476*] and cycloheptanones [471] are often oxidized to the corresponding α-dicarbonyl compounds or α-hydroxy-α-enones on reaction with SeO$_2$ in aqueous dioxane or ethanol (Schemes 193, 194). This reaction has been used as a key step in the synthesis of lycopodine (Scheme 193e) [472] as well as for the synthesis of α-diones especially in the camphor series in which the competing α,β-dehydrogenation leading to enones is disfavoured for steric reason [473, 474, 476] (Scheme 194a–c, e, f). This process permits the synthesis of phytohormone analogs [474, 476] from acetoxy camphor (Scheme 194f) but was unsuccessful for 2,2,3,3-tetramethyl cyclobutanone [481] (Scheme 193a). Polyfunctionalized ketones whose structures are related to those described above have also been oxidized by SeO$_2$ [473, 480, 482, 483]. Some of these results are collected in Schemes 195 and 196. For example α-hydroxy and α-bromoketones have been selectively oxidized on the carbon bearing the heteroatom and produced α-diketones [473, 480] (Scheme 195). The reaction is particularly efficient [480] for 60 (Scheme 195e) which cannot be oxidized [@480] with other reagents such as Cu(OAC)$_2$, BiO$_2$, PbO$_2$, Pb(OAC)$_4$, CrO$_3$, N$_2$O$_4$, DMSO/AcO$_2$, in various solvents.

---

* Ref. 475 (Scheme 194d) describes a related reaction taking place on an hydrazone (see Sect. 9.6).

9. Selenium Dioxide Oxidations

a) X = OH	Excess SeO$_2$/EtOH/reflux/2h	40%	[473]
b) "	SeO$_2$/neat/heated 0.2h	85%	[473]
c) X = Br	SeO$_2$/Ac$_2$O/135°C	00%	[473]
d) "	SeO$_2$/neat/heated 6h	55%	[473]

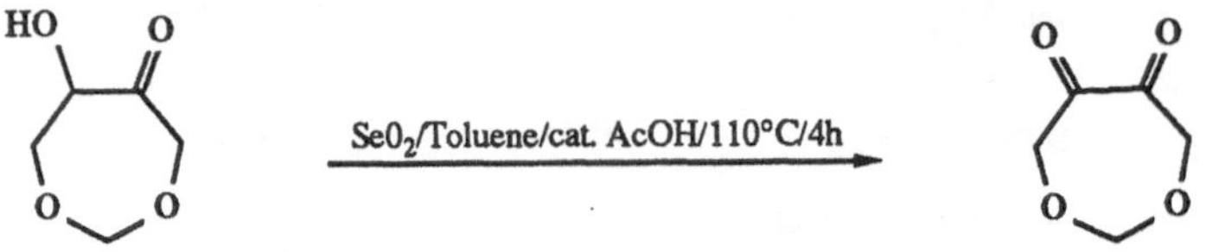

SeO$_2$/Toluene/cat. AcOH/110°C/4h

| e) | 60 | | 41% | [480] |

Scheme 195

α-Diketones bearing hydrogens on the carbon α to the carbonyl group are susceptible to further oxidation to triketones [482, 483]. These often rearrange and finally produce ring contracted α-diones [482, 483] after spontaneous extrusion of carbon monoxide (Schemes 196, 197) or to unsaturated α-diones [484–486] (Scheme 198). Therefore care has to be taken in order to avoid overoxidation during the reaction of SeO$_2$ with monoketones. It was found [483] for example that 3,3,5,5-tetramethyl cyclohexanone and 3,3,5-trimethyl-5-phenyl cyclohexanone produce, especially if acetic acid is used as the solvent, 2,2,4,4-tetramethyl cyclopentane-1,5-dione and 2,2,3-trimethyl-5-phenyl cyclopentane-1,4-dione respectively when reacted with excess of SeO$_2$ (Scheme 197b–d). In accord with the proposals presented above, it was found that 2,2,4,4-tetramethyl cyclohexane-1,6-dione ist cleanly transformed [482, 483] to 2,2,4,4-cyclopentane-1,5-dione under similar conditions (Scheme 196).

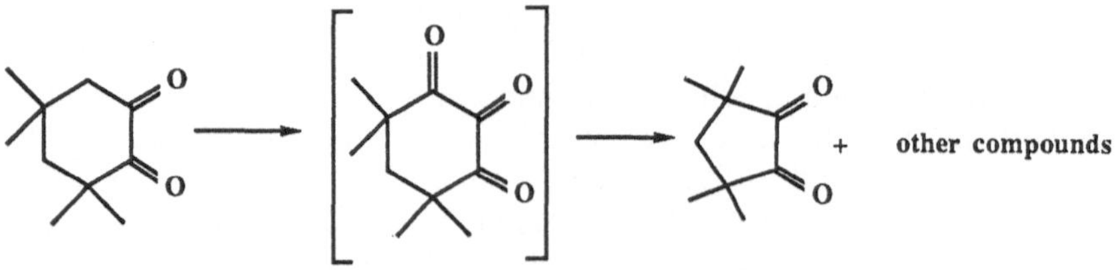

a)	1eq. SeO$_2$/Dioxane/reflux/5h	50%	[482, 483]
b)	2.5eq. SeO$_2$/Dioxane/reflux/5h	71%	[483]
c)	2eq. SeO$_2$/AcOH/1h	77%	[483]

Scheme 196

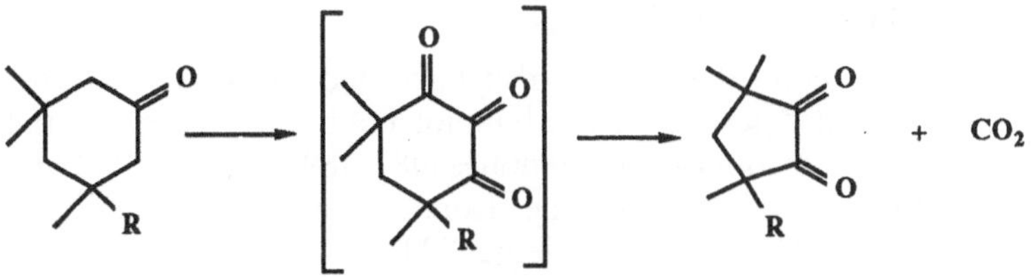

a) R = Me	2eq. SeO_2/Dioxane/reflux/20h	30%	[482,483]
b) "	3.5eq. SeO_2/AcOH/reflux/2h	87%	[483]
c) R = Ph	2.5eq. SeO_2/AcOH/reflux/2h	54%	[483]
d) "	3.5eq. SeO_2/AcOH/reflux/2h	92%	[483]

Scheme 197

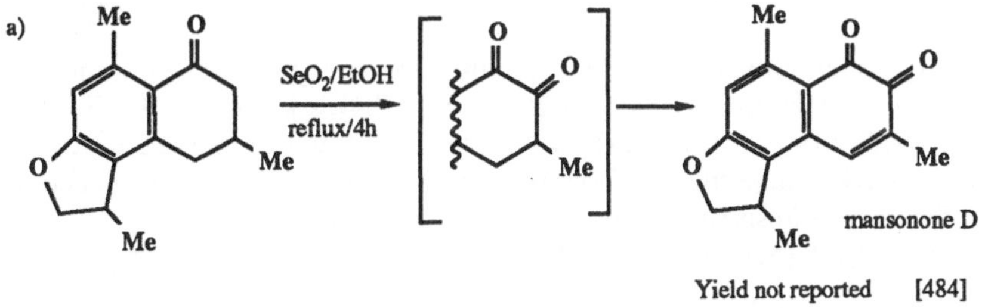
a)
SeO2/EtOH
reflux/4h
mansonone D
Yield not reported [484]

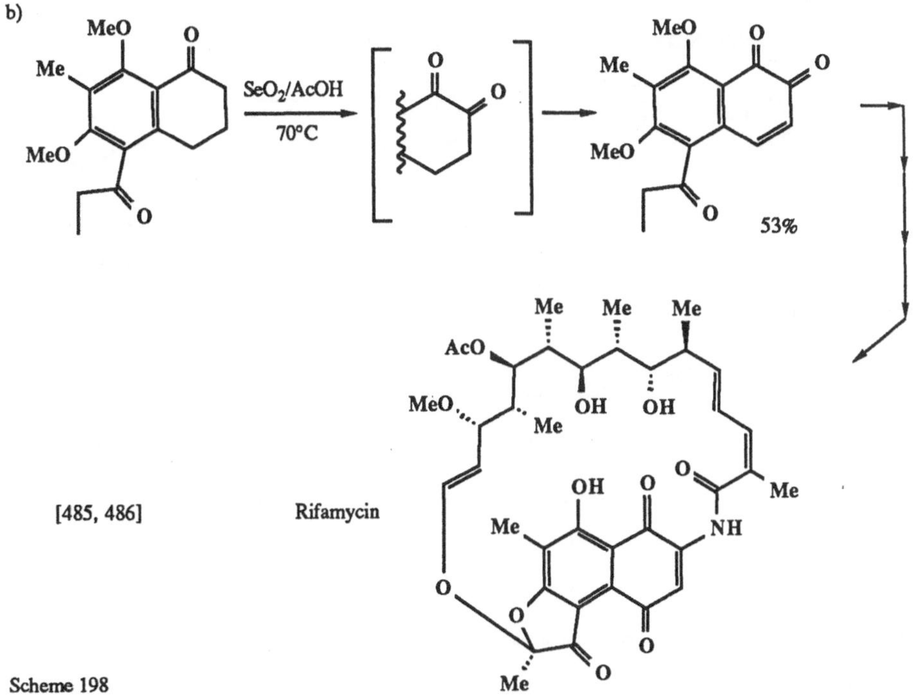
b)
SeO2/AcOH
70°C
53%
[485, 486]
Rifamycin

Scheme 198

9.3.1.3 Dehydrogenation of Carbonyl Compounds

Selenium dioxide also promotes the α,β-dehydrogenation of ketones. The reaction is particularly favoured if performed in t-butanol and has been used for the synthesis of cyclopentenones from cyclopentanones [487–489] (Scheme 199) and of cyclohexenones and α,α'-cyclohexadienones from cyclohexanones especially those fused to other cycles [425, 490–498] (Schemes 200–203). It was used as a key step in the synthesis of various natural products such as pentenomycin I (Scheme 199a) [487, 488], 6-α,9-α-difluoro-16-α-methylprednisolone [495] (Scheme 203), mayurone, thujopsene and thujopsadiene (Scheme 200) [490]. The SeO$_2$ dehydrogenation of ketones to enones proved to be superior to the DDQ dehydrogenation reaction in the case of 2-carbomethoxycyclopentanone (Scheme 199 compare entry b to c) [489] and even to the Sharpless / Reich selenoxide elimination sequence in the case of mayurone (Scheme 200) [490]. It was recently found [289, 295] that cholestan-3-one is oxidized to the corresponding 1,4-dien-3-one on reaction with catalytic amounts (0,5 equiv.) of SeO$_2$ in the presence of iodoxybenzene. The last reagent is able to regenerate SeO$_2$ but is apparently unable to overoxidize it to SeO$_3$.

The dehydrogenation reaction is a highly favoured process when another carbonyl group such as an ester [439–441] (Scheme 204), a ketone [479, 499–501] (Scheme 205–207) or a carbon-carbon double bond [499] (Scheme 205) or an aromatic ring [484–486] (Scheme 198) is present on the beta carbon. This reaction was used for the synthesis of various natural products such as steroids [*81, 479, 491, 492, 495, 502] (Schemes 202, 203, 206), especially for the synthesis of

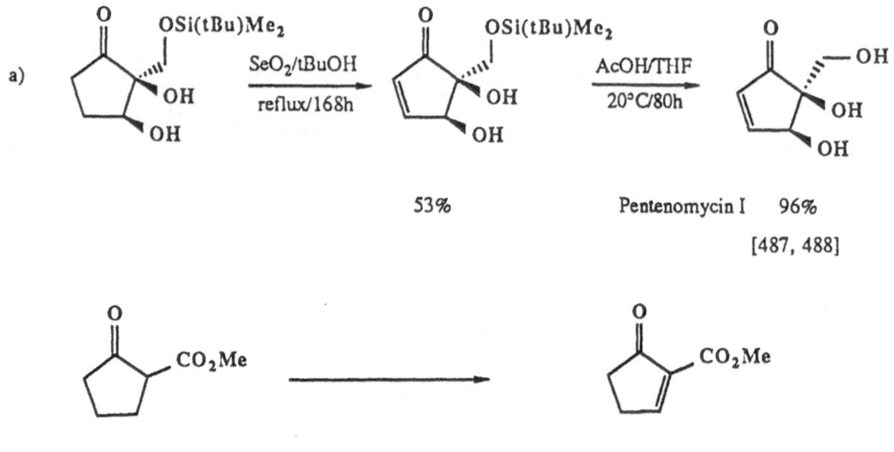

53%

Pentenomycin I   96%

[487, 488]

| b) | SeO$_2$/Dioxane*/reflux | 45% | [489] |
| c) | DDQ | 10% | [489] |

* The reaction does not work using dioxane distilled from LiAlH$_4$

Scheme 199

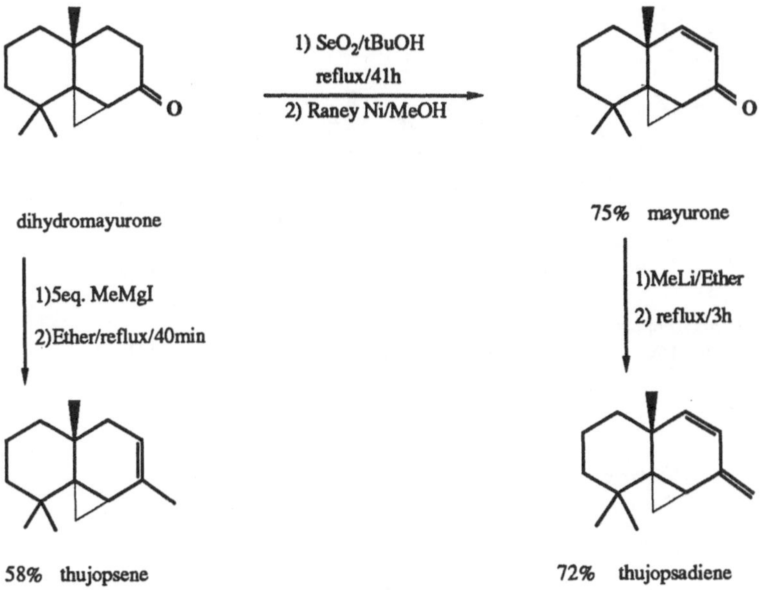

dihydromayurone

75% mayurone

1) SeO$_2$/tBuOH
   reflux/41h

2) Raney Ni/MeOH

1)5eq. MeMgI

2)Ether/reflux/40min

1)MeLi/Ether

2) reflux/3h

58% thujopsene

72% thujopsadiene

Scheme 200 [490]

a)	R=Me, X=OH	SeO$_2$/Various solvents	traces	[491]
b)	R=Me, X=OAc	ibid	70%	[491]
c)	R=Et, X=H	SeO$_2$/tBuOH,Pyridine	61%	[492]

d)

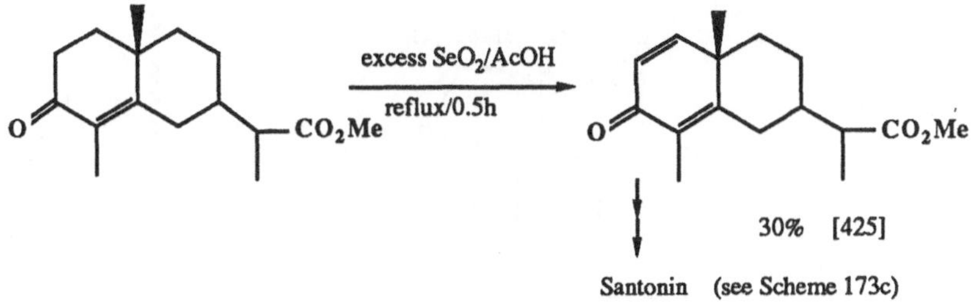

excess SeO$_2$/AcOH

reflux/0.5h

30% [425]

Santonin (see Scheme 173c)

Scheme 201

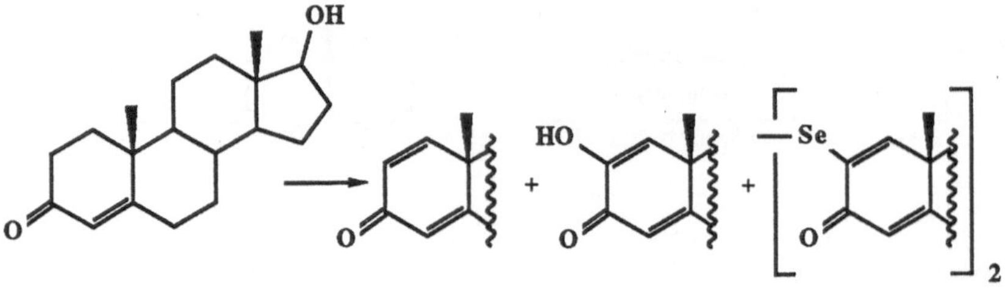

9. Selenium Dioxide Oxidations

SeO$_2$/C$_6$H$_6$-H$_2$O/reflux/5h	35%	00%	--	[493]
1 eq. SeO$_2$/tBuOH/AcOH/reflux then 0.3eq. SeO$_2$/reflux/16h	53%	2%	24%	[494]

Scheme 202

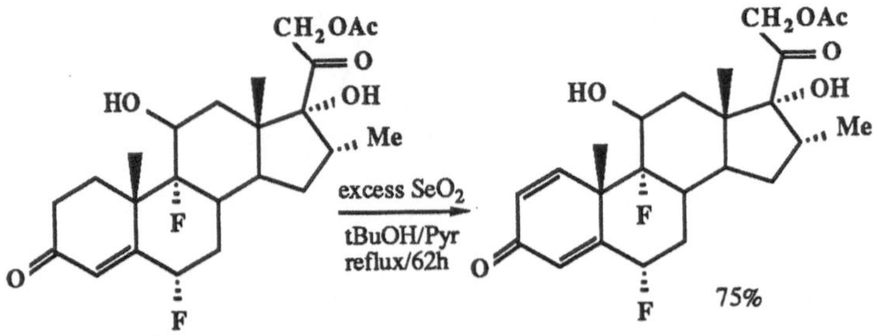

Scheme 203 [495]

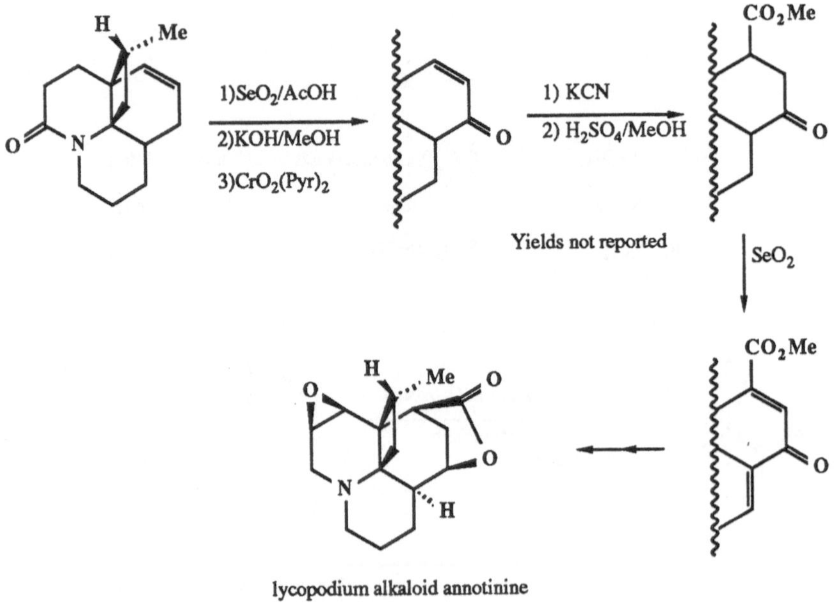

Yields not reported

lycopodium alkaloid annotinine

Scheme 204 [439, 440, 441]

152

a)

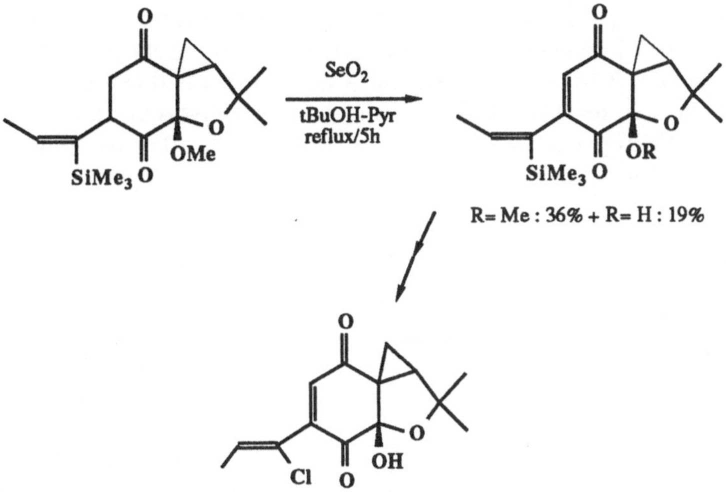
SeO2
tBuOH-Pyr
reflux/5h
OMe
SiMe3
OR
R= Me : 36% + R= H : 19%
OH
Cl

mycorrhizin A [499]

b)

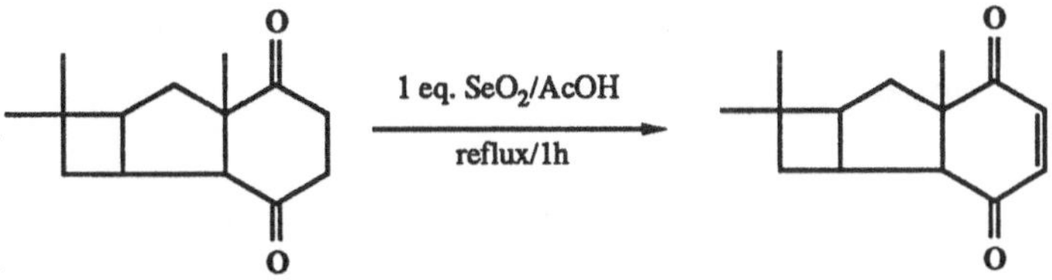
1 eq. SeO2/AcOH
reflux/1h

good yield [500]

Scheme 205

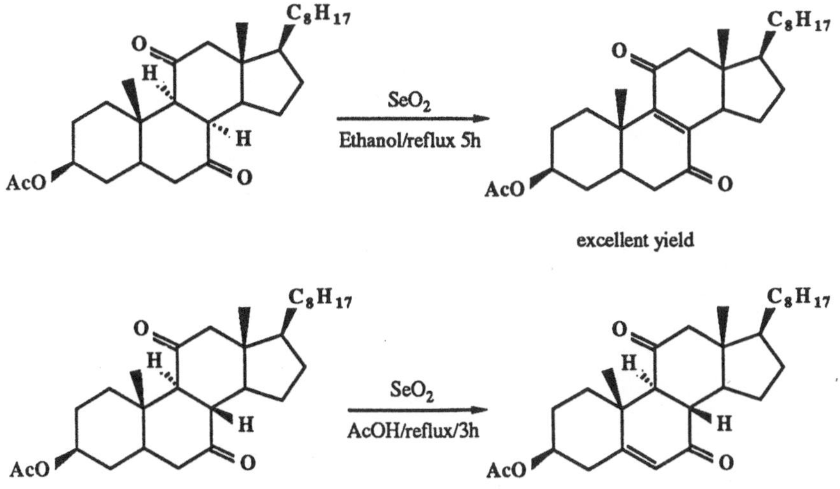
C8H17
SeO2
Ethanol/reflux 5h
AcO
excellent yield
SeO2
AcOH/reflux/3h

Yield not reported

Scheme 206 [479]

9. Selenium Dioxide Oxidations

cortisone [*81, 502] from 12-keto bile acids as well as for the preparation of mansonone D [484] (Scheme 198a), rifamycin [485, 486] (Scheme 198b), santonin [425] (Scheme 201d) and mycorrhizin A antibiotics [499] (Scheme 205a). In the case of mansonone D [484], the resistance of the dihydrobenzofuran moiety to undergo dehydrogenation with SeO_2 as well as with other oxidants such as $Pb(OAc)_4$ or chloranil has been mentionned [484].

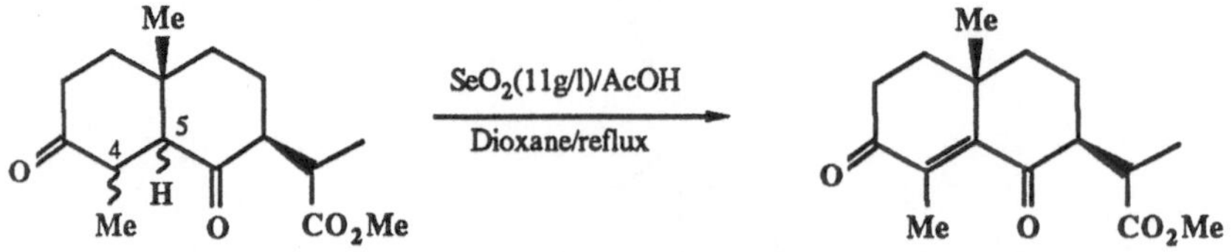

	relative configuration at C_4 - C_5	time required for	
		20% oxidation	40 % oxidation
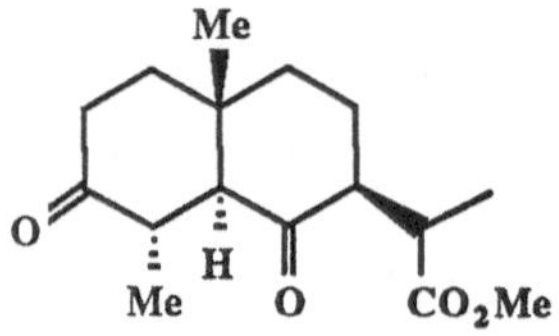	trans		4% in 120h
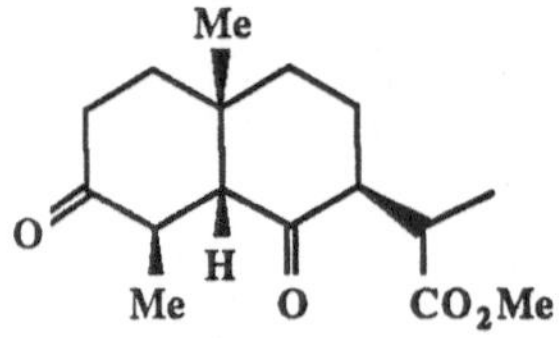	trans	32h	77h
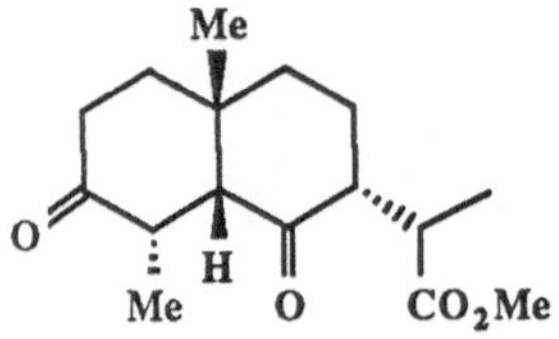	cis	20h	100h
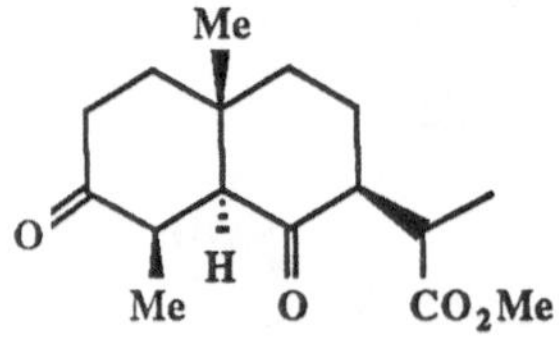	cis	8h	18h
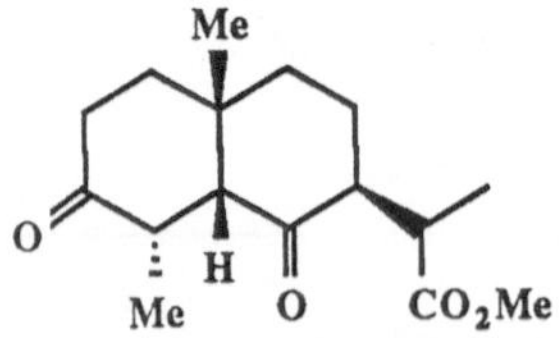	cis	4h	10h

Scheme 207 [501]

The dehydrogenation of 1,4-dicarbonyl compounds is highly favoured when the two hydrogens to be removed possess a cis relationship [479, 501, 503]. For example Barton et al. [501] found that methyl 3,6-dioxoeudesmanates possessing hydrogens in the cis relationship at the 4 and 5 positions are more readily oxidized to methyl 3,6-dioxoeudesm-4-enoates than their trans analogues (Scheme 207). It was also observed [479, 503] that the dehydrogenation of the 1,4-diketones takes a different course [479] with the cis (8,9) and the trans (8,9) isomers of 7,11-deketolanosterylacetates (Scheme 206). The cis isomer being in general the more easily dehydrogenated (see also Scheme 208 compare b to c) [479, 503].

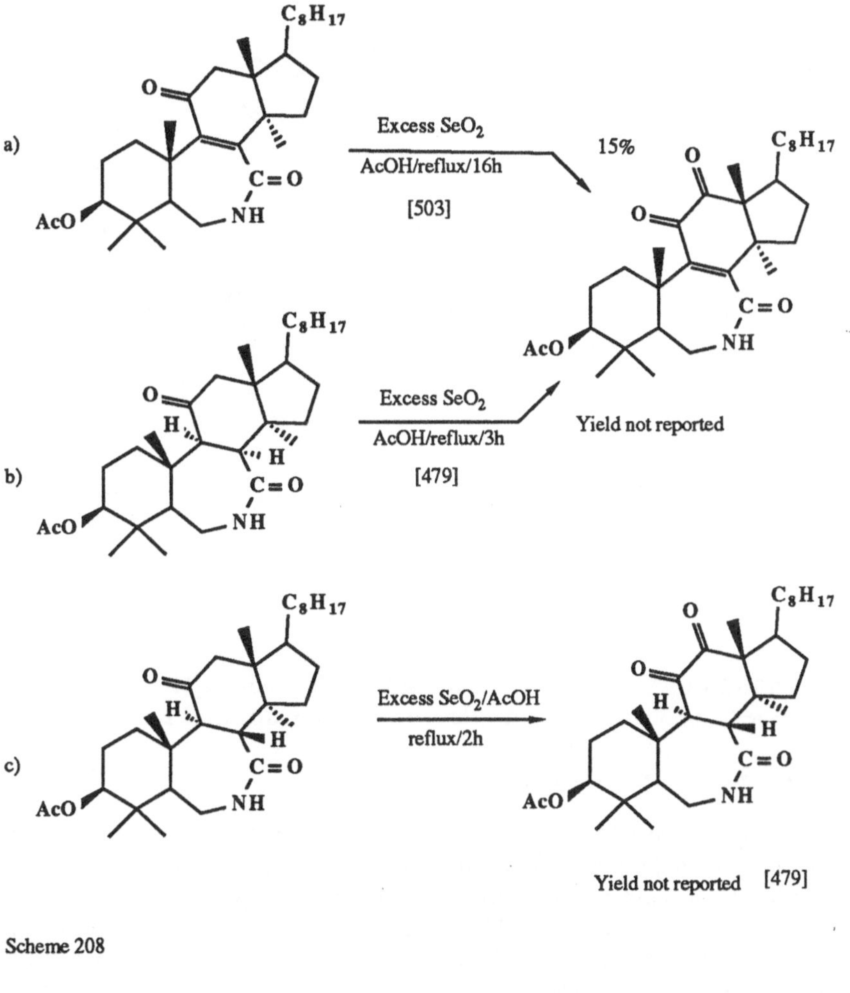

Scheme 208

9.3.1.4 Acetalisation of Carbonyl Compounds Catalyzed by SeO$_2$

SeO$_2$ interacts with acetals. For example it was found, during the synthesis of mycorrhizin A, that partial deprotection of an acetal occurs [499] (Scheme 205a).

It was also found that SeO$_2$ in methanol [504, 505] or in ethylene glycol [506] catalyses the formation of dimethyl acetals or of 1,3-dioxolans from 3-keto steroids (Scheme 209). Under similar conditions however 11-, 17- and 20-ketosteroids as well as Δ^4-3-ketosteroids do not produce the corresponding acetals.

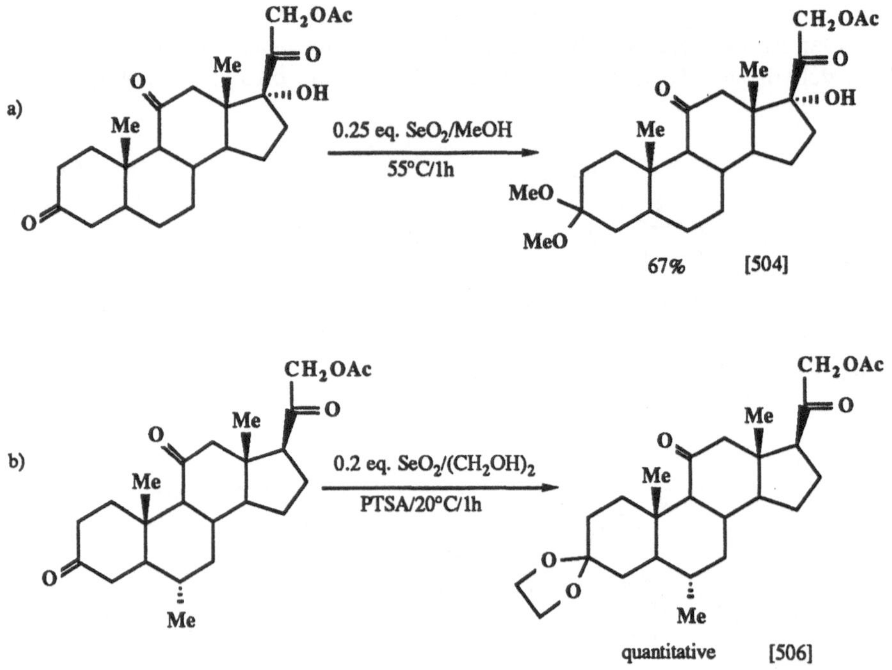

Scheme 209

9.3.2 Oxidation of Carbonyl Compounds With SeO$_2$/H$_2$O$_2$ – Synthesis of Carboxylic Acids from Aldehydes and Ketones

9.3.2.1 Reaction of SeO$_2$/H$_2$O$_2$ with Ketones

The course of the reaction of SeO$_2$ with carbonyl compounds is dramatically changed if the reaction is performed in the presence of hydrogen peroxide. Under these conditions usually no α-dicarbonyl compounds and enones but rather cycloalkane carboxylic acids [507–517] arising from an oxidative ring contraction are obtained (Schemes 210–213a, b) as well as in some cases lactones [512, 514–518] resulting from a process analogous to the Baeyer-Villiger oxidation (Schemes 213–215). The latter transformation is particularly favoured with strained ketones such as adamantanone [519] (Scheme 215).

Acyclic ketones rearrange oxidatively to carboxylic acids on reaction with SeO$_2$/H$_2$O$_2$. Thus acetone produces propionic acid [520], deoxy-benzoine leads to

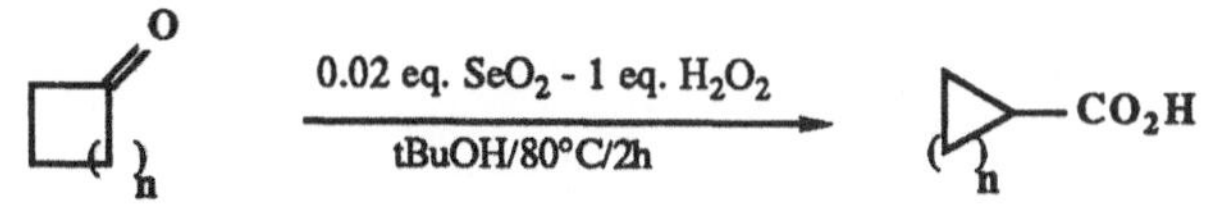

	n	Yield (%)	
a)	1	00	[507, 508]
b)	2	23	[507]
c)	3	60	[507, 508]
d)	4	35	[507]
e)	5	25	[508]
f)	6	20	[508]
g)	7	10	[508]
h)	9	30	[509]

Scheme 210

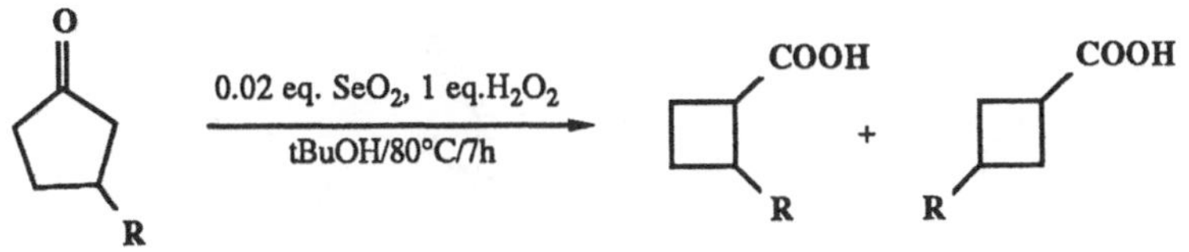

a)	R = Me	14%	13%
b)	R = tBu	24%	21%

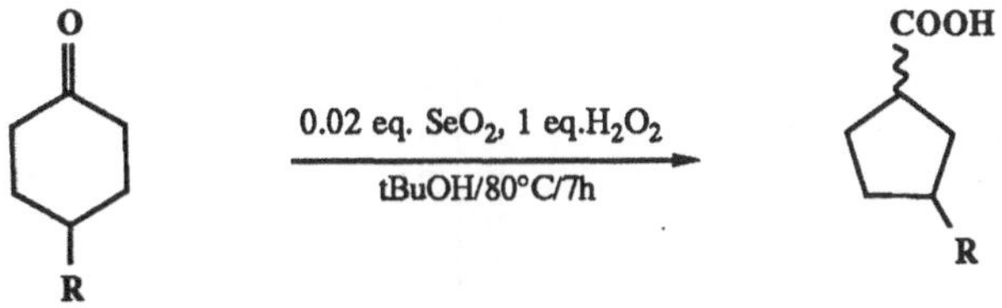

c)	R = Me	28% (cis/trans : 34/64)
d)	R = tBu	45% (cis exclusively)

Scheme 211 [510]

diphenyl acetic acid [521] and cyclopropyl methyl ketone leads to cyclopropyl acetic acid [508] (Scheme 216). Glutaric acid has been obtained from levulinic acid [522] and functionalized carboxylic acids have been produced in low yield from alkyl aryl ketones [521, 523] and from α-halogenoketones [524].

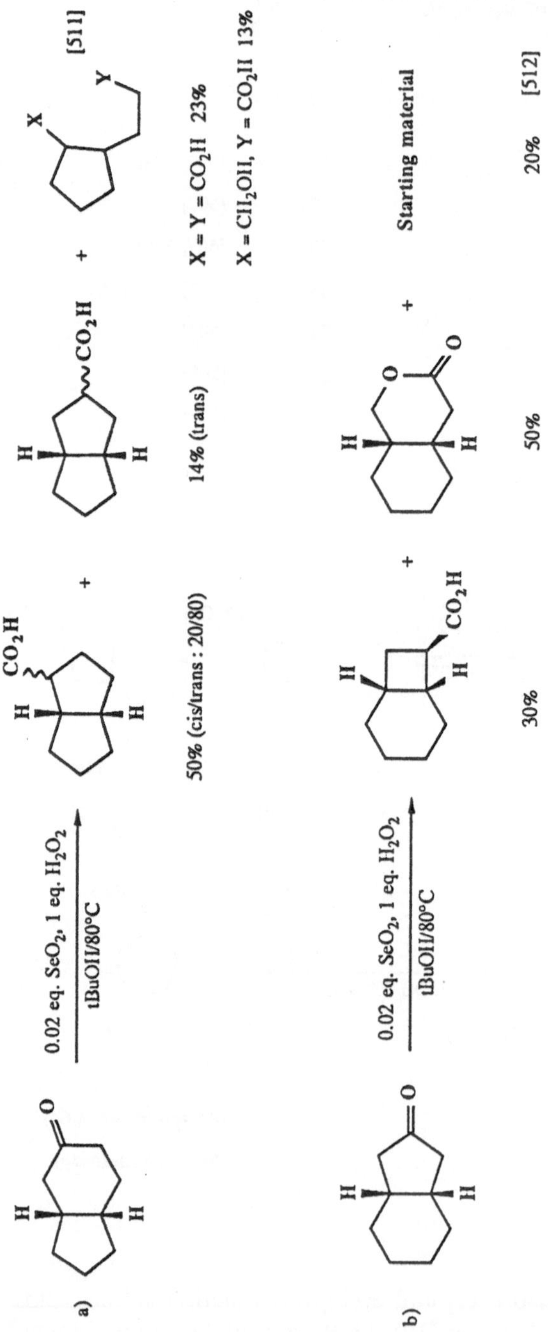
a)
0.02 eq. SeO2, 1 eq. H2O2
tBuOH/80°C
50% (cis/trans : 20/80)
14% (trans)
X = Y = CO2H 23%
X = CH2OH, Y = CO2H 13%
[511]
b)
0.02 eq. SeO2, 1 eq. H2O2
tBuOH/80°C
30%
50%
Starting material
20%
[512]

Scheme 212

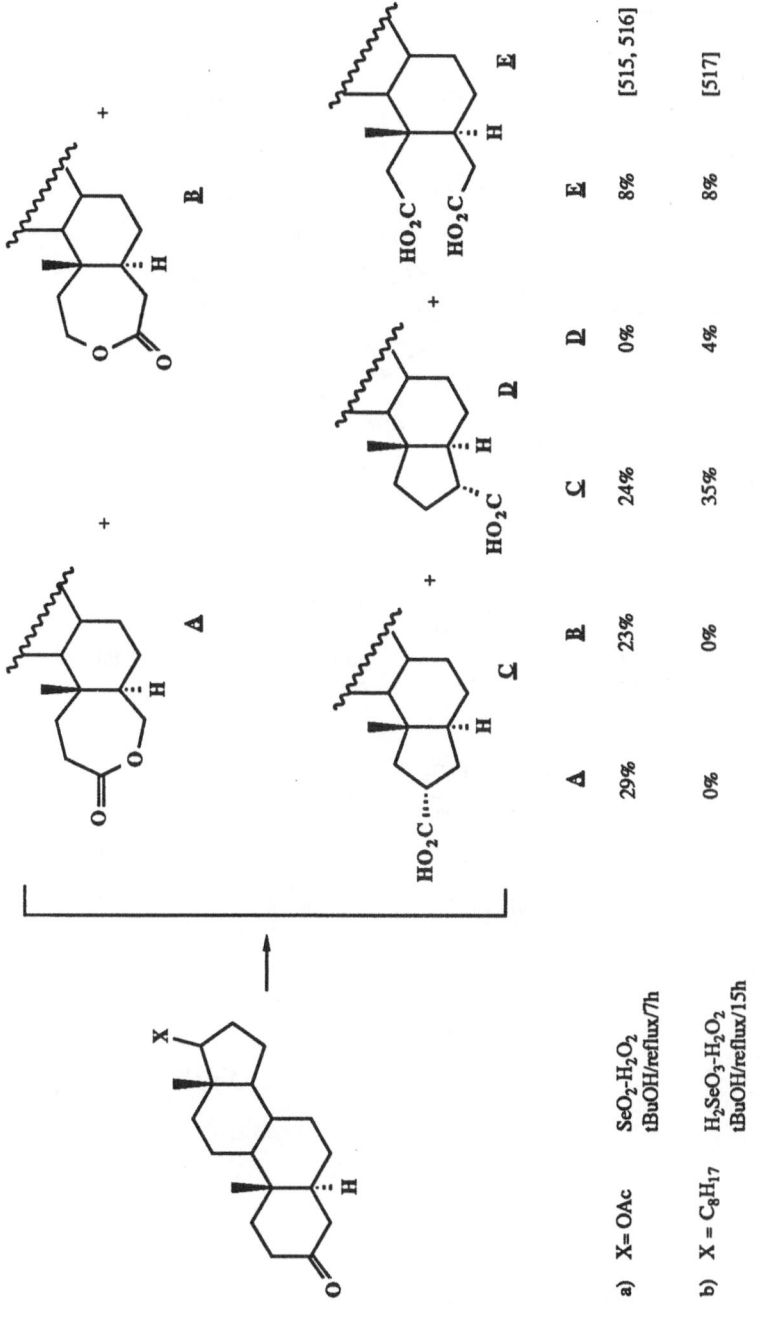

	A	B	C	D	E	
a) X= OAc	SeO$_2$-H$_2$O$_2$ tBuOH/reflux/7h	29%	23%	24%	0%	8%	[515, 516]
b) X = C$_8$H$_{17}$	H$_2$SeO$_3$-H$_2$O$_2$ tBuOH/reflux/15h	0%	0%	35%	4%	8%	[517]

Scheme 213

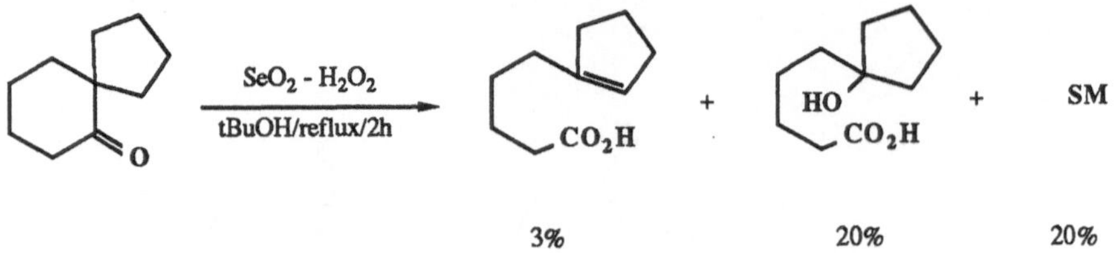

9. Selenium Dioxide Oxidations

3%　　　　　　20%　　　　　20%

Scheme 214 [518]

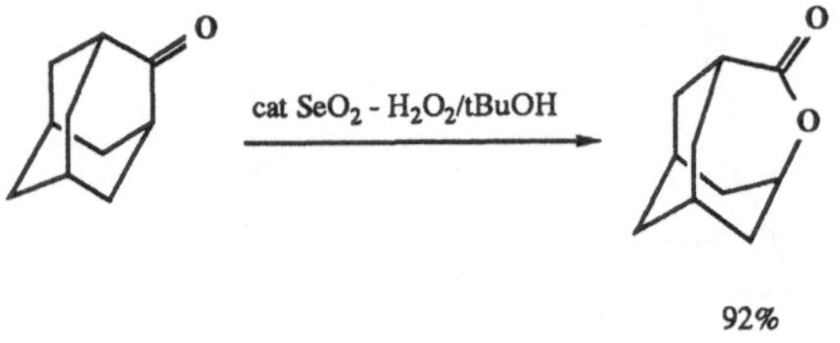

92%

Scheme 215 [519]

a) $Me-C-Me$ (with O double bond)   1)SeO$_2$ or H$_2$SeO$_4$/ 2 eq. H$_2$O$_2$/tBuOH/reflux/6h   2)KOH   3)H$_3$O$^+$ →   MeCH$_2$CO$_2$H   39% based on acetone used   [520]

b) $Ph-C-CH_2Ph$ (with O double bond)   cat SeO$_2$/2 eq. H$_2$O$_2$ tBuOH/reflux/17h   →   Ph$_2$CHCO$_2$H + PhCO$_2$H   6%   20%   [521]

c)   1 eq. H$_2$SeO$_4$ tBuOH/MeOH 70°C/24h   →   PhC—CPh (diketone, two C=O)   71%

d)   SeO$_2$ or H$_2$SeO$_4$ H$_2$O$_2$/tBuOH   →   CO$_2$H   [508]   exclusively

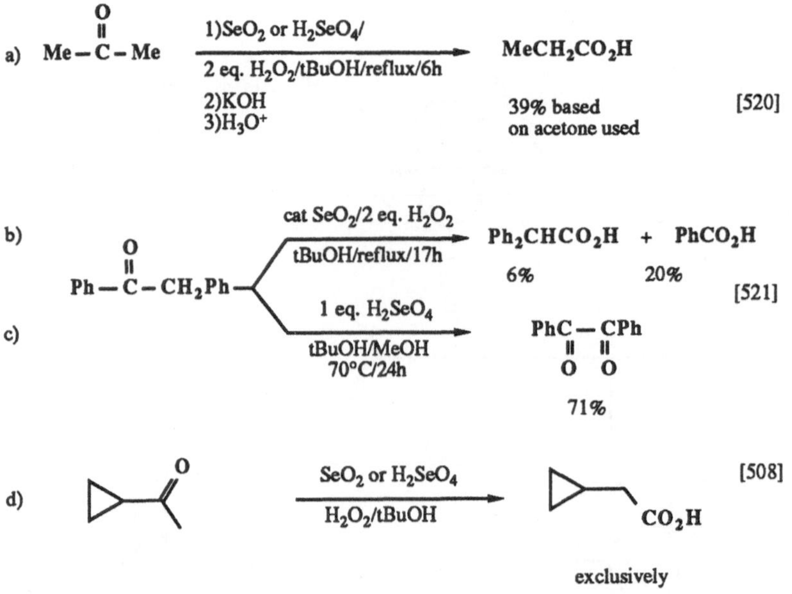

Scheme 216

The following features of the reaction are worth noting:

(i)　selenious acid alone, hydrogen peroxide alone or selenious acid plus an alkyl hydroperoxide are ineffective;

160

(ii) non enolisable ketones such as norcamphor and 2,4-diphenyl bicyclo[3.3.1] nonanone [525] do not undergo the reaction;

(iii) unsymmetrical ketones capable of enolizing in more than one way produce two different carboxylic acids [*508, 510, 511, 523] (Schemes 211a, b, 212) the ratios of which are nearly the same as those reported in the monobromination reactions [@508]. The selectivity observed with cyclopropyl methyl ketone fits into this proposal [508] (Scheme 216d);

(iv) several unsubstituted cyclanones (Scheme 210) [507, *508, 509] at the exclusion of cyclobutanones lead to ring contracted carboxylic acids which otherwise require chlorination of the ketone and Favorski rearrangement to be formed. The reaction is particularly efficient with cyclohexanones [507, 511], which produce cyclopentane carboxylic acids (Schemes 210c, 211c, d). Cyclopentanones [507, 510, 512], lead to cyclobutane carboxylic acids (Schemes 210b, 211 a, b) which are not available by the Favorski route [507].

(v) the rearrangement takes place with several alkyl substituted cycloalkanones [509–512] (Schemes 210, 211) including bicyclic ones [512, 513] (Scheme 212b) and various 3-ketosteroids [514–517, 526, 527] (Scheme 213) especially 3-cholestanone which leads to 2-α-A-norcholestane carboxylic acid [517, 526, 527]. This oxidation, besides its preparative value, also furnishes important information on the stereochemistry of the reaction;

(vi) 2-alkylcyclohexanones lead to products resulting from a Baeyer-Villiger type oxidation [518, 519] (Scheme 214, 215) rather than to the ring contracted carboxylic acids. SeO$_2$/H$_2$O$_2$ has been also reacted with the steroidal enone **61** shown in Scheme 217 and effects in one pot a series of remarkable reactions which finally lead [528] to the lactone **62** (Scheme 217).

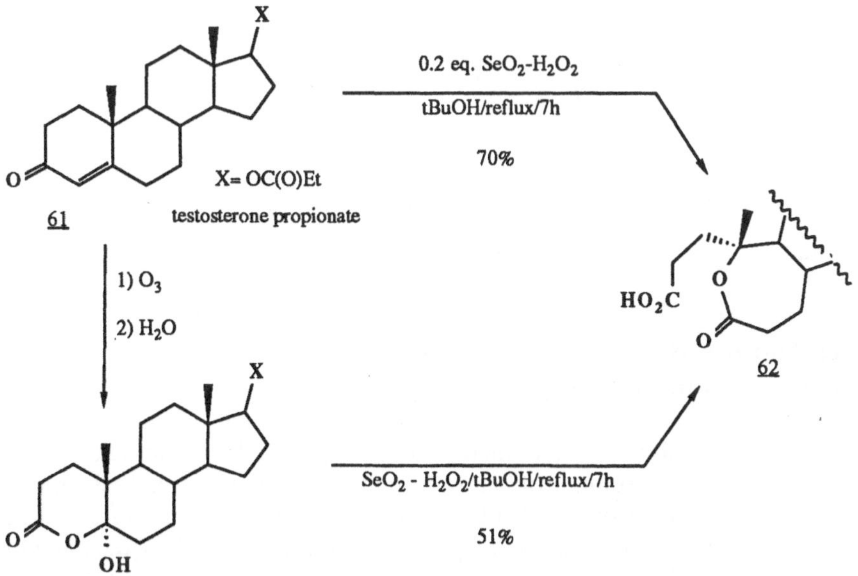

Scheme 217 [528]

9.3.2.2 Reaction of SeO_2/H_2O_2 with Aldehydes

Although aldehydes are usually oxidized [463] in the α position by SeO_2 (Scheme 192a, b), oxidation leading to carboxylic acids exclusively occurs if the reaction is performed with 1 equivalent of hydrogen peroxide [528] in the presence of catalytic amount of SeO_2 (Scheme 218). Similar reaction takes place with enals and produces [529, 530] α,β-unsaturated acids or their esters depending upon the solvent used (Scheme 219).

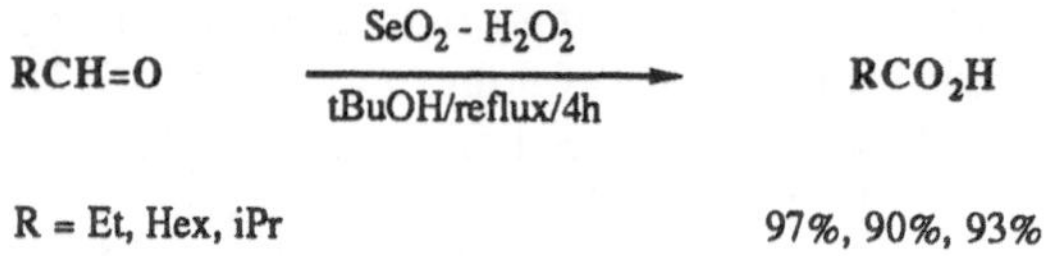
RCH=O
SeO2 - H2O2
tBuOH/reflux/4h
RCO2H
R = Et, Hex, iPr
97%, 90%, 93%

Scheme 218 [529]

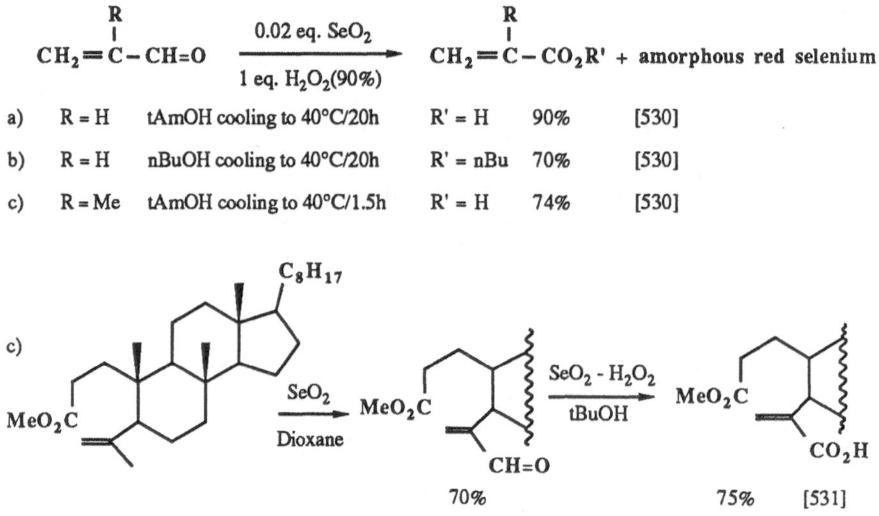
CH2=C(R)-CH=O
0.02 eq. SeO2
1 eq. H2O2(90%)
CH2=C(R)-CO2R' + amorphous red selenium
a) R = H tAmOH cooling to 40°C/20h R' = H 90% [530]
b) R = H nBuOH cooling to 40°C/20h R' = nBu 70% [530]
c) R = Me tAmOH cooling to 40°C/1.5h R' = H 74% [530]
c)
C8H17
MeO2C
SeO2
Dioxane
MeO2C
CH=O
70%
SeO2 - H2O2
tBuOH
MeO2C
CO2H
75%
[531]

Scheme 219

9.4 Oxidation of the Alkyl Chain of Aromatic and Heteroaromatic Compounds

Except in very special cases which will be discussed later in this section, SeO_2 does not react on the ring of aromatic or heteroaromatic compounds. Alkyl substituted derivatives are however oxidized, often under quite drastic conditions, at the benzylic position producing aromatic or heteroaromatic carbonyl compounds (Schemes 169, 220–224) [408, 409, 532–540] or aryl or heteroaromatic substituted olefins (Schemes 170e, 225, 226) [411, 532, 541, 542].

9.4.1 Synthesis of Aryl Carbonyl Compounds

Methyl substituted aromatic or heteroaromatic compounds can be selectively oxidized to the corresponding aromatic or heteroaromatic aldehydes or acids depending especially upon the number of equivalents of SeO_2, the solvent and the temperature used. Excess of SeO_2, use of pyridine as the solvent and high temperature favour the formation of the acid [532, 533, 535], (Schemes 221a, b, c, f, compare f to e, 223a, b) whereas molar equivalent of SeO_2 in dioxane usually leads to aldehydes [534, 535, 537] (Scheme 221 compare e to f; Schemes 221d, 222, 223c).

The reaction has been sucessfully applied *inter alias* for the oxidation of alkyl benzenes or naphthalenes [532] (Scheme 220), of alkyl pyridines [533, 536] (Scheme 221a–c, h, i), of alkyl quinoleines [533–535, 537, 546] (Scheme 221g) of alkyl pyrimidines [535–538] and their oxides [535] (Scheme 223) and of isochromane to isochroman-1-one (Scheme 224) [539, 540]. Chromane however is not oxidized [539].

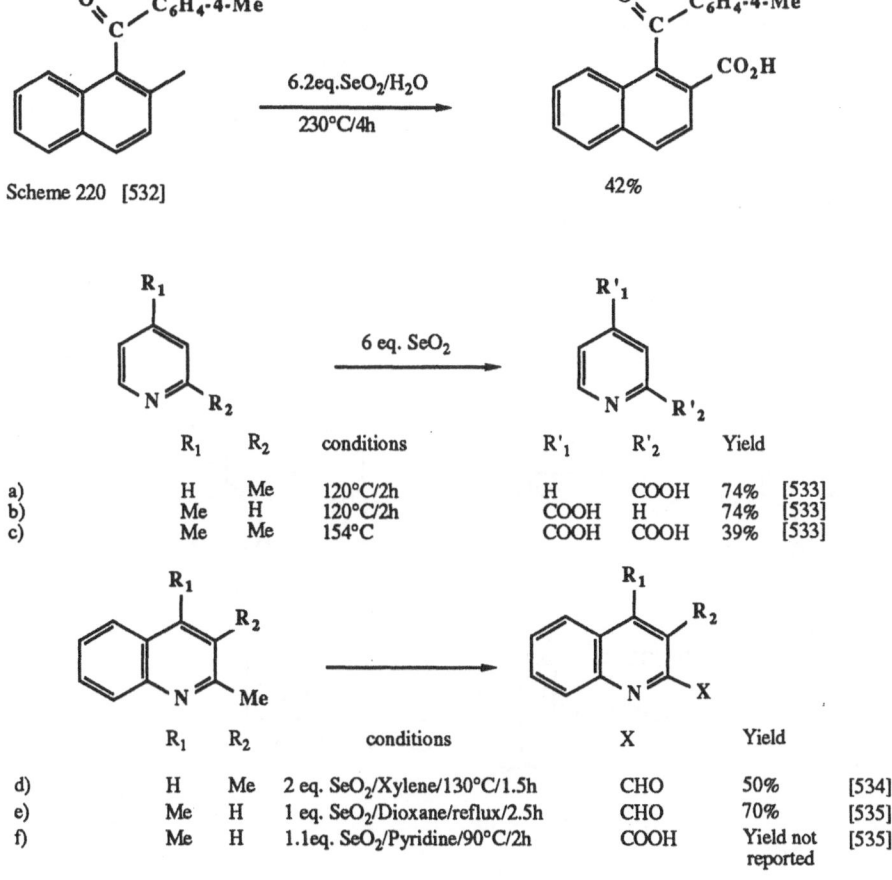

Scheme 220 [532]

42%

	R_1	R_2	conditions	R'_1	R'_2	Yield	
a)	H	Me	120°C/2h	H	COOH	74%	[533]
b)	Me	H	120°C/2h	COOH	H	74%	[533]
c)	Me	Me	154°C	COOH	COOH	39%	[533]

	R_1	R_2	conditions	X	Yield	
d)	H	Me	2 eq. SeO_2/Xylene/130°C/1.5h	CHO	50%	[534]
e)	Me	H	1 eq. SeO_2/Dioxane/reflux/2.5h	CHO	70%	[535]
f)	Me	H	1.1eq. SeO_2/Pyridine/90°C/2h	COOH	Yield not reported	[535]

Scheme 221

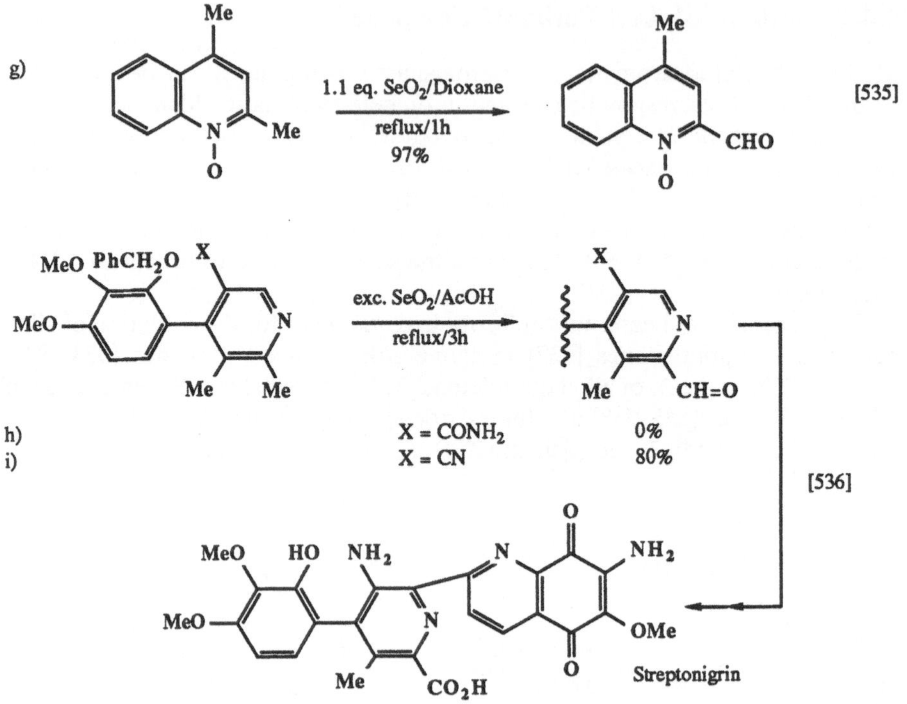

9. Selenium Dioxide Oxidations

g)

1.1 eq. SeO$_2$/Dioxane
reflux/1h
97%

[535]

exc. SeO$_2$/AcOH
reflux/3h

h)
i)

X = CONH$_2$ 0%
X = CN   80%

[536]

Streptonigrin

Scheme 221 (contd.)

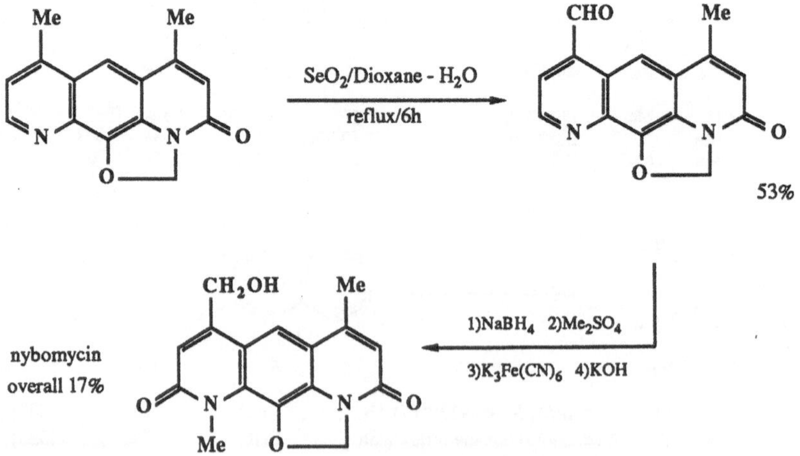

SeO$_2$/Dioxane - H$_2$O
reflux/6h

53%

nybomycin
overall 17%

1)NaBH$_4$ 2)Me$_2$SO$_4$
3)K$_3$Fe(CN)$_6$ 4)KOH

Scheme 222 [537]

164

9.4 Oxidation of the Alkyl Chain of Aromatic and Heteroaromatic Compounds

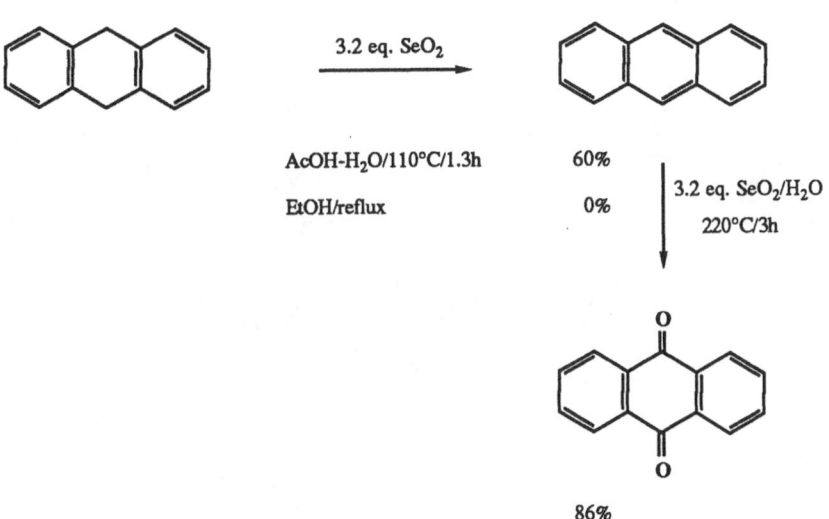

	R	Conditions	X	Yield (%)	
a)	H	SeO₂/Pyridine/reflux/5h	CO₂H	65	[538]
b)	Me	SeO₂/Pyridine/reflux/5h	CO₂H	26	[538]
c)	Ph	SeO₂/Dioxane/reflux/5h	CHO	90	[535]

Scheme 223

1. eq. SeO₂/Xylene/reflux/12h   85%   [539]

1.2 eq. SeO₂/Neat/160°C/2h   65%   [540]

Scheme 224

3.2 eq. SeO₂

AcOH-H₂O/110°C/1.3h   60%

EtOH/reflux   0%

3.2 eq. SeO₂/H₂O
220°C/3h

86%

Scheme 225 [532]

165

Scheme 226

a)

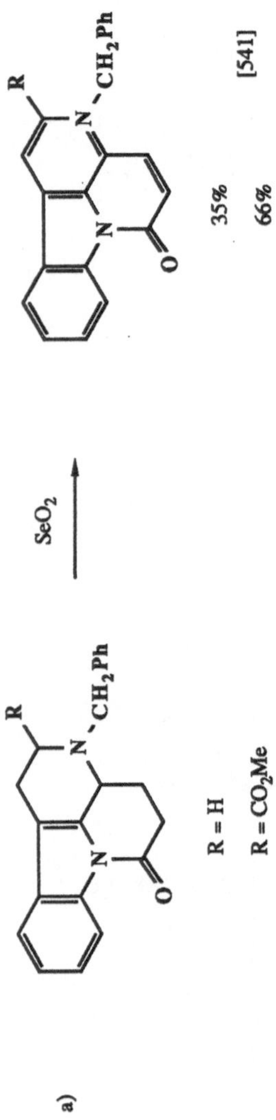

R = H
R = CO₂Me

SeO₂

35%
66%

[541]

b)

Dihydromocimycin

Mocimycin

[542]

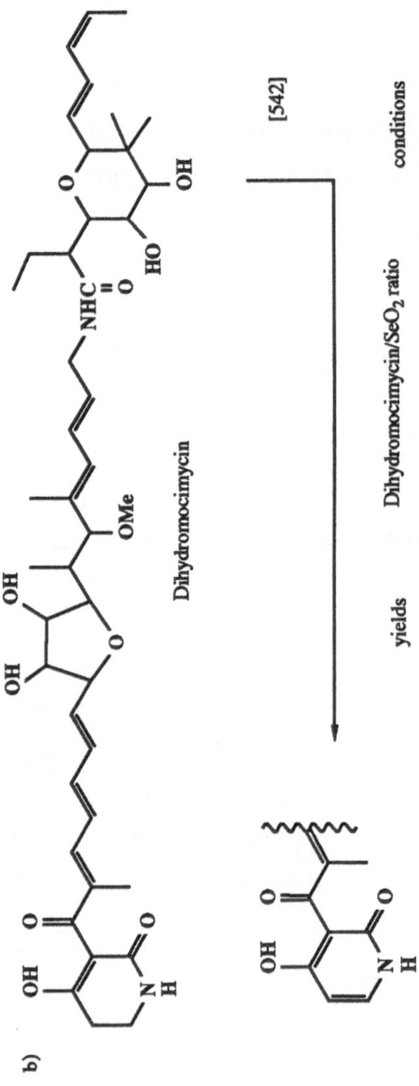

yields	Dihydromocimycin/SeO₂ ratio	conditions
42%	3:1	HMPA/H₂O/90°C/2h
79%	1.2:1	HMPA/95°C/1h
55%	3.3:1	tAmOH/96°/3h
71%	3.2:1	tAmOH-HMPA/96°C/3h

This reaction was used for the partial synthesis of antitumor antibiotic streptonigrin (Scheme 221i) [536, 543, 544] and for the synthesis of nybomycin antibiotic (Scheme 222) [537]. The last synthesis requires a multistep sequence to construct the hydroxymethyl quinolone moiety from the quinoline ring (Scheme 222) [537] rather than the more direct route which involves the oxidation of a methyl quinolone moiety (Scheme 227 compare c to b and a) [537, 545, 546] which requires very drastic conditions (neat, 175 °C) to occur (Scheme 227a, b) [537, 545]. The reactions are usually easier (i) with N-oxides than with the original heterocycles [535] (ii) when a strongly electron withdrawing group is present on the aromatic ring in para position to the methyl group to be oxidized

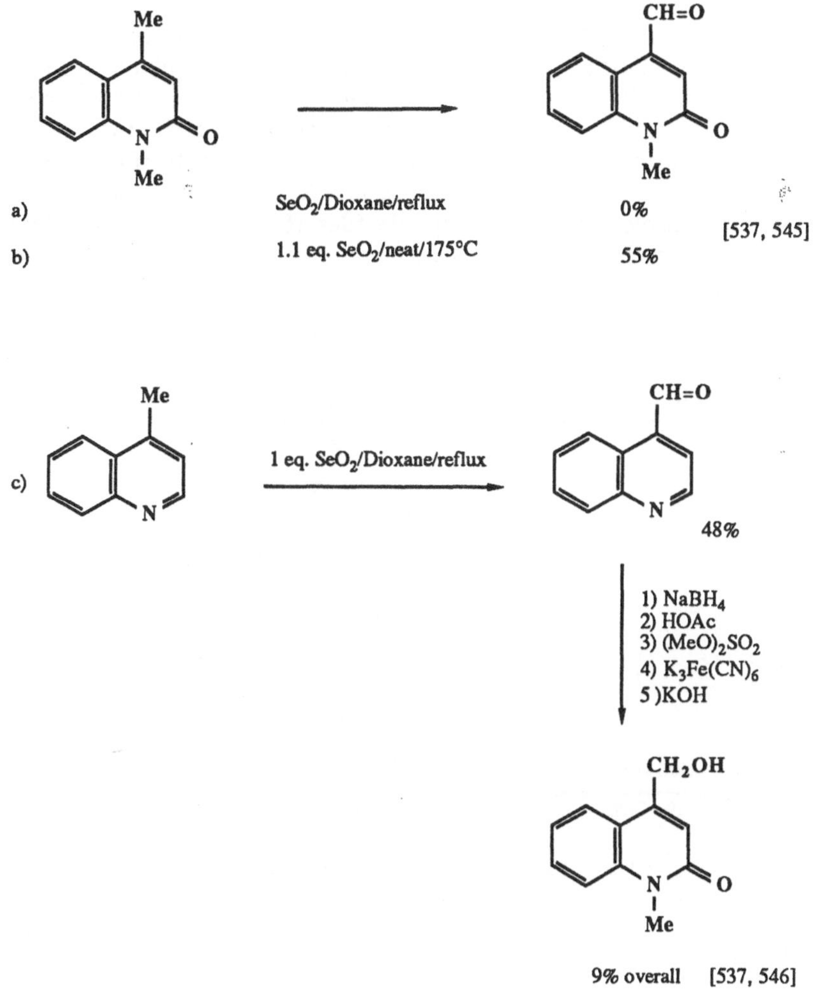

Scheme 227

9. Selenium Dioxide Oxidations

(Scheme 221 compare h to i). When different alkyl groups are attached to the aromatic ring the reaction can be performed selectively on one of them. Thus 2,4-dimethylquinoline [535] and 2,4-dimethylpyrimidines [535, 538] can be selectively oxidized at the 2 and 4 positions respectively.

The Sharpless procedure proved particularly efficient for the oxidation of alkyl indoles to acyl indoles (SeO_2-$tBuO_2H$, CH_2Cl_2, 20 °C, 236 h, Scheme 169a). Interestingly however the reaction does not proceed in the absence of t-BuO_2H (Scheme 169b). These compounds are difficult to prepare by acylation of indoles or by oxidation of alkyl indoles with other oxidants such as dicyanodichloroquinone (DDQ) [408] (Scheme 169c). The mild conditions used for this transformation even allow the synthesis of acyl indoles from alkyl indoles bearing a piperidino group which remains untouched. It must be recalled that the latter moiety is oxidized to a pyridinic moiety when the reaction is carried out with SeO_2 in dioxane [408] (Scheme 169d).

As mentionned earlier in this section SeO_2 does not usually react on the aromatic or heteroaromatic rings, although pyridines [533] and quinolines [533] form complexes with SeO_2. Therefore pyridine and xylene have been often used as solvent in SeO_2 oxidation reactions. Other reactions which involve the participation of the aromatic ring have been reported. Thus, in the presence of SeO_2, 2,6-dimethyl phenol couples oxidatively in low yields [547], anisole and heterosubstituted anisoles produce [547] selenides (Scheme 228) and finally tris(p-tolyl) selenonium chloride is formed on reaction of toluene with SeO_2 but in the presence of aluminium trichloride [548] (Scheme 229).

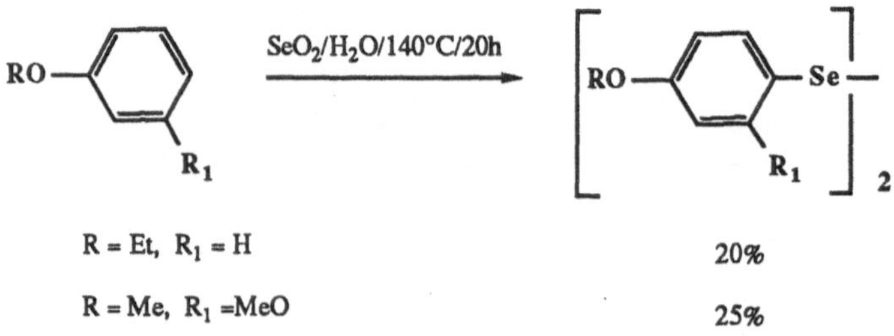

R = Et, R_1 = H   20%

R = Me, R_1 = MeO   25%

Scheme 228 [547]

77%

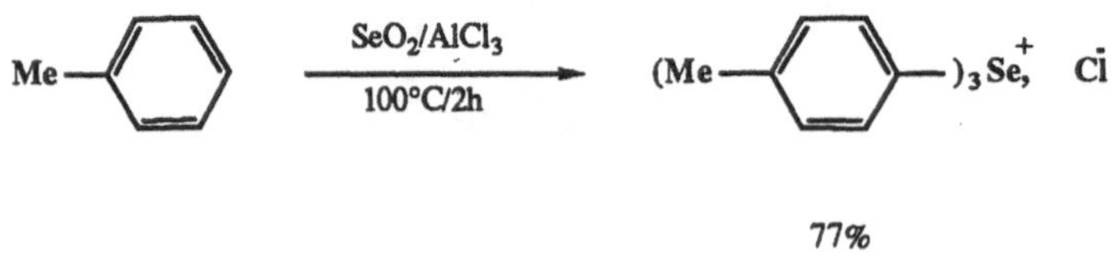

Scheme 229 [548]

168

9.4.2 Aromatisation of Cyclic and Heterocyclic Compounds

SeO$_2$ promotes the dehydrogenation of cyclic and polycyclic compounds (Scheme 170e) [411, *549] especially the heterocyclic ones and allows the synthesis of aromatic and heteroaromatic compounds. The reaction has been *inter alias* used for the synthesis of mocimycin from dihydromocimycin [542] (Scheme 226b) and for the synthesis of indole alkaloids (Schemes 169 and 226a) [408, 409, 541].

9.5 Oxidation of Benzylic and Allylic Alcohols

Aliphatic alcohols do not react with SeO$_2$ [427]. For example ethanol is often used as the solvent in SeO$_2$ promoted oxidation of organic compounds [*54, *60, *63, *81, *84, *89]. It is however oxidized [427] to glyoxal in very low yield (5%) under drastic conditions (SeO$_2$, neat, 230 °C). Benzylic alcohols [361, 550] on the other hand are more efficiently oxidized to carbonyl compounds (Scheme 230) or are

ArCH$_2$OH $\longrightarrow$ ArCH=O

a) Ar = Ph   SeO$_2$/C$_6$H$_6$/100°C/10h   83%   [550]

b) Ar = Ph   *   93%   [361]

c) Ar = 4-X-Ph; X = Me, MeO, Cl, NO$_2$ *   89-92% [361]

d) Ar = Ph   SeO$_2$/tBuOH   0%   [361]

e)

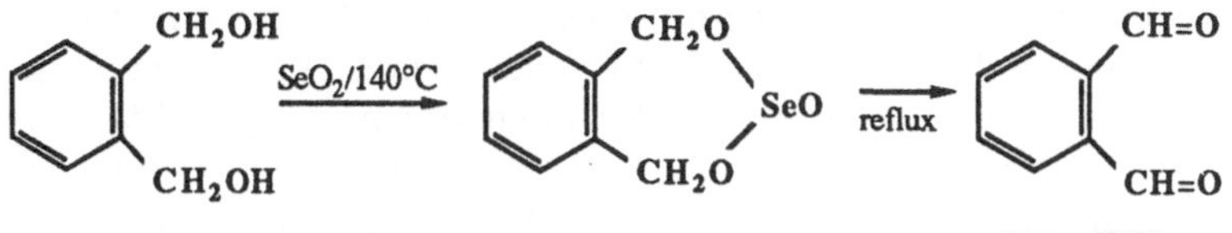

66%   68%   [550]

* cat Se or cat SeO$_2$ /1.5eq.(4-MeOPh)$_2$SeO/Dioxane/reflux/12h

Scheme 230

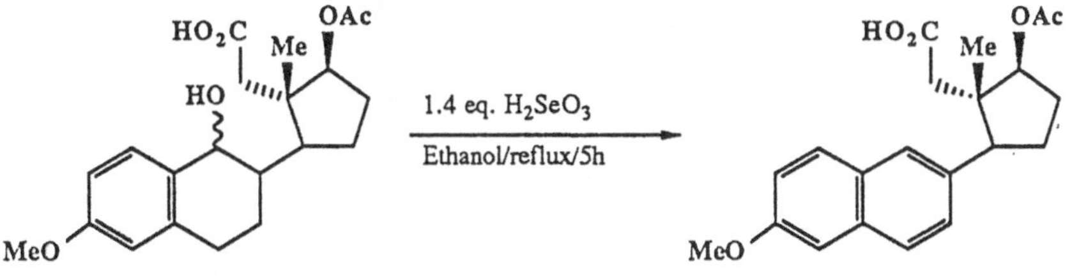

1.4 eq. H$_2$SeO$_3$

Ethanol/reflux/5h

72%

Scheme 231 [551]

169

transformed to aromatic compounds [551] especieally in the case of α-aryl cyclohexanols (Scheme 231). The first method [550] used in the case of benzyl alcohol takes place under drastic conditions (SeO$_2$, neat, 10 h). Recently however better results have been obtained [361] by refluxing the benzyl alcohol in dioxane with catalytic amounts of metallic selenium or SeO$_2$ but in the presence of 1,5 molar equiv. of bis(p-methoxy phenyl)selenoxide as an extremely mild and selective oxidizing agent able to regenerate SeO$_2$. Surprisingly it was noticed [361] that the Sharpless method which uses t-butyl hydroperoxide as the co-oxidant, and which proved particularly valuable in the allylic oxidation of olefins was unsuitable for the oxidation of benzyl alcohols.

Oxidation of allyl alcohols also takes place on reaction with SeO$_2$. Aldehydes are obtained [427] in the case of α-methyl allyl alcohol or 3-phenyl-2-propene-1-ol (Scheme 176a, b), but oxidation on the allylic methyl group leading to dials seems to be observed [522] when 2-methyl-2-hexene-4-ol or 2-methyl-2-octene-6-ol are reacted with SeO$_2$ in acetic acid/acetic anhydride mixture.

9.6 Oxidation of Hydrazones, Imines, Oximes and of Semicarbazones

Hydrazones [475], and oximes [553] behave similarly to the corresponding carbonyl compounds towards SeO$_2$ and are usually oxidized in α position. Imino compounds bearing an α-carbonyl group are formed [475, 553] often in low yield, especially when the starting materials lack β hydrogens (Scheme 232) or when the formation of α carbon-carbon double bonds is disfavoured [475] for steric reasons (Scheme 194d) whereas the latter compounds are usually formed in the other cases [475, 554, 555] (Scheme 233).

Aldoximes with a free hydroxyl group on the nitrogen behave differently since they are transformed [556] in very high yield to nitriles (Scheme 234). The reaction takes place in chloroform with stoichiometric or catalytic (5%) amounts of SeO$_2$ at room temperature for alkyl derivatives [556] and at reflux for aryl analogues [556]. In the latter case, water must be continuously removed [556]. This method was further developped [@557] as a one-flask procedure involving the addition of SeO$_2$ to an in situ generated oxime (Scheme 235). The oxidation reaction can be performed with only 10% SeO$_2$ [557].

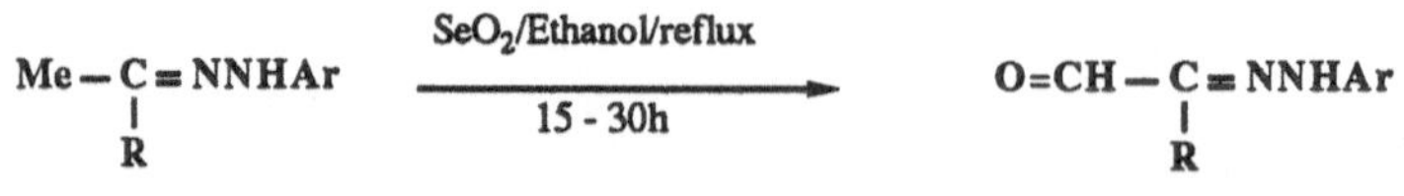

$$\text{Me}-\underset{\underset{R}{|}}{C}=\text{NNHAr} \quad \xrightarrow[\text{15 - 30h}]{\text{SeO}_2/\text{Ethanol/reflux}} \quad \text{O=CH}-\underset{\underset{R}{|}}{C}=\text{NNHAr}$$

R	Ar	
Me	4-NO$_2$-C$_6$H$_4$	10% *
Me	3,4-(NO$_2$)$_2$C$_6$H$_3$	40% *
4-XC$_6$H$_4$ X = F, Cl, Br	4-NO$_2$C$_6$H$_4$	80% *

* Yield in crude products

Scheme 232   [475]

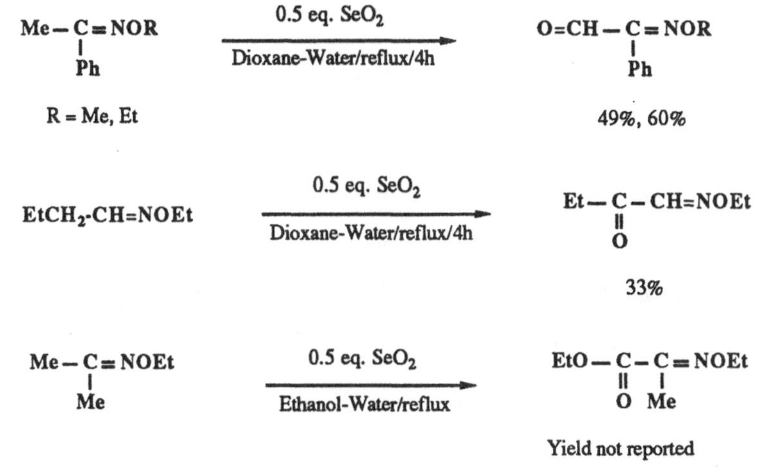

$$Me-\underset{\underset{Ph}{|}}{C}=NOR \quad \xrightarrow[\text{Dioxane-Water/reflux/4h}]{0.5 \text{ eq. } SeO_2} \quad O=CH-\underset{\underset{Ph}{|}}{C}=NOR$$

R = Me, Et

49%, 60%

$$EtCH_2\text{-}CH=NOEt \quad \xrightarrow[\text{Dioxane-Water/reflux/4h}]{0.5 \text{ eq. } SeO_2} \quad Et-\underset{\underset{O}{\|}}{C}-CH=NOEt$$

33%

$$Me-\underset{\underset{Me}{|}}{C}=NOEt \quad \xrightarrow[\text{Ethanol-Water/reflux}]{0.5 \text{ eq. } SeO_2} \quad EtO-\underset{\underset{O}{\|}}{C}-\underset{\underset{Me}{|}}{C}=NOEt$$

Yield not reported

Scheme 232 (contd.)   [553]

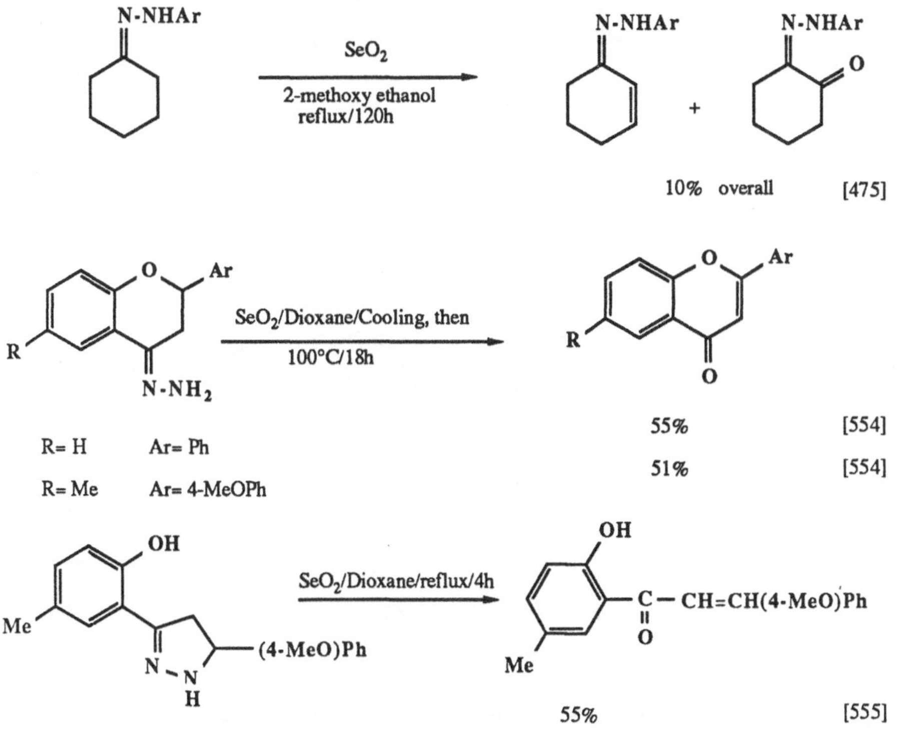

SeO$_2$

2-methoxy ethanol
reflux/120h

+

10% overall   [475]

SeO$_2$/Dioxane/Cooling, then

100°C/18h

R= H   Ar= Ph

R= Me   Ar= 4-MeOPh

55%   [554]

51%   [554]

SeO$_2$/Dioxane/reflux/4h

55%   [555]

Scheme 233

171

9. Selenium Dioxide Oxidations

R-CH=NOH ────────────────────→ **RC≡N**

R		Yields (%)
Ph	1 eq. SeO$_2$/CHCl$_3$/reflux/2h	100
Ph	0.05 eq. SeO$_2$/CHCl$_3$/reflux/2h	86
nHeptyl	1 eq. SeO$_2$/CHCl$_3$/20°C/3h	68
nHeptyl	0.5 eq. SeO$_2$/CHCl$_3$/20°C/3h	59
nHeptyl	0.05 eq. SeO$_2$/CHCl$_3$/reflux/3h	74

Scheme 234 [556]

$$\text{PrCH=O} \xrightarrow[\text{2) 1 eq. SeO}_2/20°C]{\text{1) H}_2\text{NOH/DMF/70°C/2h}} \text{PrC≡N}$$

80%

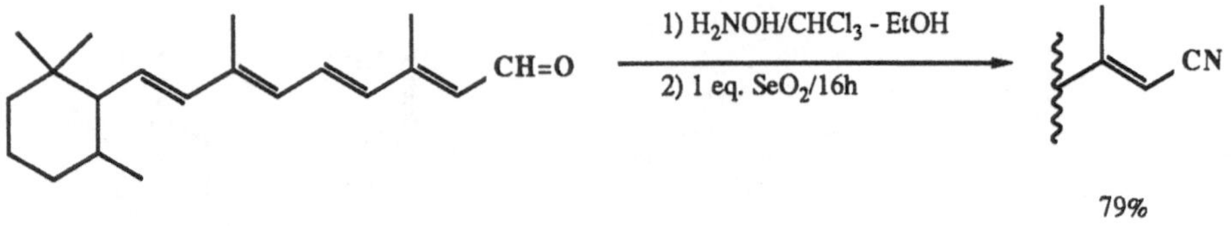

1) H$_2$NOH/CHCl$_3$ - EtOH
2) 1 eq. SeO$_2$/16h

79%

Scheme 235 [557]

$$\underset{\overset{|}{\text{CH}_2\text{R}}}{\text{Ar}-\text{C}=\text{N}}-\text{NH}-\overset{\overset{\text{O}}{\|}}{\text{C}}-\text{NH}_2 \xrightarrow[\text{100°C/few minutes}]{\text{1 eq. SeO}_2/\text{AcOH}} \underset{\underset{\text{N}}{\text{N}}\diagdown\diagup \text{Se}}{\text{Ar}-\text{C}=\text{C}-\text{R}} \xrightarrow{\text{reflux}} \text{ArC≡CR}$$

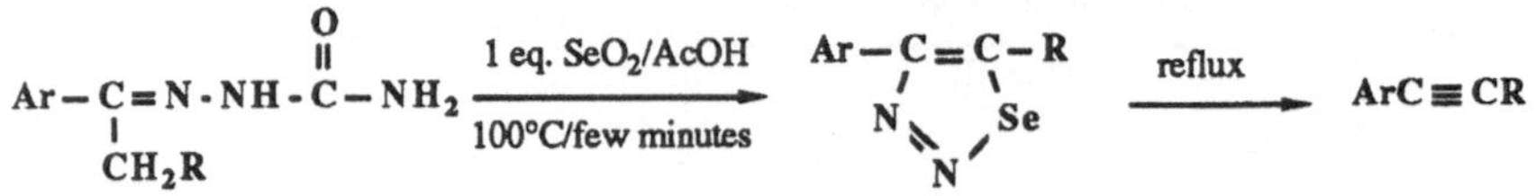

	Ar	R			
a)	Ph	H	70%	84%	[559]
b)	4-ClC$_6$H$_4$	H	65%	80%	[559]
c)	Ph	Me	68%	...	[558]
d)	Ph	Ph	... *	67%	[559]

* This compound is isolated in 61% yield if the reaction is instead carried out in dioxane

Scheme 236

172

The reaction of SeO_2 with semicarbazones [558–566] (Schemes 236–240) and with bis-semicarbazones [567] takes a different course and produces 1,2,3-selenadiazoles in good yield. Its postulated mechanism [558], disclosed in the Scheme 241, is in good agreement with that proposed [446] for selenium dioxide oxidation of ketones (Scheme 191). The semicarbazone of deoxybenzoin [559] (Scheme 236d) and the guanilhydrazone of acetophenone [568] (Scheme 242)

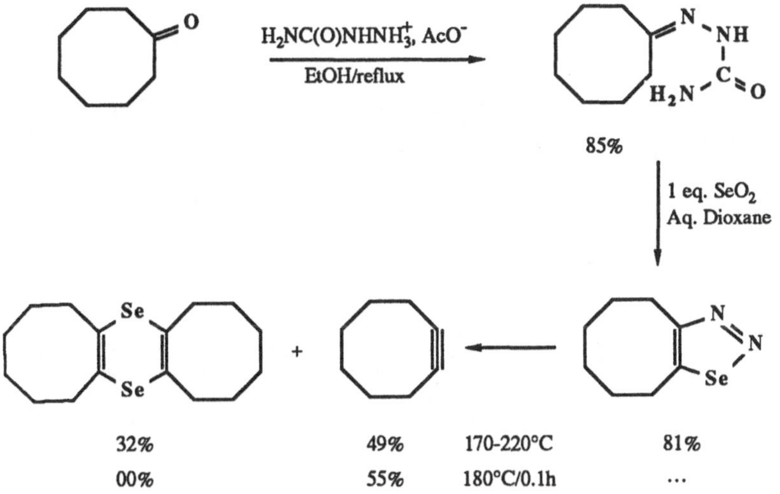
O
H2NC(O)NHNH3+, AcO-
EtOH/reflux
N
NH
C
H2N
O
85%
1 eq. SeO2
Aq. Dioxane
Se
Se
+
N
N
Se
32%
49%
170-220°C
81%
00%
55%
180°C/0.1h
...

Scheme 237 [560, 561, 562]

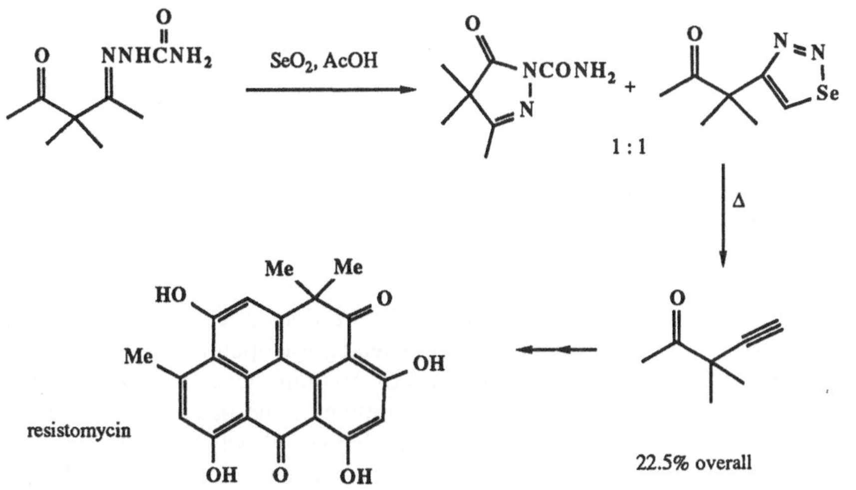
O
O
NNHCNH2
SeO2, AcOH
O
NCONH2
N
+
O
N=N
Se
1 : 1
Δ
Me
Me
HO
O
Me
OH
resistomycin
OH
O
OH
O
22.5% overall

Scheme 238 [563]

behave somewhat differently since diphenyl acetylene [559] is directly formed[1] in the first case whereas a triazine is obtained in the second example. The selenadiazoles formed were found valuable starting materials for the synthesis of acetylenic compounds (Schemes 236–239) [559–561, 563, 567]. The reaction occurs by heating the product to 200 °C [559–561, 563, 564, 566, 567, 569–574] or on addition of butyllithium [564, 575, 576]. It was successfully used for the preparation of the strained cyclooctyne [560–562] (Scheme 237), as a key step in the synthesis of resistomycine antibiotic from ethyl acetoacetate [563] (Scheme 238) and for the synthesis of tribenzo dehydrooctatetraene [564] (Scheme 239).

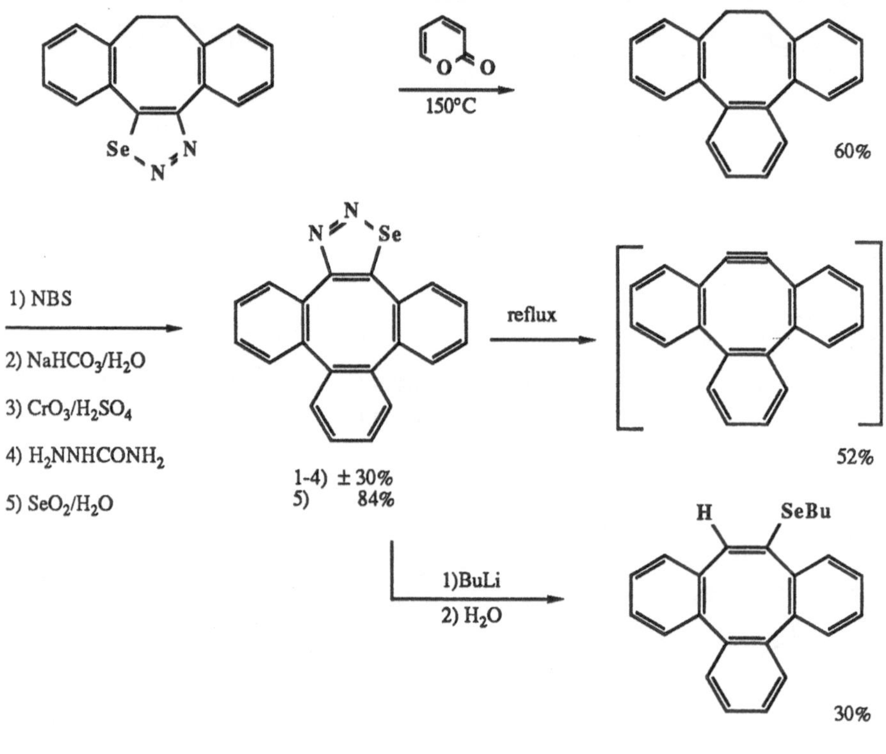

Scheme 239 [564]

As expected cyclopenteno-1,2,3-selenadiazole does not lead to the highly strained cyclopentyne on heating but gives instead dicyclopenteno-1,4-diselenine in 27% yield [562–565] (Scheme 240). Flash or vacuum thermolysis at 500 °C of the cyclopenteno-1,2,3-selenadiazole and trapping of the products in argon at 12 °K gave trimethyleneselenoketene [565] (Scheme 240) whereas ethylene and propadieneselone are simultaneously produced if the reaction is performed [565]

---

[1] This spontaneous elimination of nitrogen takes place in acetic acid (the usual solvent for these reactions), whereas it can be avoided if the reaction is performed in dioxane.

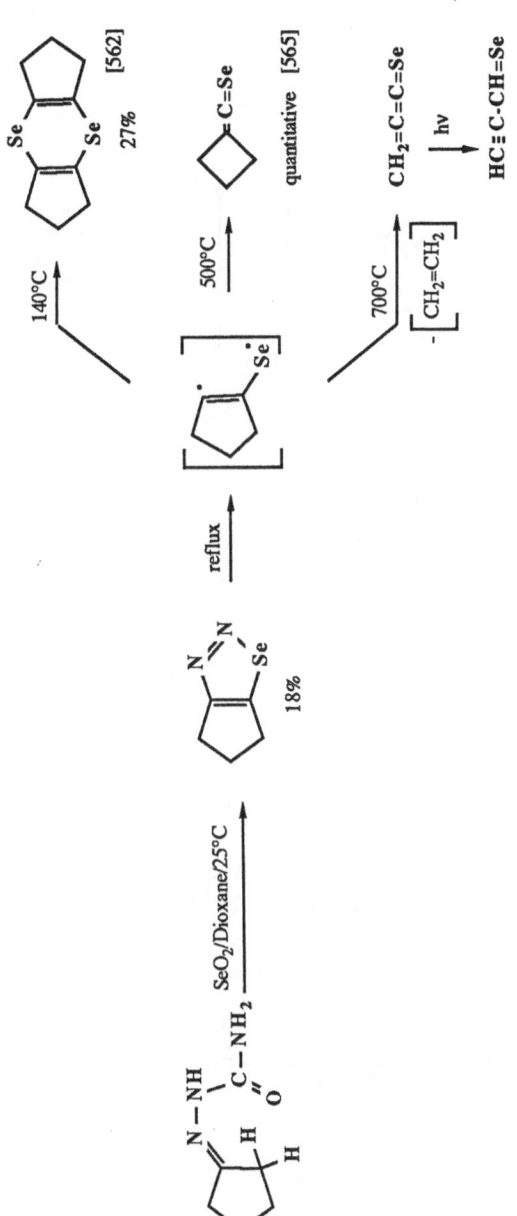

Scheme 240

175

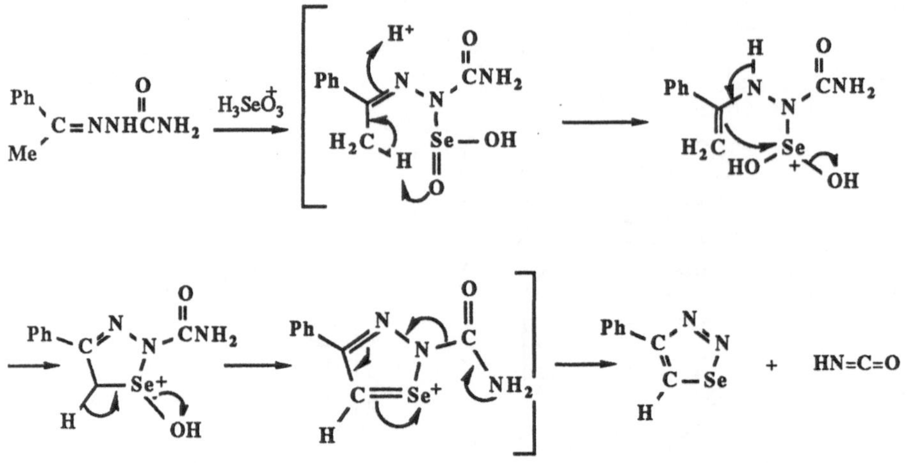

9. Selenium Dioxide Oxidations

Scheme 241 [558]

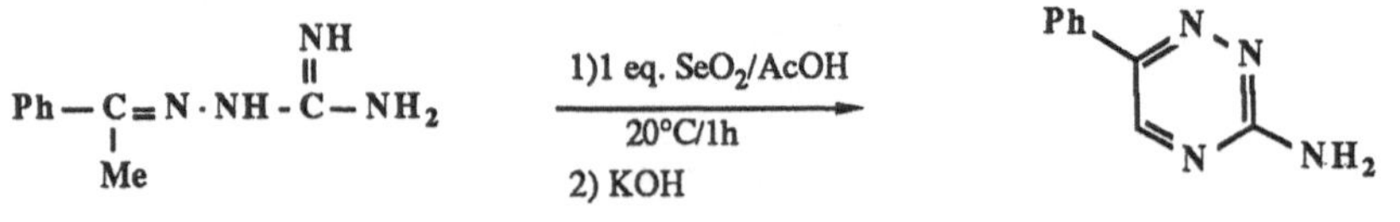

Scheme 242 [568]

at 700 °C. The latter compound leads [565] to selenopropynal on further irradiation (> 220 nm) of the matrix at 12 °K.

If the thermolysis is performed in an excess of CS_2 (3 h, 160 °C), cyclopenta-1,3-selenathiolane-2-thione is formed [*115, 577] (Scheme 243). This may be a

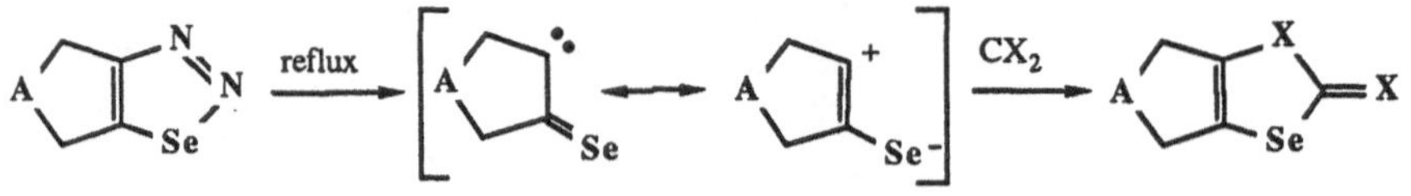

A	X			
CH$_2$	S	160°C/3h	53%	[577]
S*	Se	110°C/2h	35%	[578, 580]

* Obtained in 18% yield from the corresponding semicarbazone and SeO$_2$ in glacial AcOH

Scheme 243

useful starting material for the synthesis of tetraheterofulvalenes [*115], important partners in organic metals [*115]. Similar behaviour of other cyclopenta-1,2,3-selenadiazoles [578, 580] (Scheme 243) and 1,2,3-benzoselenadiazole [577, 579] towards carbon disulfide [577, 579] and carbon diselenide [578–580] have also been reported.

9.7 Conversion of Thio- and Selenocarbonyl Compounds to Carbonyl Compounds

Selenium dioxide is able to transform thio- and selenocarbonyl compounds to carbonyl compounds [322]. Similar reactions have been shown to occur with selenoxides (Sect. 4.1.3) or with BSA (Sect. 6.10.1). But except for the special case of 5-α-cholestan-3-β-selenobenzoate which is transformed in high yield to the corresponding benzoate on reaction with SeO$_2$ for 3 days [322] (Scheme 136e), the reaction takes very long times and often produces a mixture of products. We therefore suggest the use of BSA for this purpose [322] as a superior reagent (Scheme 136 compare entry e to c).

9.8 Reactions of Nitroalkanes: Synthesis of N-hydroxycarboxamides and Nitriles

Primary nitroalkanes react with SeO$_2$ (1 equiv.) in the presence of triethyl amine and produce N-hydroxy carboxamides [581] in 63–76% yield (Scheme 244). The presence of an amine as well as the 1:1:2 ratio of nitroalkane/SeO$_2$/triethyl amine are crucial for the success of the reaction. The nitroalkane is for example recovered unchanged if triethyl amine is omitted [581] and the corresponding nitrile (Scheme 244) is formed if a 1:2:2 ratio of reactants is used [581].

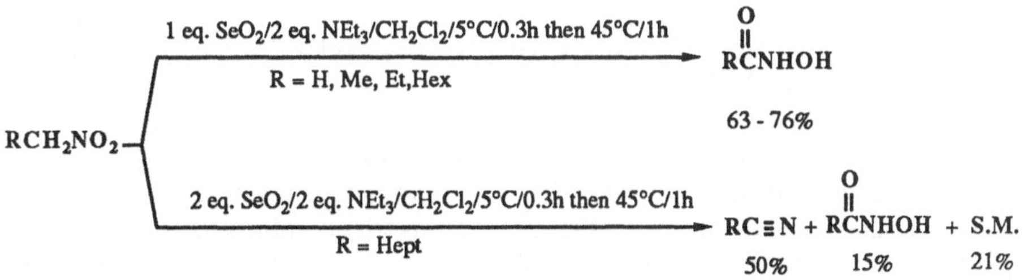

Scheme 244 [581]

9.9 Synthesis of Olefins from Phosphorus Ylides and Diazoalkanes

Phosphorus ylides [582] and diphenyl diazomethane [583] have been dimerised oxidatively to the corresponding olefins on reaction with SeO$_2$ (Schemes 245,

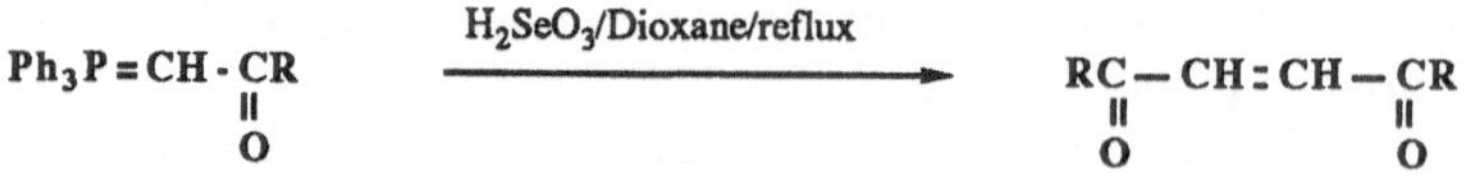

9. Selenium Dioxide Oxidations

$$Ph_3P=CH-CR \xrightarrow{\text{H}_2\text{SeO}_3/\text{Dioxane/reflux}} RC-CH=CH-CR$$

R	Reaction time	Yield (%)
Ph	6.5h	87
4-NO$_2$C$_6$H$_4$	8h	70
OEt	4h	50

Scheme 245 [582]

$$Ph_2C=N_2 \longrightarrow Ph_2C=CPh_2 \ + \ Ph_2C=O \ + \ \text{other products}$$

1 eq. SeO$_2$/CS$_2$/reflux/72h	40%	49%
0.7 eq. SeO$_2$/CS$_2$/20°C/48h	59%	34%
O$_2$/CS$_2$/20°C/200h	60%	13%
SO$_2$	08%	19%

Scheme 246 [583]

246). The reactivity of diphenyl diazomethane towards SeO$_2$ and SO$_2$ was compared at that occasion [583].

9.10 Reaction of SeO$_2$ with Phosphines, Arsines, Stibines and Related Compounds

Phosphines, arsines and stibines react [584] with SeO$_2$ and usually give the corresponding oxides. Trialkoxyarsines and stibines produce selenites [585] in modest to good yields (Scheme 247a, b) whereas trialkyl phosphites are oxidized [585] to trialkyl phosphates (Scheme 247c) in very good yield at the condition that 0.5 molar equivalent of SeO$_2$ is used. Lowering the amount of SeO$_2$ lowers the yield of the phosphate in favour of the phosphoroselenoates [585] [(RO)$_3$P = Se] arising from the reaction of selenium with unreacted phosphite [585]. These have been transformed to the phosphates on further reaction with SeO$_2$ [585] (Scheme 247d). The reactivity of various trivalent phosphorus compounds towards SeO$_2$ has been compared [585] and was found to decrease in the order (MeO)$_3$P > BuP(OEt)$_2$ < P(OEt)$_3$. Under closely related conditions arsonites [RAs(OR$_1$)$_2$] have been transformed [586] to arsonates [RAs(O)(OR$_1$)$_2$].

$$(RO)_3X \xrightarrow{\text{1.5 eq. SeO}_2/\text{C}_6\text{H}_6} (RO)_2Se{=}O \quad + \quad X_2O_3$$

a) X=As   R = Me, Et, nPr, iPr   100°C/48h   83, 62, 65, 51%

b) X=Sb   R = Et, nPr, iPr   100°C/72h   33, 34, 41%

$$(RO)_3P \xrightarrow[\text{exothermic, then 20°C/12h}]{\text{0.5 eq. SeO}_2/\text{C}_6\text{H}_6} (RO)_3P{=}O \quad + \quad Se$$

c)   R = Me, Et   84, 82%

$$(EtO)_3P{=}Se \xrightarrow[\text{conditions not described}]{\text{0.5 eq. SeO}_2/\text{C}_6\text{H}_6} (EtO)_3P{=}O \quad + \quad Se$$

d)

                                                     96%

Scheme 247 [585]

9.11 Reaction with Hydrazines

Few hydrazines have been reacted with SeO_2. Hydrazine, the parent compound is oxidized to nitrogen [587], whereas a series of mono aryl hydrazines are transformed to the coresponding diazonium salts [588] (Scheme 248a) on reaction with SeO_2. N,N-Diphenyl hydrazine produces diphenyl amine [588] (Scheme 248b) and N-alkyl-N'-tosyl hydrazines lead (Scheme 248c), probably via a radical pathway [589], to alkane sulphinic esters (in yields ranging from 40 to 80%) along with some olefins (14–18%).

The oxidation of hydrazines to azo derivatives has been proposed as a key step in these transformations [589]. Nevertheless the azo system was never isolated even in the favourable case of tosyl hydrazines [589].

a) $PhNHNH_2$, HCl $\xrightarrow{\text{1 eq. SeO}_2/\text{H}_2\text{O}}$ $PhN_2^+ \ Cl^- \ + \ Se°$

                                                    60%   [588]

b) $Ph_2N{-}NH_2$ $\xrightarrow{\text{1 eq. SeO}_2/\text{H}_2\text{O}}$ $Ph_2NH \ + \ Se°$

                                                    94%   [588]

c) $4{-}MeC_6H_4SO_2NHNHCH_2CH_2R \xrightarrow[\text{20°C/2h}]{\text{1 eq. SeO}_2/\text{THF}} N_2 \ + \ 4{-}MeC_6H_4SO_3H$

    R= $C_{14}H_{29}$, cHex $+ \ 4{-}MeC_6H_4S(O)OCH_2CH_2R$ (40-80%)

                                                $+ \ RCH{=}CH_2$ (14-18%)   [589]

Scheme 248

9.12 Reaction with Organometallics

It was recently found that SeO_2 reacts with Grignard reagents [590], alkyl lithiums [590] and trialkyl boranes [591] to produce selenides. It was also described [308] that arylmercury chlorides produce arylseleninic acids on reaction with SeO_2. The reaction was successfully applied to the synthesis of polystyrene bound benzeneseleninic acid [308] (Scheme 249), precursor of the corresponding perseleninic acid. The latter proved to be a valuable reagent [308] able to oxidize [308] olefins to diols, to promote the oxidation [308] of allyl- and benzylalcohols and of phenolic systems and to effect [308] the Baeyer-Villiger reaction on ketones (Chapter 8).

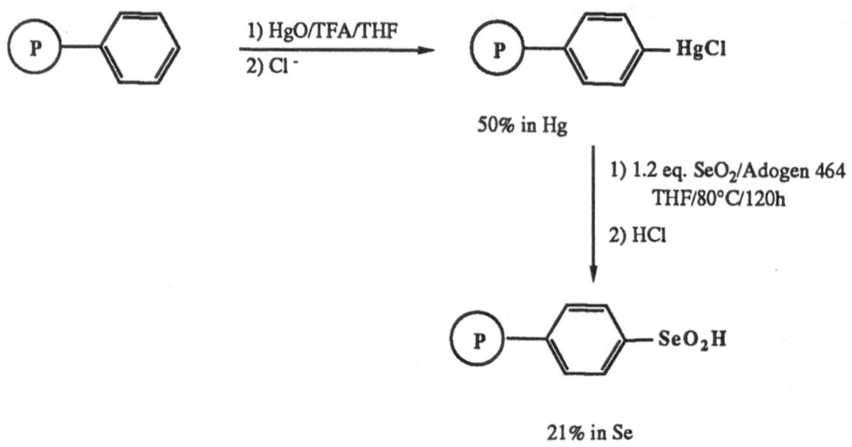

Scheme 249 [308]

180

Chapter 10

Reactions of Imidoselenium Compounds

10.1 With Alkenes and Alkynes: Synthesis of Allylic and Propargylic Amines

In 1976 Sharpless reported [592] that β-pinene afforded the corresponding allylic amine when reacted in methylene chloride whith selenium diimides **63** and **64** (Schemes 250, 251) obtained from selenium tetrachloride and t-butyl amine (2

SeCl₄ $\xrightarrow{\text{2 RNH}_2}$ RN=Se=NR

R = tBu	<u>63</u>
R = Ts	<u>64</u>

TsNClNa $\xrightarrow[\text{CH}_2\text{Cl}_2]{\text{Se}}$ <u>65</u>

Scheme 250 [592]

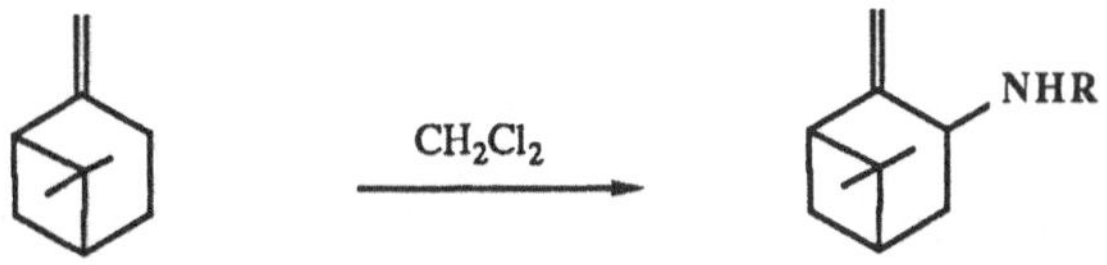

$\xrightarrow{\text{CH}_2\text{Cl}_2}$

R = tBu	2 eq.	<u>63</u>	62%
R = Ts	2 eq.	<u>64</u>	82%
R = Ts	0.63 eq.	<u>65</u>	82%

Scheme 251 [592]

equiv.) or p-toluene sulfonamide (2 equiv.) respectively in the presence of an amine base (4 equiv.). Even better results have been observed [592] using reagent **65** (0.63 to 0.83 mol. equiv. around 20 °C. Scheme 250), although **64** and **65** should be structurally identical. Under these conditions side reactions leading to multiple allylic amination or to the corresponding diene are minimized [592].

10. Reaction of Imidoselenium Compounds

Differently substitued olefins have been successfully transformed [592] to the corresponding N-tosyl allyl amines on reaction with **65** (Schemes 251, 252) whereas acetylenic compounds (Scheme 253) produced N-tosyl propargylic

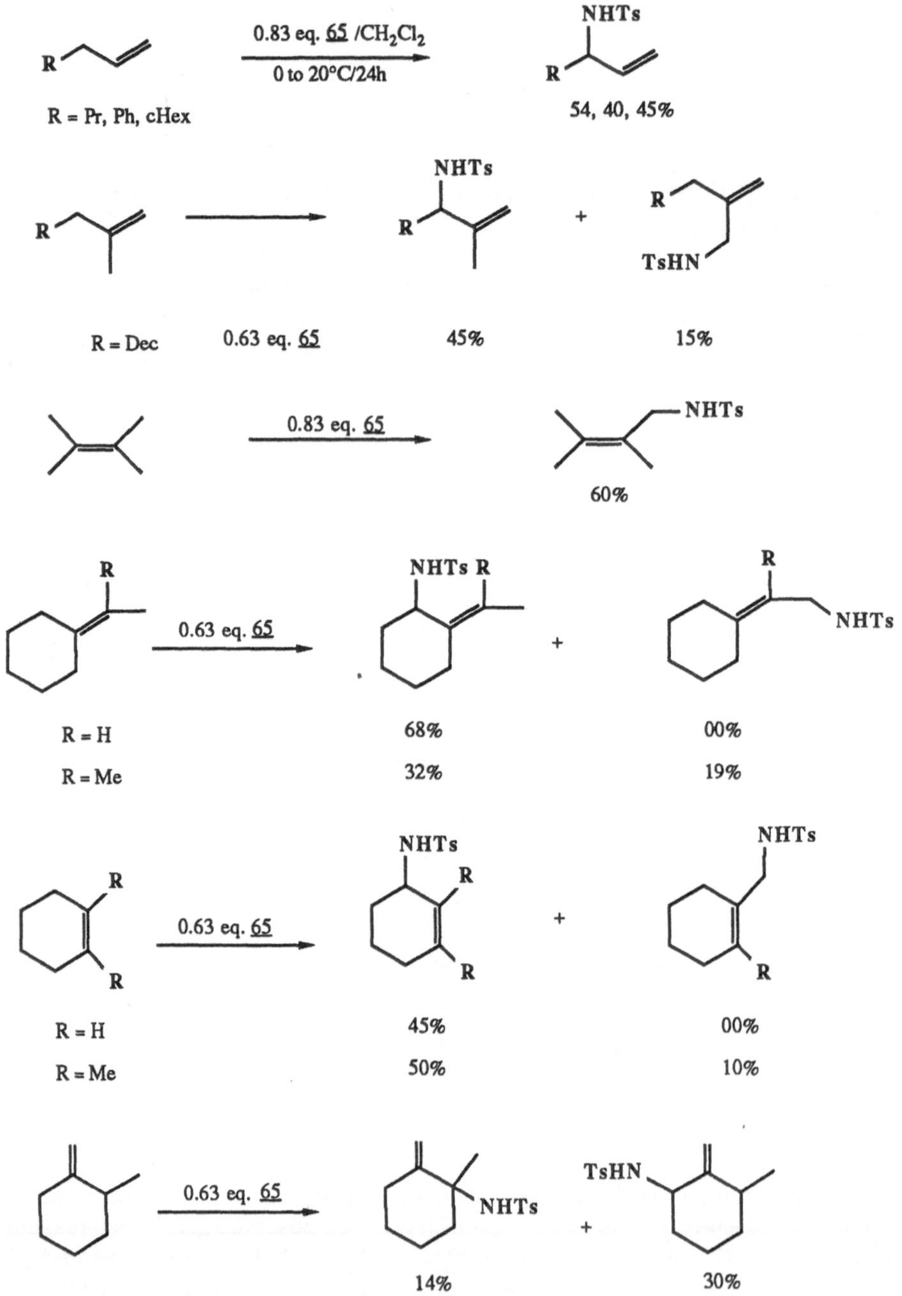

R = Pr, Ph, cHex

54, 40, 45%

R = Dec   0.63 eq. **65**   45%   15%

0.83 eq. **65**   60%

R = H   68%   00%
R = Me   32%   19%

R = H   45%   00%
R = Me   50%   10%

14%   30%

Scheme 252 [592]

182

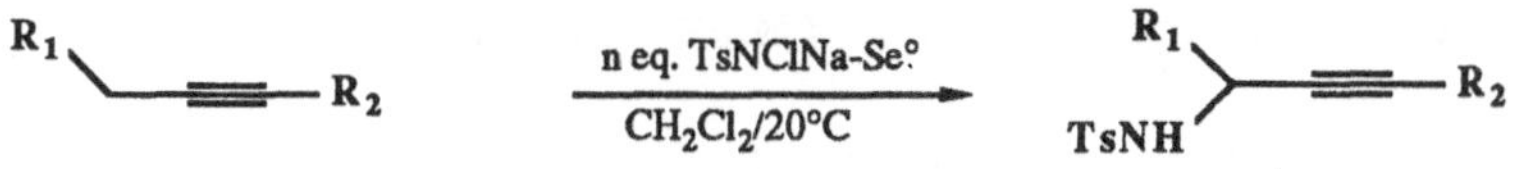

a) R_1 = Hept; R_2 = H;   n = 1.25   23%

b) R_1, R_2 = Pr ;   n = 0.63   51%

Scheme 253 [592]

amines but in much lower yields [592]. The positional selectivity for these
aminations is reminiscent of that described by Guillemonat [*63, *81, *82, *84,
*85, *89, *358, *372] for SeO$_2$ allylic oxygenation but striking differences are
observed with cyclic olefins. The reactions reported above are related to the ones
implying (i) the analogous bis(p-toluenesulfonyl)sulfodiimide [593, 594] which
also produces allylic N-tosylamines from olefins in similar or even better yields
and to (ii) selenium dioxide oxidations which lead to allyl alcohols which have
been discussed in Chapter 9. It is therefore not surprising that a mechanism
involving an ene reaction followed by a [2,3] sigmatropic shift, reminiscent of the
one already described for the oxidation of olefins (Scheme 177) [344, 432], has
been proposed to rationalize these interesting transformations (Scheme 254)
[592, 593].

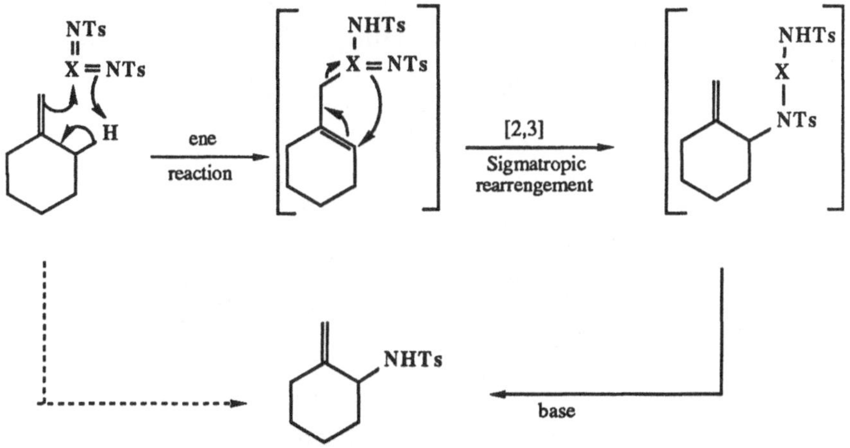

a)   X = Se   45%   [592]

b)   X = S   84%   [593]

Scheme 254

10.2 With Dienes: Synthesis of *Cis* Diaminoalkanes

Reagent **65** also reacts with dienes and produces α, β-disulfonamides [595] (Scheme 255). In the case of cyclohexadiene and cyclopentadiene the sulfonamide groups are introduced cis to each other [595]. With dienes possessing differently substituted carbon-carbon double bonds the less substituted one is usually diaminated [595] (Scheme 255d, e). A speculative mechanism taking into account the fact that the best results are obtained when the reaction is performed in the presence of equimolecular amounts of p-toluenesulfonamide is presented

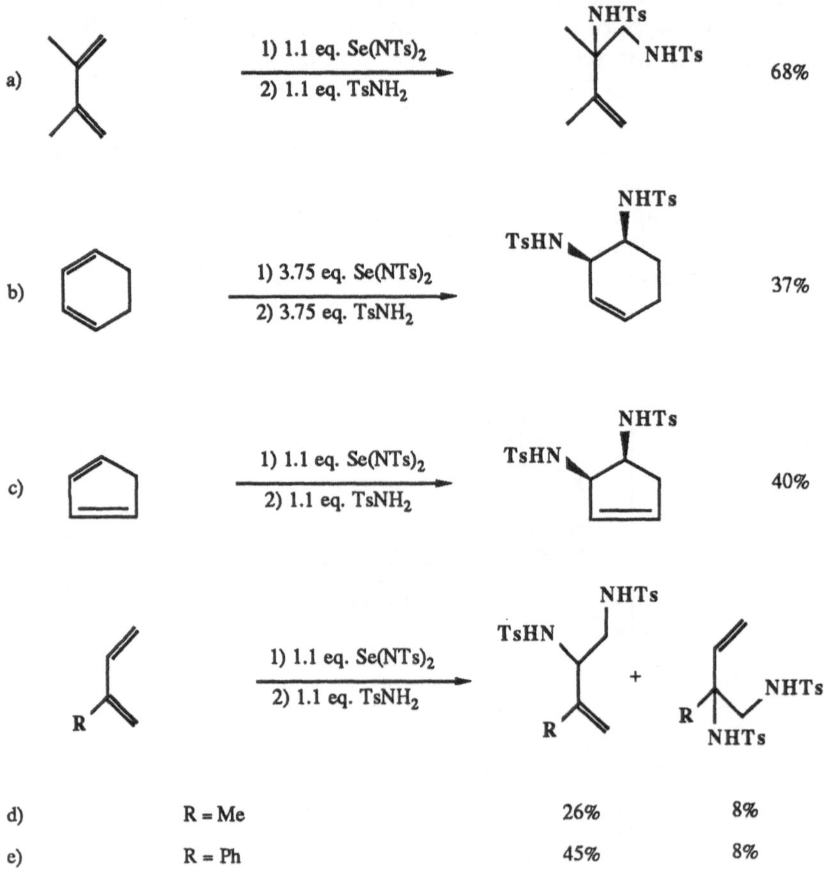

Scheme 255 [595]

in the Scheme 256. Finally the reagent obtained from diphenyl diselenide and chloramine-T and containing probably the selenimido moiety behaves differently [596] since it adds accross the carbon-carbon double bond of olefins and produces phenylseleno seleninamidines (Scheme 257). Attempts at isolating products led

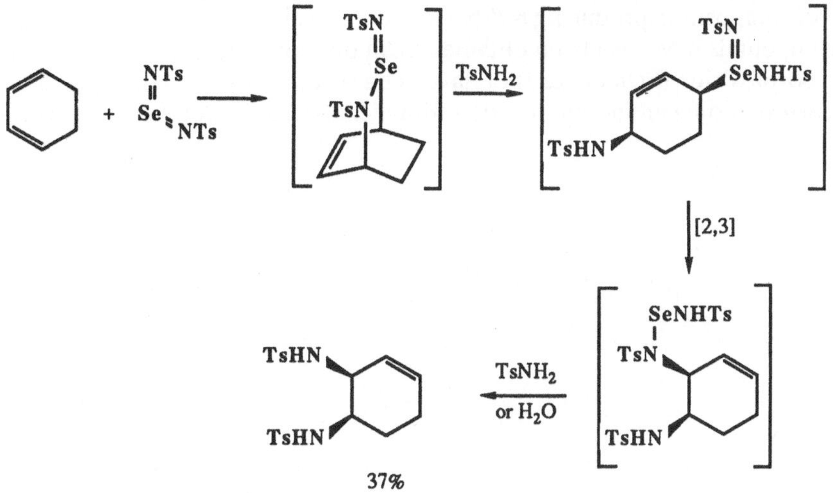

Scheme 256 [595]

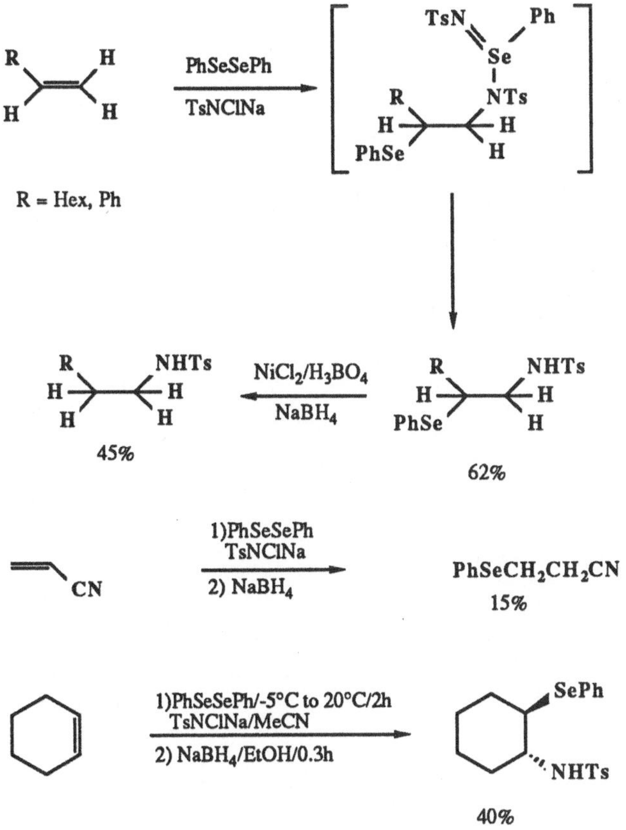

R = Hex, Ph

PhSeCH₂CH₂CN
15%

Scheme 257 [595]

to their decomposition producing β,β-N-(tosylamino) alkyl phenyl selenides. The latter compounds have also been obtained [596] on reduction of the adduct with sodium borohydride (Scheme 257) or have been successfully deselenylated [596] to the corresponding amines on reaction with nickel boride [596] (Scheme 257a).

Chapter 11

Reactions Involving Selenium Oxychloride and Selenium Tetrahalides

11.1 Reactivity of Selenium Oxychloride

Selenium oxychloride in benzene or ether permits the synthesis of α-chloroketones from ketones [597] (Scheme 258). The reaction is exothermic and proceeds through dichloroorganoselenium compounds which decompose to α-chloroketones at their boiling point or after several hours in boiling benzene.

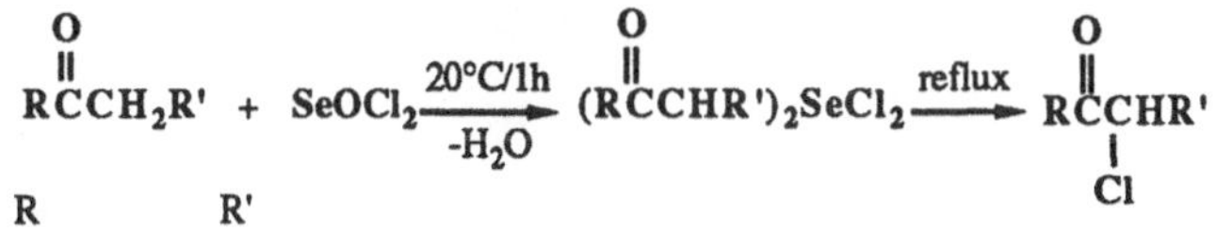

$$
\underset{\text{O}}{\overset{\text{O}}{\text{R}\overset{\|}{\text{C}}\text{CH}_2\text{R}'}} + \text{SeOCl}_2 \xrightarrow[-\text{H}_2\text{O}]{20°\text{C}/1\text{h}} (\text{R}\overset{\text{O}}{\overset{\|}{\text{C}}}\text{CHR}')_2\text{SeCl}_2 \xrightarrow{\text{reflux}} \underset{\text{Cl}}{\overset{\text{O}}{\text{R}\overset{\|}{\text{C}}\text{CHR}'}}
$$

R	R'	
Ph	H	54%
- (CH$_2$)$_4$-		41%
Ph	Me	42%

Scheme 258 [597]

11.2 Reactivity of Selenium Tetrafluoride

Selenium tetrafluoride [598], a liquid (bp 106 °C, mp −10 °C) soluble in halogenated solvents, and its pyridine complex [598] proved to be convenient fluorinating agents particularly suitable for the replacement of hydroxyl and carbonyl groups by fluorine atom(s). The reactions are generally carried out at room temperature and at atmospheric pressure, in usual laboratory glass equipment, under conditions milder than those involved when the sulfur analogue [*598, 599] (SF$_4$) is used.

11.2.1 Synthesis of Geminal Difluoroalkanes from Aldehydes and Ketones

On reaction with SeF$_4$ aldehydes, ketones and N,N-dimethyl benzamide are transformed, in yields ranging from 65 to 100%, to the corresponding geminal difluorides [598, 600] (Scheme 259).

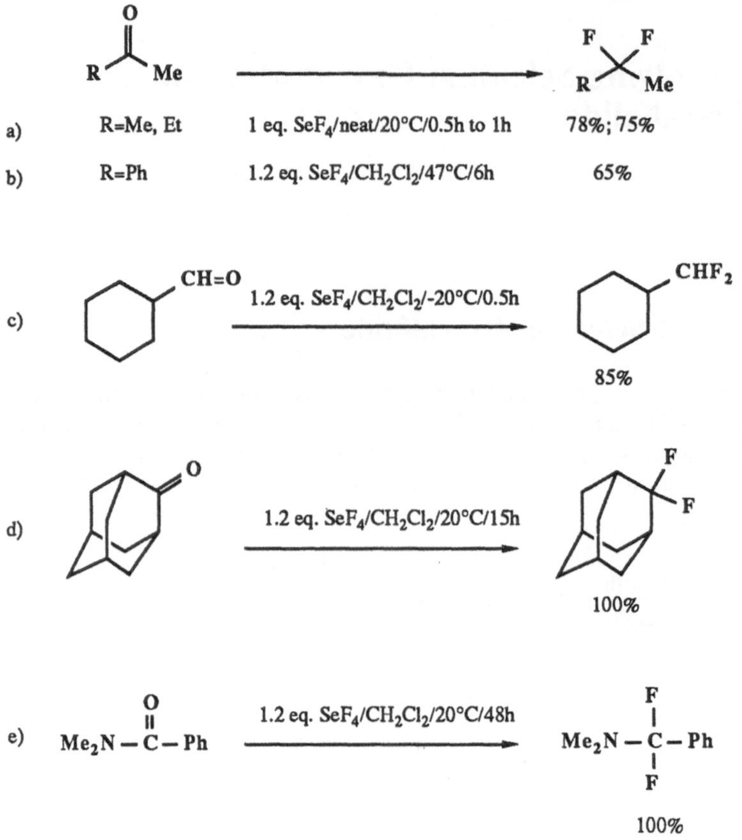

a) R=Me, Et   1 eq. SeF$_4$/neat/20°C/0.5h to 1h   78%; 75%

b) R=Ph   1.2 eq. SeF$_4$/CH$_2$Cl$_2$/47°C/6h   65%

c) 1.2 eq. SeF$_4$/CH$_2$Cl$_2$/-20°C/0.5h   85%

d) 1.2 eq. SeF$_4$/CH$_2$Cl$_2$/20°C/15h   100%

e) Me$_2$N – C – Ph   1.2 eq. SeF$_4$/CH$_2$Cl$_2$/20°C/48h   Me$_2$N – C – Ph   100%

Scheme 259 [598]

11.2.2 Synthesis of Alkylfluorides from Alcohols

SeF$_4$ also promotes the formation of alkylfluorides [598] from alcohols (Schemes 260, 261) and of acyl fluorides [598] from carboxylic acids or anhydrides (Scheme 262). Best results are obtained when the reactions are performed with the SeF$_4$-Pyridine complex [598] (readily available on reaction of SeF$_4$ with pyridine) since the pyridine liberated in the medium permits the trapping of the HF concomitantly formed.

Alkoxyselenium trifluorides (ROSeF$_3$) first formed [598] in the case of alcohols decompose [598] later on heating to the corresponding alkyl fluorides which are formed without rearrangement (except with susceptible systems such as cyclopropyl carbinyl and isobutyl alcohols) [598] (Scheme 261). It is interesting to notice that α-hydroxyketones are selectively tansformed [598] to α-fluoroketones (Scheme 260d) under these conditions.

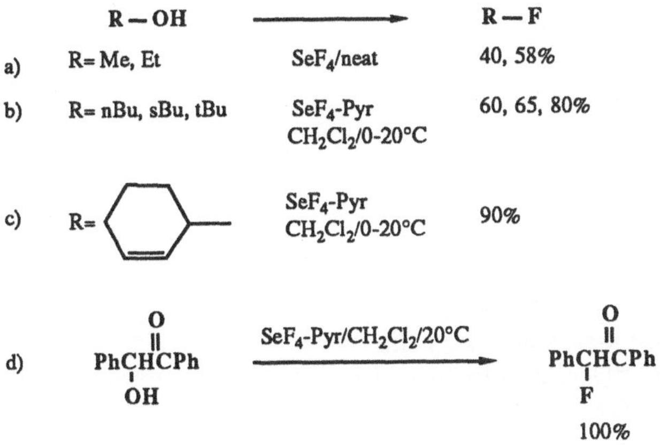

$$R-OH \quad \xrightarrow{\hspace{3cm}} \quad R-F$$

a) R= Me, Et $\quad$ SeF$_4$/neat $\quad$ 40, 58%

b) R= nBu, sBu, tBu $\quad$ SeF$_4$-Pyr $\quad$ 60, 65, 80%
CH$_2$Cl$_2$/0-20°C

c) R= $\quad$ SeF$_4$-Pyr $\quad$ 90%
CH$_2$Cl$_2$/0-20°C

d) PhCHCPh $\quad$ $\xrightarrow{\text{SeF}_4\text{-Pyr/CH}_2\text{Cl}_2/20°C}$ $\quad$ PhCHCPh
(with O above, OH below) (with O above, F below)
100%

Scheme 260 [598]

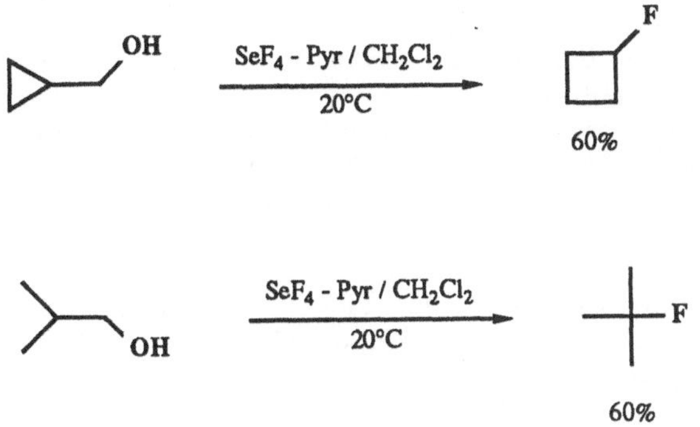

$\xrightarrow{\text{SeF}_4 - \text{Pyr / CH}_2\text{Cl}_2}{20°C}$ 60%

$\xrightarrow{\text{SeF}_4 - \text{Pyr / CH}_2\text{Cl}_2}{20°C}$ 60%

Scheme 261 [598]

RCO$_2$H $\quad$ $\xrightarrow{\text{SeF}_4\text{-Pyr/CH}_2\text{Cl}_2}{20°C}$ $\quad$ RCOF

H, Me, Et, cHex, Ph $\qquad\qquad$ 25, 80, 85, 90, 95%

(EtC)$_2$O $\quad$ $\xrightarrow{\text{SeF}_4\text{-Pyr/CH}_2\text{Cl}_2}{20°C}$ $\quad$ 2 EtCF

(with O above (EtC)$_2$O) (with O above EtCF)
85%

Scheme 262 [598]

11.3 Reactivity of Selenium Tetrachloride

Selenium tetrachloride was described in a patent [601] to selectively oxidize propene to allyl chloride.

References

1. Berzelius JJ: Acad Handl Stockholm **39,** 13 (1818)
2. Rankama K, Sahama TG: "Geochemistry" quoted in T Moeller's "Inorganic Chemistry" John Wiley, New York, 30 (1952)
3. Turekian KK, Wedepohl KH: Geol Soc Amer Bull **72,** 175 (1961)
4. Andrews ED, Hartley WJ: Grant AB NZ Vet J **16,** 3 (1968)
5. Hartley WJ, Grant AB: Fed Proc **20,** 679 (1961)
6. Lakin HW: Trace Elements in the Environment, Am Chem Soc 96 (1973), See also Chem Abstr 68: 51917r, (1968)
7. Louderback T: Mineral Industries Bulletin **18,** 3 (1975)
8. Oldfield JE: Nat Acad Sci **1,** 57 (1974)
9. Shrift A: Organic Selenium Compounds: Their Chemistry and Biology (Klayman DL, Günther WHH Ed) John Wiley and Sons, New York, 763 (1973)
10. Rosenfeld I, Beath OA: Selenium Geobotany, Biochemistry, Toxicity and Nutrition, Academic Press, New York, (1964)
11. Frost DV, Ingvoldstad D: Chem Script **8A,** 96 (1975)
12. Shamberger RJ: Biochemistry of Selenium, Plenum Press, New York, (1983)
13. Trelease SF Di Somma AA, Jacobs AL: Science **132,** 618 (1960)
14. Hurd-Karner AM: J Agr Res **54,** 601 (1937)
15. Martin JL: Organic Selenium Compounds: Their Chemistry and Biology (Klayman DL, Günther WHH Ed) John Wiley and Sons, New York, 663 (1973)
16. Martin JL: Proceedings 3rd Int Symp Selenium and Tellurium Comp (9–12 July Metz, France) (Cagniant D, Kirsch G Ed) Metz, 133 (1979)
17. Scott ML: Organic Selenium Compounds: Their Chemistry and Biology (Klayman DL, Günther WHH Ed) John Wiley and Sons, New York, 629 (1973)
18. Schwarz K, Pathak KD: Chem Script **8A,** 85 (1975)
19. Underwood EJ: Trace Elements in Human and Animal Nutrition, Academic Press, New York, (1971)
20. Diplock AT: Vit Horm **32,** 445 (1974)
21. Moxon AL: Science, **88,** 81 (1938)
22. Levander OA, Agrett LC: Toxic Appl Pharmacol **14,** 308 (1969)
23. Schrauzer GN, McGinness JE, Kuehn K: Proceedings 3rd Int Symp Selenium and Tellurium Comp (9–12 July Metz, France) (Cagniant D, Kirsch G Ed) 173 (1979)
24. Stadman TC: Ann Rev Biochem **49,** 93 (1980)
25. Stadman TC, Dilworth GL: Proceedings 3rd Int Symp Selenium and Tellurium Comp (9–12 July Metz, France) (Cagniant D, Kirsch G Ed) 117 (1979)
26. Ganther HE: Chem Script **8A,** 79 (1975)
27. Hoekstra WG: Fed Proc **34,** 2083 (1975)
28. Scott ML, Noguchi T, Coombs GF: Vit Horm **32,** 429 (1974)
29. Rotruck JT, Pope AL, Gauther HC, Swanson AB, Hafeman DG, Hoekstra WG: Science **179,** 588 (1973)
30. Flohe L, Gunzler WA, Schock HH: FEBS Lett **32,** 132 (1973)

31. Wendel A, Sies H: Proceedings 3rd Int Symp Selenium and Tellurium Comp (9–12 July Metz, France) (Cagniant D, Kirsch G Ed) 157 (1979)
32. Ladenstein R, Epp O, Huber R: Proceedings 3rd Int Symp Selenium and Tellurium Comp (9–12 July Metz, France) (Cagniant D, Kirsch G Ed) 147 (1979)
33. Flohe L, Eisele B, Wendel A: Hoppe-Seylers Z Physiol Chem 352, 151, (1971), See also Chem Abstr 74: 107494u, (1971)
34. Oh SH, Ganther HE, Hoeskstra WG: Biochemistry **13**, 1825 (1974)
35. Awasthi YC, Beutlar E, Srivastava SK: J Biol Chem **250**, 5144 (1975)
36. Nakamura W, Hosoda S, Hayashi K: Biochim: Biophys Acta **358**, 251 (1974)
37. Walter R, Roy J: Organic Selenium Compounds: Their chemistry and Biology (Klayman LD, Günther WHH Ed) John Wiley and Sons, New York, 601 (1973)
38. Shapiro JR: Organic Selenium Compounds: Their Chemistry and Biology (Klayman DL, Günther WHH Ed) John Wiley and Sons, New York, 693 (1973)
39. Cerwenta EA, Cooper C: Archives of Environmental Health, **3**, 189 (1961)
40. Glover JR: Trans Ass Ind Med Officers **4**, 94 (1954)
41. Glover JR: Proceedings of the Symp on Selenium – Tellurium Env, Ind Health Found (Pittsburgh, Pensylvania), U.S.A., 279 (1976)
42. Klayman DL: Organic Selenium Compounds: Their Chemistry and Biology (Klayman DL, Günther WHH Ed) John Wiley and Sons, New York, 727 (1973)
43. Hekin H, Lehne RK: Belgian Patent N° 644.855, 15 (1964), Chem Abstr 63: P8117h, (1965)
44. Matsui M, Kitahara T, Takagi K, Watanabe I, Tamokami S: Jpn Kokkai Tokkyo Koho 79 55,512, 1979, Chem Abstr 91: 157940h Ed, (1979)
45. Desiron JL, Menzies R: Société Générale des Minerais (Bruxelles, Belgium). Personal Communication, (1985)
46. Lange B: Photoelements and their Applications, Reinhold Publishing Co, New York, (1938)
47. Chu JYC: Proceedings 3rd Int Symp Selenium and Tellurium Comp (9–12 July Metz, France), 351 (1979)
48. Tsaritsyi MA, Zakharenko NI: Steklo i Keram 22, 25 (1965), Chem Abstr 63: 2715c, (1965)
49. Tsaritsyi MA, Zakharenko NI: Steklo i Keram 19, 16 (1962), Chem Abstr 57: 16149d, (1962)
50. Crystal RG: Organic Selenium Compounds: Their Chemistry and Biology (Klayman DL, Günther WHH Ed) John Wiley and Sons, New York, 13 (1973)
51. Dharmatti SS, Weaver HE: Phys Rev **86**, 259 (1952)
52. Duddeck H, Wagner P, Gegner S: Tetrahedron Lett **26**, 1205 (1985)
53. Lardon MA: Organic Selenium Compounds: Their Chemistry and Biology (Klayman DL, Günther WHH Ed) John Wiley and Sons, New York, 933 (1973)
54. Klayman DL, Gunther WHH: Organic Selenium Compounds: Their Chemistry and Biology (Klayman DL, Günther WHH Ed) John Wiley and Sons, New York, (1973)
55. Wittig G, Fritz H: Justus Liebigs Ann Chem **577**, 39 (1952)
56. Hellwinkel D, Fahrbach G: Justus Liebigs Ann Chem **715**, 68 (1968)
57. Allred AL: J Inorg Nucl Chem **17**, 215 (1961)
58. Allred AL, Rochow EG: J Inorg Nucl Chem **5**, 264 (1958)
59. Pritchard HO, Skinner HA: Chem Rev **55**, 745 (1955)
60. Rheinboldt H: Houben-Weyl, Methoden der Organischen Chemie (Müller E Ed) G. Thieme Verlag Stuttgart, **IX**, 917 (1967)
61. Krief A: Tetrahedron **36**, 2531 (1980)
62. Clive DLJ: Tetrahedron **34**, 1049 (1978)
63. Reich HJ: Oxidation in Organic Chemistry (Trahanovsky WS Ed) Academic Press, New York, **C**, 1 (1978)

64. Reich HJ: Acc Chem Res **12,** 22 (1979)
65. Krief A, Dumont W, Denis JN, Evrard G, Norberg B: J Chem Soc Chem Commun 569 (1985)
66. Krief A: The Chemistry of Selenium and Tellurium Compounds (Patai S and Rappoport Z Ed) John Wiley and Sons, Chichester, **Vol. 2,** (1987)
67. Krief A, Hevesi L: Janssen Chimica Acta **2,** 3 (1984)
68. Hevesi L: The Chemistry of Selenium and Tellurium Compounds (Patai S and Rappoport Z Ed) John Wiley and Sons, Chichester, **Vol. 1,** 307 (1986)
69. Reich HJ, Cohen ML: J Org Chem **44,** 3148 (1979)
70. Encyclopedia of Electrochemistry of the Elements, (Bard AJ Ed) Marcel Dekker, New York, **4,** 280 (1975)
71. Encyclopedia of Electrochemistry of the Elements (Bard AJ Ed) Marcel Dekker, New York, **4,** 363 (1975)
72. Nakamura N, Sekido E: Talanta **17,** 515 (1970)
73. Sugiura Y, Hojo Y, Tamai Y, Tanaka H: J Am Chem Soc **98,** 2339 (1976)
74. Huber RE, Criddle RS: Arch Biochem Biophys **122,** 164 (1967)
75. Fagioli F, Pulidori F, Bighi C, DeBattisti A: Gazz Chim Ital **104,** 639 (1974)
76. Pearson RG, Sobel H, Songstad J: J Am Chem Soc **90,** 319 (1968)
77. Guanti G, Dell'erba C, Spinelli D: Gazz Chim Ital **100,** 184 (1970)
78. Nicolaou KC, Petasis NA: Selenium in Natural Products Synthesis, CIS, Inc., Philadelphia, (1984)
79. Mayor Y: Chimie et Industrie, **43,** 188 (1940)
80. Waitkins GR, Clark CW: Chem Rev **36,** 235 (1945)
81. Rabjohn N: Organic Reactions, John Wiley and Sons, New York, **5,** 331 (1949)
82. Campbell TW, Walker HG, Coppinger GM: Chem Rev **50,** 279 (1952)
83. Gosselck J: Angew Chem **75,** 831 (1963)
84. Trachtenberg EN: Oxidation, Techniques and Applications in Organic Synthesis (Augustine RL Ed) Marcel Dekker, New York, 119 (1969)
85. Jerussi RA: Selective Organic Transformations (Thyagarajan BS Ed) John Wiley, New York, **I,** 301 (1970)
86. Organic Selenium and Tellurium Chemistry Annals of the New York Academy of Sciences (Okamoto Y, Gunther WHH Ed) **192,** (1972)
87. Zingaro RW, Copper W: Selenium, Van Nostrand-Reinhold, Princeton, 200 (1974)
88. Sharpless KB, Gordon KM, Lauer RF, Singer SP, Young MW: Chem Script **8A,** 9 (1975)
89. Rabjohn N: Organic Reactions, John Wiley and Sons, New York, **24,** 261 (1976)
90. Schmid GH, Garratt DG: The Chemistry of Double-bonded Functional Groups Part 2, Suppl A (Patai S Ed) J Wiley and Sons, New York, 855 (1977)
91. Bulka E: The Chemistry of Cyanates and Their Thio Derivatives, Part I (Patai S Ed) John Wiley and Sons, Chichester, 887 (1977)
92. Barton DHR, Ley SV: Further Perspectives in Organic Chemistry, Ciba Foundation Symposium, Elsevier, 53 (1978)
93. Clive DLJ: Aldrichimica Acta **11,** 43 (1978)
94. Magnus PD: Comprehensive Organic Chemistry (Barton DHR and Ollis WD Ed) Pergamon Press, Oxford, **3,** 491 (1979)
95. Comasseto JV, Ferreira JTB, Marcuzzo do Canto M: Quimica Nova, 58 (1979)
96. Sharpless KB, Verhoeven TR: Aldrichimica Acta **12,** 63 (1979)
97. Ley SV: Annual Reports **77,** 233 (1980)
98. Nicolaou KC: Tetrahedron **37,** 4097 (1981)
99. Pennanen SI: Kemia-Kemi 275 (1981)
100. Pennanen SI: Kemia-Kemi 501 (1981)
101. Witczak ZJ, Whistler RL: Heterocycles **19,** 1719 (1982)

102. Comasseto JV: J Organomet Chem **253,** 131 (1983)
103. Witczak ZJ: Nucleosides Nucleotides **2,** 295 (1983)
104. Liotta D: Acc Chem Res **17,** 28 (1984)
105. Krief A: Topics in Current Chemistry (De Meijere A Ed) Springer-Verlag, Berlin, **135,** 1 (1987)
106. Reid DH: Specialist Periodical Reports: Organic Compounds of Sulfur, Selenium and Tellurium, The Chemical Society, London, **1** (1970)
107. Reid DH: Specialist Periodical Reports: Organic Compounds of Sulfur, Selenium and Tellurium, The Chemical Society, London, **2** (1973)
108. Reid DH: Specialist Periodical Reports: Organic Compounds of Sulfur, Selenium and Tellurium, The Chemical Society, London, **3** (1975)
109. Hogg DR: Specialist Periodical Reports: Organic Compounds of Sulfur, Selenium and Tellurium, The Chemical Society, London, **4** (1977)
110. Hogg DR: Specialist Periodical Reports: Organic Compounds of Sulfur, Selenium and Tellurium, The Chemical Society, London, **5** (1979), ISBN 0 85186 620 4
111. Hogg DR: Specialist Periodical Reports: Organic Compounds of Sulfur, Selenium and Tellurium, The Chemical Society, London, **6** (1981)
112. Katritzky AR, Rees CW: Comprehensive Heterocyclic Chemistry, Pergamon Press, Oxford (1984)
113. Patai S, Rappoport Z: The Chemistry of Organic Selenium and Tellurium Compounds, John Wiley, Chichester, Vol. 1 (1986)
114. Patai S, Rappoport Z: The Chemistry of Organic Selenium and Tellurium Compounds, Vol. 2, John Wiley and Sons, Chichester (1987)
115. Krief A: Tetrahedron **42,** 1209 (1986)
116. Bagnall KW: Selenium, Tellurium and Polonium in Comprehensive Inorganic Chemistry, Pergamon Press, Oxford, **Vol. 2,** 935 (1973)
117. Simanek V, Klasek A: Tetrahedron Lett 3039 (1969)
118. Ahmad R, Saá JM, Cava MP: J Org Chem **42,** 1228 (1977)
119. Tiecco M, Testaferri L, Tingoli M, Chianelli D, Montanucci M: J Org Chem **48,** 4289 (1983)
120. Evers M, Christiaens L: Tetrahedron Lett **24,** 377 (1983)
121. Tiecco M, Testaferri L, Tingoli M, Chianelli D, Montanucci M: Synth Commun **13,** 617 (1983)
122. Ahmed Z, Cava MP: J Am Chem Soc **105,** 682 (1983)
123. Liotta D, Sunay U, Santiesteban H, Markiewicz W: Org Chem **46,** 2605 (1981)
124. McMurry JE: Org React (Dauben WG Ed) John Wiley and Sons, New York, **24,** 187 (1976)
125. Liotta D, Markiewicz W, Santiesteban H: Tetrahedron Lett 4365 (1977)
126. Detty MR: Tetrahedron Lett 5087 (1978)
127. Liotta D, Santiesteban H: Tetrahedron Lett 4369 (1977)
128. Scarborough RM, Smith AB: Tetrahedron Lett 4361 (1977)
129. Sindelar K, Metysova J, Protiva M: Czech Chem Commun **34,** 3801 (1969)
130. Agenas L-B: Ark Kemi **24,** 415 (1965)
131. Scarborough RM, Toder BH, Smith AB: J Am Chem Soc **102,** 3904 (1980)
132. Günther WHH: J Org Chem **31,** 1202 (1966)
133. Kelly TR, Dali HM, Tsang WG: Tetrahedron Lett 3859 (1977)
134. Hevesi L: Tetrahedron Lett 3025 (1979)
135. Seshadri R, Pegg WJ, Israel M: J Org Chem **46,** 2596 (1981)
136. Zima G, Barnum C, Liotta D: J Org Chem **45,** 2736 (1980)
137. Liotta D, Saindane M, Barnum C, Zima G: Tetrahedron **41,** 4881 (1985)
138. Prince M, Bremer BW, Brenner W: J Org Chem **31,** 4292 (1966)
139. Prince M, Bremer BW: J Org Chem **32,** 1655 (1967)

References

140. Sevrin M, Denis JN, Krief A: Tetrahedron Lett **21,** 1877 (1980)
141. Van Es T: JS Afr Chem Inst **21,** 82 (1968)
142. Van Ende D, Krief A: Tetrahedron Lett 2709 (1975)
143. Ho T-L: Synth Commun **8,** 301 (1978)
144. Behan JM, Johnstone RAW, Wright MJ: J Chem Soc Perkin Trans I, 1216 (1975)
145. Van Es T: Carbohydrate Res **5,** 282 (1967)
146. Clive DLJ, Denyer CV Derrick LJ: J Chem Soc Chem Commun 253 (1973)
147. Chan TH, Finkenbine JR: Tetrahedron Lett 2091 (1974)
148. Detty MR: J Org Chem **44,** 4528 (1979)
149. Woods TS, Klayman DL: J Org Chem **39,** 3716 (1974)
150. Fujimori K, Yoshimoto H, Oae S: Tetrahedron Lett **21,** 3385 (1980)
151. Wako Pure Chem Ind, Jpn Kokai Tokkyo Koho JP 8228,011, Chem Abstr 96: 216846k, (1982)
152. Perkins MJ, Smith BV, Turner ES: J Chem Soc Chem Commun 977 (1980)
153. Kambe N, Kondo K, Murai S, Sonoda N: Angew Chem Int Ed Engl **19,** 1008 (1980)
154. Kametani T, Aizawa M, Kurobe H, Matsuura T, Nemoto H, Fukumoto K: Chem Pharm Bull (Tokyo) **30,** 1493 (1982)
155. Fujimori K, Yoshimoto H, Oae S: Tetrahedron Lett 4397 (1979)
156. Miyata T, Kondo K, Murai S, Hirashima T, Sonoda N: Angew Chem Int Ed Engl **19,** 1008 (1980)
157. James FG, Perkins MJ, Porta O, Smith BV: J Chem Soc Chem Commun 131 (1977)
158. Morgan GT, Burstall FH: J Chem Soc 2197 (1929)
159. Reich HJ, Renga JM, Reich IL: J Am Chem Soc **97,** 5434 (1975)
160. Culvenor CCJ, Davies W, Heath NS: J Chem Soc, 278 (1949)
161. Callear AB, Tyerman WJR: Trans Faraday Soc, **62,** 2760 (1966)
162. Tyerman WJR, O'Callaghan WB, Kebarle P, Strausz OP, Gunning HE: J Am Chem Soc **88,** 4277 (1966)
163. Corey EJ, Marfat A, Falck JR, Albright JO: J Am Chem Soc **102,** 1433 (1980)
164. Mathey F, Muller G CR: Acad Sci Paris, Ser C, **281,** 881 (1975)
165. Clive DLJ, Menchen SM: J Org Chem **45,** 2347 (1980)
166. Kudelska W, Michalska M: Tetrahedron Lett **22,** 2989 (1981)
167. Calo V, Lopez L, Mincuzzi A, Pesce G: Synthesis, 200 (1976)
168. Paulsen H, Heiker FR, Feldmann J, Heyns K: Synth Commun 636 (1980)
169. Ogawa A, Miyake JI, Murai S, Sonoda N: Tetrahedron Lett **26,** 669 (1985)
170. Ogawa A, Miyake JI, Karasaki Y, Murai S, Sonoda N: J Org Chem **50,** 384 (1985)
171. Perkins MJ, Smith BV, Terem B, Turner ES: J Chem Research 341 (1979)
172. Clive DLJ, Menchen SM: J Chem Soc Chem Commun 168 (1979)
173. Clive DLJ, Kiel WA, Menchen SM, Wong CK: J Chem Soc Chem Commun 657 (1977)
174. Detty MR, Seidler MD J Org Chem **47,** 1354 (1982)
175. Sonoda N, Kondo K, Nagano K, Kambe N, Morimoto F: Angew Chem Int Ed Engl **19,** 308 (1980)
176. Laing IG: Rodd's Chemistry of the Carbon Compounds (Coffey, S Ed) Elsevier, Amsterdam, **IIIC,** 59 (1973)
177. Emerson WS: Organic Reactions (Adams R Ed) John Wiley and Sons, New York, **4,** 174 (1948)
178. Scheithauer M, Mayer R: Z Chem **6,** 375 (1966)
179. Chakravarti RN, Robinson R: J Chem Soc 78 (1947)
180. Kataev EG, Gabdrakhmov FG: Zh Obshch Khim **37,** 772 (1967)
181. Morimoto S: J Chem Soc Japan **75,** 557 (1954)
182. Cravador A, Krief A, Hevesi L: J Chem Soc Chem Commun 451 (1980)

183. Clarembeau M, Cravador A, Dumont W, Hevesi L, Krief A, Lucchetti J, Van Ende D: Tetrahedron **41,** 4793 (1985)
184. Goldsmith DJ, Liotta DC, Volmer M, Hoekstra W, Waykole L: Tetrahedron **41,** 4873 (1985)
185. Hobé M: Mémoire de licence, Facultés ND de la Paix, Namur, (1986)
186. Cravador A, Krief A: Unpublished Results (1986)
187. Georges B, Hevesi L: Unpublished Results (1986)
188. Back TG, Collins S, Law KW: Tetrahedron Lett **25,** 1689 (1984)
189. Back TG, Kerr RG: Tetrahedron, **41,** 4759 (1985)
190. Tayim HA, Bailar JC: J Am Chem Soc **89,** 4330 (1967)
191a. Maccarone E, Mamo A, Perrini G, Torre M: J Chem Soc Perkin Trans II 324 (1981)
191b. Maccarone E, Mamo A, Scarlata G, Torre: M Tetrahedron **34,** 3531 (1978)
191c. Scarlata G, Torre M: J Heterocycl Chem **13,** 1193 (1976)
191d. Fitzpatrick JD, Orchin M: J Org Chem **22,** 1177 (1957)
192. Fitzpatrick JD, Orchin M: J Am Chem Soc **79,** 4765 (1957)
193. Ucciani E, Piantoni R, Naudet M: Chem Phys Lipids 240 (1968)
194. Gunstone FD, Ismael IA: Chem Phys Lipids 264 (1967)
195. Teeter HM, Bell EW, O'Donnell JL, Danzig MJ, Cowan JC: Amer Oil Chem Soc J **35,** 238 (1958)
196. Libert H, Schmid L: Monatsh **98,** 19 (1967)
197. Ruzicka L: Helv Chim Acta **19,** 419 (1936)
198. Buchta E, Kallert W: Justus Liebigs Ann Chem **576,** 1 (1952)
199. Cocker W, Cross BE, Edward JT, Jenkinson DS, McCormick J: J Chem Soc 2355 (1953)
200. Clemo GR, Dickenson HG: J Chem Soc 735 (1935)
201. Gensler WJ, Sherman GM: J Am Chem Soc **81,** 5217 (1959)
202. Barker RL, Clemo GR: J Chem Soc 1277 (1940)
203. Sen-Gupta SC: Current Sci **6,** 295 (1936)
204. Ruzicka L: Helv Chim Acta **17,** 442 (1934)
205. Ruzicka L, Peyer E: Helv Chim Acta **18,** 676 (1935)
206. Clemo GR, Dickenson HG: J Chem Soc 255 (1937)
207. Gravel D, Gauthier J: Tetrahedron Lett 5489 (1968)
208. Diels O, Gädke W, Kording P: Justus Liebigs Ann Chem **459,** 1 (1927)
209. Diels O, Karstens A: Justus Liebigs Ann Chem **478,** 129 (1930)
210. Bisarya SC, Nayak UR, Dev S: Tetrahedron Lett 2323 (1969)
211. Diels O: Chem Ber **69A,** 195 (1936)
212. Charonnat R, Girard M: Bull Soc Chim Fr 209 (1949)
213. Mesta CK, Paknikar SK, Bhattacharyya SC: Chem Commun 584 (1968)
214. Gruber W, Proske G: Monatsh **82,** 255 (1951)
215. Dieterle H, Salomon A: Arch Pharm **270,** 495 (1932)
216. Banerjee DK, Ramadas SR, Ramani G: Tetrahedron Lett 5311 (1966)
217. Ruzicka L, Van Veen AG: Rec Trav Chim Pays-Bas **48,** 1018 (1929)
218. Brunner O, Hofer H, Stein R: Monatsh **61,** 293 (1932)
219. Thiele W, Trautmann G: Chem Ber **68,** 2245 (1935)
220. Ohta Y, Sakai T, Hirose Y: Tetrahedron Lett 6365 (1966)
221. Castille A: Ann Real Acad Farm **32,** 121 (1966)
222. Pentegova VA, Kashtanova NK, Rezvulahin AI, Kolipova EI: Khim Prirodn Soedin Akad Nauk SSR, **2,** 239 (1966)
223. Lammens H, Verzele M: Bull Soc Chim Belg **77,** 497 (1968)
224. Carman RM, Craig W Aust J: Chem **24,** 361 (1971)
225. Cuthbertson E, Gall JH, MacNicol DD: Tetrahedron Lett 3204 (1977)
226. Klem RE, Skinner HF, Walba H, Isensee RW: J Heterocycl Chem **7,** 403 (1970)

References

227. Yokoyama M, Kotake M: J Chem Soc Japan **56,** 336 (1935)
228. Dorée C, Petrow VA: J Chem Soc 1391 (1935)
229. Yokoyama M, Kotake M: Bull Chem Soc Japan, **10,** 138 (1935)
230. Kondo K, Sonoda N, Sakurai H: Bull Chem Soc Japan **48,** 108 (1975)
231. Sonoda N, Yasukara T, Kondo K, Ikeda T, Tsutsumi S: J Am Chem Soc **93,** 6344 (1971)
232. Kondo K, Sonoda N, Yoshida K, Koishi M, Tsutsumi S: Chem Lett 401 (1972)
233. Kondo K, Yokoyama S, Miyoshi N, Murai S, Sonoda N: Angew Chem Int Ed Engl **18,** 700 (1979)
234. Kondo K, Yokoyama S, Miyoshi N, Murai S, Sonoda N: Angew Chem Int Ed Engl **18,** 692 (1979)
235. Kondo K, Sonoda N, Tsutsumi S: Chem Lett 373 (1972)
236. Koch P: Tetrahedron Lett 2087 (1975)
237. Kondo K, Sonoda N, Tsutsumi S: J Chem Soc Chem Commun, 307 (1972)
238. Kondo K, Sonoda N, Sakurai H: J Chem Soc Chem Commun 853 (1973)
239. Koch F, Perroti E: Tetrahedron Lett 2899 (1974)
240. Sonoda N, Yamamoto G, Natsukawa K, Kondo K, Murai S: Tetrahedron Lett 1969 (1975)
241. Kondo K, Sonoda N, Sakurai H: Chem Lett 1429 (1974)
242. Kondo K, Sonoda N, Tsutsumi S: Tetrahedron Lett 4885 (1971)
243. Kondo K, Sonoda N, Sakurai H: Tetrahedron Lett 803 (1974)
244. Kondo K, Sonoda N, Sakurai H: J Chem Soc Chem Commun 160 (1974)
245. McCoy JJ, Zajacek JG, Fuger KE: US Pat 3,989,755 (1976), Chem Abstr 86: 106052j, (1977)
246. Kondo K, Murai S, Sonoda N: Tetrahedron Lett 3727 (1977)
247. Ogawa A, Kambe N, Murai S, Sonoda N: Tetrahedron **41,** 4813 (1985)
248. Ogawa A, Kondo K, Murai S, Sonoda N: J Chem Soc Chem Commun 1283 (1982)
249. Barnard D, Woodbridge DT: Chem Ind (London), 1603 (1959)
250. Krief A: Aldrich Technical Information 191 (1981)
251. Denis JN, Krief A: Tetrahedron Lett 3995 (1979)
252. Denis JN, Krief A: J Chem Soc Chem Commun 229 (1983)
253. Ogura F, Yamaguchi H, Otsubo T, Tanaka H: Bull Chem Soc Japan **55,** 641 (1982)
254. Poje M, Balenovic K: Bull Sci Acad Youg Sect A, **20,** 1 (1975)
255. Naddaka VI, Gar'kin VP, Minkin VI: Zh Org Khim **12,** 2481 (1976)
256. Balenovic K, Lazic R, Polak V, Stern P: Bull Sci Acad Youg Sect A, **17,** 147 (1972)
257. Bregant N, Perina I, Balenovic K: Bull Sci Acad Youg Sect A, **17,** 148 (1972)
258. Wallace TJ: J Am Chem Soc **86,** 2018 (1964)
259. Perina I, Bregant N, Balenovic K: Bull Sci Acad Youg Sect A **18,** 4 (1973)
260. Balenovic K, Bregant N, Perina I: Synthesis 172 (1973)
261. Perina I, Bregant N, Balenovic K: Bull Sci Acad Youg Sect A, **18,** 3 (1973)
262. Marino JP, Schwartz A: Tetrahedron Lett 3253 (1979)
263. Mikolajczyk M, Luczak J: J Org Chem **43,** 2132 (1978)
264. Tamagaki S, Hatanaka I, Kozuka S: Bull Chem Soc Japan **50,** 3421 (1977)
265. Kinoshita T, Sato S, Tamura C: Bull Chem Soc Japan **49,** 2236 (1976)
266. Abatjoglou AG, Bryant DR: Tetrahedron Lett **22,** 2051 (1981)
267. Abatjoglou AG: U.S. Pat 4,278,517, Chem Abstr 95: 132284s Ed, (1981)
268. Hevesi L, Krief A: Angew Chem Int Ed Engl **15,** 381 (1976)
269. Takaki K, Yasumura M, Negoro K: J Org Chem **48,** 54 (1983)
270. Corey EJ, Kim CU: J Am Chem Soc **94,** 7586 (1972)
271. Baudat R, Petrzilka M: Helv Chim Acta **62,** 1406 (1979)
272. Lucchetti J, Krief A: CR Acad Sci Ser C, **288,** 537 (1979)
273. Horn VV, Paetezold R: Z Anorg Allg Chem **398,** 186 (1973)

274. Marino JP, Larsen RD: J Am Chem Soc **103,** 4642 (1981)
275. Hori T, Sharpless KB: J Org Chem **44,** 4204 (1979)
276. Hori T, Sharpless KB: J Org Chem **44,** 4208 (1979)
277. Francisco CG, Freire R, Hernandez R, Salazar JA, Suarez E: Tetrahedron Lett **25,** 1621 (1984)
278. Takaku H, Shimida Y, Nakajima Y, Hata T: Nucleic Acids Research, **3,** 1233 (1976)
279. Barton DHR, Brewster AG, Ley SV, Read CM, Rosenfeld MN: J Chem Soc Perkin Trans I, 1473 (1981)
280. Sukumaran KB, Harvey RG: J Org Chem **45,** 4407 (1980)
281. Sukumaran KB, Harvey RG: J Am Chem Soc **101,** 1353 (1979)
282. Barton DHR, Ley SV, Magnus PD, Rosenfeld MN: J Chem Soc Perkin Trans I, 567 (1977)
283. Barton DHR, Magnus PD, Rosenfeld MN: J Chem Soc Chem Commun 301 (1975)
284. Barton DHR, Brewster AG, Ley SV, Rosenfeld MN: J Chem Soc Chem Commun 985 (1976)
285. Jeffs PW, Lynn DG: Tetrahedron Lett 1617 (1978)
286. Barton DHR, Brewster AG, Ley SV, Rosenfeld MN: J Chem Soc Chem Commun 147 (1977)
287. Barton DHR, Lester DJ, Ley SV: J Chem Soc Chem Commun 130 (1978)
288. Barton DHR, Lester DJ, Ley SV: J Chem Soc Perkin Trans I, 2209 (1980)
289. Barton DHR, Morzycki JW, Motherwell WB, Ley SV: J Chem Soc Chem Commun, 1044 (1981)
290. Yamakawa K, Satoh T, Ohba N, Sakaguchi R: Chem Lett 763 (1979)
291. Yamakawa K, Satoh T, Takita S :Heterocycles, 259 (1982)
292. Yamakawa K, Satoh T, Ohba N, Sakaguchi R, Takita S, Tamura N: Tetrahedron **37,** 473 (1981)
293. Khoi N, Polonsky J: Helv Chim Acta **64,** 1540 (1981)
294. Woodward RB, Cava MP, Ollis WD, Hunger A, Daeniker HU, Schenker K: Tetrahedron **19,** 247 (1963)
295. Barton DHR, Godfrey CRA, Morzycki JW, Motherwell WB, Ley SV: J Chem Soc Perkin Trans I, 1947 (1982)
296. Barton DHR, Hui RAHF, Lester DJ, Ley SV: Tetrahedron Lett 331 (1979)
297. Barton DHR, Hui RAH, Ley SV, Williams DJ: J Chem Soc Perkin Trans I, 1919 (1982)
298. Back TG: J Chem Soc Chem Commun 278 (1978)
299. Back TG: J Org Chem **46,** 1442 (1981)
300. Back TG, Ibrahim N: Tetrahedron Lett 4931 (1979)
301. Back TG, Ibrahim N, McPhee DJ: J Org Chem **47,** 3283 (1982)
302. Sharpless KB, Lauer RF, Teranishi AY: J Am Chem Soc **95,** 6137 (1973)
303. Reich HJ, Reich IL, Renga JM: J Am Chem Soc **95,** 5813 (1973)
304. Barton DHR, Brewster AG, Hui RAHF, Lester DJ, Ley SV, Back TG: J Chem Soc Chem Commun 952 (1978)
305. Shimizu M, Kuwajima I: Tetrahedron Lett 2801 (1979)
306. Shimizu M, Urabe H, Kuwajima I: Tetrahedron Lett **22,** 2183 (1981)
307. Kuwajima I, Shimizu M, Urabe H: J Org Chem, **47,** 837 (1982)
308. Taylor RT, Flood LA: J Org Chem **48,** 5160 (1983)
309. Lucchetti J, Krief A: CR Acad Sci Paris, Ser C, **289,** 287 (1979)
310. Kice JL, Lee TWS: J Am Chem Soc **100,** 5094 (1978)
311. Rheinbolt H, Giesbrecht E: Chem Ber **88,** 1037 (1955)
312. Faehl LG, Kice JL: J Org Chem **44,** 2357 (1979)
313. Barton DHR, Lusinchi X, Milliet P: Tetrahedron **41,** 4727 (1985)
314. Czarny MR: J Chem Soc Chem Commun 81 (1976)

References

315. Barton DHR, Lusinchi X, Milliet P: Tetrahedron Lett **23,** 4949 (1982)
316. Ninomiya I, Hashimoto C, Kiguchi T, Barton DHR, Lusinchi X, Milliet P: Tetrahedron Lett **26,** 4187 (1985)
317. Ninomiya I, Kiguchi T, Hashimoto C, Barton DHR, Lusinchi X, Milliet P: Tetrahedron Lett **26,** 4183 (1985)
318. Barton DHR, Billion A and Boivin J: Tetrahedron Lett **26,** 1229 (1985)
319. Barton DHR, Lester DJ, Ley SV: J Chem Soc Perkin Trans I, 1212 (1980)
320. Back TS, Collins S, Kerr RG: J Org Chem **46,** 1564 (1981)
321. Barton DHR, Hui RAH, Ley SV: J Chem Soc Perkin Trans I, 2179 (1982)
322. Cussans NJ, Ley SV, Barton DHR: J Chem Soc Perkin Trans I, 1650 (1980)
323. Barton DHR, Cussans NJ, Ley SV: J Chem Soc Chem Commun, 393 (1978)
324. Barton DHR, Lester DJ, Ley SV: J Chem Soc Chem Commun, 445 (1977)
325. Barton DHR, Cussans NJ, Ley SV: J Chem Soc Chem Commun, 751 (1977)
326. Barton DHR, Bielska MT, Cardoso JM, Cussans NJ, Ley SV: J Chem Soc Perkin Trans I, 1840 (1981)
327. Cussans NJ, Ley SV, Barton DHR: J Chem Soc Perkin Trans I, 1654 (1980)
328. Burton A, Hevesi L, Dumont W, Cravador A, Krief A: Synthesis, 877 (1979)
329. Barrett AGM, Read RW, Barton DHR: J Chem Soc Perkin Trans I, 2191 (1980)
330. Cravador A, Krief A: J Chem Soc Chem Commun 951 (1980)
331. Shimizu M, Takeda R, Kuwajima I: Tetrahedron Lett 3461 (1979)
332. Shimizu M, Takeda R, Kuwajima I: Bull Chem Soc Japan **54,** 3510 (1981)
333. Magnus P, Cooke F, Sarkar T: Organometallics, **1,** 562 (1982)
334. Sosnovsky G, Krogh JA: Z Naturforsch **34b,** 511 (1979)
335. Czarny MR: Synth Commun **6,** 285 (1976)
336. Grieco PA, Yokoyama Y, Gilman S, Nishizawa M: J Org Chem **42,** 2034 (1977)
337. Kametani T, Nemoto H, Fukumoto K: Heterocycles **6,** 1365 (1977)
338. Hori T, Sharpless KB: J Org Chem **43,** 1689 (1978)
339. Reich HJ, Chow F, Peake SL: Synthesis 299 (1978)
340. Williams JR, Leber JD: Synthesis 427 (1977)
341. Kametani T, Nemoto H, Fukumoto K: Bioorg Chem **7,** 215 (1978)
342. Reich HJ, Renga JM, Reich IL: J Org Chem **39,** 2133 (1974)
343. Reich HJ: J Org Chem **40,** 2570 (1975)
344. Sharpless KB, Lauer RF: J Am Chem Soc **94,** 7154 (1972)
345. Sharpless KB, Lauer RF: J Am Chem Soc **95,** 2697 (1973)
346. Reich HJ, Wollowitz S, Trend JE, Chow F, Wendelborn DF: J Org Chem **43,** 1697 (1978)
347. Sharpless KB, Michaelson RC: J Am Chem Soc **95,** 6136 (1973)
348. Tanaka S, Yamamoto H, Nozaki H, Sharpless KB, Michaelson RC, and Cutting JD: J Am Chem Soc **96,** 5254 (1974)
349. Henbest HB, Wilson RAL: J Chem Soc 1958 (1957)
350. Nsunda KM, Hevesi L: Tetrahedron Lett **25,** 4441 (1984)
351. Grieco PA, Yokoyama Y, Gilman S, Ohfune Y: J Chem Soc Chem Commun 870 (1977)
352. Baschiardes G, Dumont W, Krief A: Unpublished Results, (1986)
353. Nicolaou KC, Magolda RL, Barnette WE: J Chem Soc Chem Commun 375 (1978)
354. Nicolaou KC, Magolda RL, Sipio WJ, Barnette WE, Lysenko Z, Joullie MM: J Am Chem Soc **102,** 3784 (1980)
355. Nicolaou KC, Barnette WE, Magolda RL: J Am Chem Soc **103,** 3486 (1981)
356. Nicolaou KC, Claremon DA, Papahatjis DP, Magolda RL: J Am Chem Soc **103,** 6969 (1981)
357. Edwards MP, Ley SV, Lister SG, Palmer BD: J Chem Soc Chem Commun 630 (1983)

358. House HO: Modern Synthetic Reactions, Benjamin WA, New York, (1972)
359. Hirayama SI: Chem Rev Jap **5,** 134 (1939)
360. Stein G: Angew Chem **54,** 146 (1941)
361. Ogura F, Otsubo T, Ariyoshi K, Yamaguchi H: Chem Lett **12,** 1833 (1983)
362. Javaid KA, Sonoda N, Tsutsumi S: Tetrahedron Lett 4439 (1969)
363. Javaid KA, Sonoda N, Tsutsumi S: Bull Chem Soc Japan **43,** 3475 (1970)
364. Javaid KA, Sonoda N, Tsutsumi S: Bull Chem Soc Japan **42,** 2056 (1969)
365. Sonoda N, Yamamoto Y, Murai S, Tsutsumi S: Chem Lett 229 (1972)
366. Jephcote VJ and Thomas EJ: Tetrahedron Lett **26,** 5327 (1985)
367. Mugdan M, Young DP: J Chem Soc 2988 (1949)
368. Sumimoto M, Suzuki T, Kondo T: Agr Biol Chem **38,** 1061 (1974)
369. Seguin P: CR Acad Sci, Paris Ser C **216,** 667 (1943)
370. Umbreit MA, Sharpless KB: J Am Chem Soc **99,** 5526 (1977)
371. Warpehoski MA, Chabaud B, Sharpless KB: J Org Chem **47,** 2897 (1982)
372. Fieser LF, Fieser M: Reagents for Organic Synthesis, Wiley Interscience, New York, **I,** 994 (1967)
373. Guillemonat A: Ann Chim **11,** 143 (1939)
374. Arno M, Garcia B, Pedro JR, Seoane E: Tetrahedron, **40,** 5243 (1984)
375. Sathe VM, Chakravarti KK, Kadival MV, Bhattacharyya SC: Ind J Chem **4,** 393 (1966)
376. Snider BB, Duncia JV: J Org Chem **45,** 3461 (1980)
377. Buchi G, Wüest H: Helv Chim Acta **50,** 2440 (1967)
378. Camps F, Coll J, Parente A: Synthesis 215 (1978)
379. Paaren ME, De Luca HF, Schnoes HK: J Org Chem **45,** 3253 (1980)
380. Haruna M, Ito K: J Chem Soc Chem Commun 483 (1981)
381. Trachtenberg EN, Carver JR: J Org Chem **35,** 1646 (1970)
382. Furlenmeier A, Fürst A, Langemann A, Waldvogel G, Hocks P, Kerb U, Wiechert R: Experientia, **22,** 573 (1966)
383. Sum FW, Weiler L: J Am Chem Soc **101,** 4401 (1979)
384. Bhalerao UT, Plattner JJ, Rapoport H: J Am Chem Soc **92,** 3429 (1970)
385. Schulte-Elte KH, Gadola M, Ohloff G: Helv Chim Acta **56,** 2028 (1973)
386. Buchi G, Wüest H: J Org Chem **34,** 857 (1969)
387. Bhalerao UT, Rapoport H: J Am Chem Soc **93,** 4835 (1971)
388. Taylor WG: J Org Chem **44,** 1020 (1979)
389. Kumonaka T, Kanai Y, Yanagiya M, Matsumoto T: Chem Lett 1715 (1982)
390. Chhabra BR, Hayano K, Ohtsuka T, Shirahama H, Matsumoto T: Chem Lett 1703 (1981)
391. Bhalerao UT, Rapoport H: J Am Chem Soc **93,** 5311 (1971)
392. Trost BM, Verhoeven TR: J Am Chem Soc **100,** 3435 (1978)
393. Schmuff NR, Trost BM: J Org Chem **48,** 1404 (1983)
394. Kishi Y, Aratani M, Fukuyama T, Nakatsubo F, Goto T, Inoue S, Tanino H, Sugiura S, Kakoi H: J Am Chem Soc **94,** 9217 (1972)
395. Takayanagi H, Nishino C: J Chem Ecol **8,** 883 (1982)
396. Quinn JM: J Chem Eng Data **9,** 389 (1964)
397. Coxon JM, Dansted E, Hartshorn MP cited in Fieser LF and Fieser M, John Wiley Intersciences, New York, **II,** 362 (1969)
398. Coxon JM, Dansted E, Hartshorn MP: Org Synth, **56,** 25 (1977)
399. Matsumoto T, Imai S, Yuki S: Bull Chem Soc Japan **54,** 1448 (1981)
400. Francis MJ, Grant PK, Low KS, Weavers RT: Tetrahedron **32,** 95 (1976)
401. Tanaka K, Uchiyama F, Sakamoto K, Inubushi Y: J Am Chem Soc **104,** 4965 (1982)
402. Fried J, Heim S, Etheredge SJ, Sunder-Plassmann P, Santhanakrishnan TS, Himizu JI, Lin CH: J Chem Soc Chem Commun 634 (1968)

References

403. Martin SF, Desai SR, Phillips GW, Miller AC: J Am Chem Soc **102,** 3294 (1980)
404. Pan H-L, Cole C-A, Fletcher TL: Synthesis 813 (1980)
405. Matsui M, Yamada Y: Agr Biol Chem **29,** 956 (1965)
406. Matsui M, Yamada Y: Agr Biol Chem **27,** 373 (1963)
407. Shirahama H, Hayano K, Arora GS, Ohtsuka T, Murata Y, Matsumoto T: Chem Lett 1417 (1982)
408. Campos O, Cook JM: Tetrahedron Lett 1025 (1979)
409. Cain M, Campos O, Guzman F, Cook JM: J Am Chem Soc **105,** 907 (1983)
410. Callow RK: J Chem Soc 462 (1936)
411. Johnson WS, Ackerman J, Eastham JF, De Walt HA: J Am Chem Soc **78,** 6302 (1956)
412. Wender PA, Wolanin DJ: J Org Chem **50,** 4418 (1985)
413. Tang C, Rapoport H: J Am Chem Soc **94,** 8615 (1972)
414. Stephenson LM, Speth DR: J Org Chem **44,** 4683 (1979)
415. Greenlee ML: J Am Chem Soc **103,** 2425 (1981)
416. Doyle P, Maclean IR, Parker W, Raphael RA: Proc Chem Soc 239 (1963)
417. Danieli N, Mazur Y, Sondheimer F: Tetrahedron Lett 1281 (1962)
418. Danieli N, Mazur Y, Sondheimer F: J Am Chem Soc **84,** 875 (1962)
419. Danieli N, Mazur Y, Sondheimer F: Tetrahedron **22,** 3189 (1966)
420. Danieli N, Mazur Y, Sondheimer F: Tetrahedron **23,** 715 (1967)
421. Danieli N, Mazur Y, Sondheimer F: Tetrahedron **23,** 509 (1967)
422. Tankard MH, Whitehurst JS: Tetrahedron **30,** 451 (1974)
423. Curley RW, Ticoras CJ: J Org Chem **51,** 256 (1986)
424. Evans DA, Sims CL: Tetrahedron Lett 4691 (1973)
425. Abe Y, Harukawa T, Ishikawa H, Miki T, Sumi M, Toga T: J Am Chem Soc **78,** 1422 (1956)
426. Kariyone K, Yazawa H: Tetrahedron Lett 2885 (1970)
427. Kautter KT: Ger Pat 634,501, Chem Abstr 31: 420 (1937)
428. Elliott M, Janes NF, Pulman DA: J Chem Soc Perkin Trans I, 2470 (1974)
429. Woggon WD, Ruther R, Egli H: J Chem Soc Chem Commun 706 (1980)
430. Arigoni D, Godtfredsen S, Obrecht JP: Cited in our ref 429
431. Wiberg KB, Nielsen SD: J Org Chem **29,** 3353 (1964)
432. Arigoni D, Vasella A, Sharpless KB, Jensen HP: J Am Chem Soc **95,** 7917 (1973)
433. Djerassi C: Steroid Reactions. An Outline for Organic Chemists, Holden Day Inc., San Francisco (1963)
434. Pettit GR, Kamano Y, Inoue M, Komeichi Y, Nassimbeni LR, Niven ML: J Org Chem **47,** 1503 (1982)
435. Rosenheim O, Starling WW: J Chem Soc 377 (1937)
436. Buchi G, Pickenhagen W, Wüest H: J Org Chem **37,** 4192 (1972)
437. Toth JO, Luu B, Ourisson G: Tetrahedron Lett **24,** 1081 (1983)
438. Thomas AF: J Chem Soc Chem Commun 947 (1967)
439. Wiesner K, Jirkovsky J, Fishman M, Williams CA: J Tetrahedron Lett 1523 (1967)
440. Wiesner K, Jirkovsky J: Tetrahedron Lett 2077 (1967)
441. Wiesner K, Poon L: Tetrahedron Lett 4937 (1967)
442. Seguin P, Delepine M: CR Acad Sci Paris, Ser C, **216,** 667 (1943)
443. Itakura J, Tanaka H, Ito H: Bull Chem Soc Japan **42,** 1604 (1969)
444. Sonoda N, Tsutsumi S: Bull Chem Soc Japan **38,** 958 (1965)
445. Backer HJ, Strating J: Rec Trav Chim Pays-Bas **53,** 1113 (1934)
446. Mock WL, McCausland JH: Tetrahedron Lett 391 (1968)
447. Radlick P: J Org Chem **29,** 960 (1964)
448. Buchi G, Pickenhagen W, Wüest H: J Org Chem **37,** 4192 (1972)
449. Chabaud B, Sharpless KB: J Org Chem **44,** 4202 (1979)

450. Danieli N, Mazur Y, Sondheimer F: Tetrahedron Lett 310 (1961)
451. Suzuki E, Hamajima K, Inoue S: Synthesis 192 (1975)
452. Buchi G, Foulkes DM, Kurono M, Mitchell GF: J Am Chem Soc **88,** 4534 (1966)
453. Buchi G, Foulkes DM, Kurono M, Mitchell GF, Schneider RS: J Am Chem Soc **89,** 6745 (1967)
454. Huang DL, Weng T, Chen FC: J Heterocycl Chem **7,** 1189 (1970)
455. Mahal HS, Venkataraman K: J Chem Soc 569 (1936)
456. Matsuura S, Kuni T, Matsuura A: Chem Pharm Bull (Tokyo) **21,** 2757 (1973)
457. Hauser FM, Rhee RP: J Am Chem Soc **101,** 1628 (1979)
458. Truchet R: CR Acad Sci, Paris, Ser C **196,** 706 (1933)
459. Truchet R: CR Acad Sci, Paris, Ser C **196,** 1613 (1933)
460. Postowsky JJ, Lugowkin BP: Chem Ber **68,** 852 (1935)
461. Riley HL: British Pat 354,798 (1934), Chem Abstr 29: 5370(9) (1935)
462. Riley HL: U.S. Pat 1,955,890, Chem Abstr 28: 4067 (1934)
463. Riley HL, Morley JF, Friend NAC: J Chem Soc 1875 (1932)
464. Sharpless KB, Gordon KM J: Am Chem Soc **98,** 300 (1976)
465. Muller R: Chem Ber **66,** 1668 (1933)
466. Corey EJ, Schaefer JP: J Am Chem Soc **82,** 918 (1960)
467. Schaefer P, Corey EJ: J Org Chem **24,** 1825 (1959)
468. Woodward RB, Cava MP, Ollis WD, Hunger A, Daeniker HV, Schenker K: J Am Chem Soc **76,** 4749 (1954)
469. Sato K, Suzuki S, Kojima Y: J Org Chem **32,** 339 (1967)
470. Hach CC, Banks CV, Diehl H: Org Synth Coll Vol **IV,** 229 (1963)
471. Vander Haar RW, Voter RC, Banks CV: J Org Chem **14,** 836 (1949)
472. Ayer WA, Bowman WR, Joseph TC, Smith P: J Am Chem Soc **90,** 1648 (1968)
473. Vene J: CR Acad Sci, Paris, Ser C, **216,** 772 (1943)
474. Marquet A, Dvolaitzky M, Arigoni D: Bull Soc Chim Fr 2956 (1966)
475. Venien F, Mandrier C: CR Acad Sci, Paris, Ser C, **270,** 845 (1970)
476. Allen MS, Lamb N, Money T, Salisbury P: J Chem Soc Chem Commun 112 (1979)
477. Polonski T: J Chem Soc Perkin Trans I 305 (1983)
478. Astin S, De V Moulds L, Riley HL: J Chem Soc 901 (1935)
479. Barnes CS, Barton DHR: J Chem Soc 1419 (1953)
480. Miller MW: Tetrahedron Lett 2545 (1969)
481. Gream GE, Worthley S: Tetrahedron Lett 3319 (1968)
482. Simonen T, Laitalainen T: Finn Chem Lett 144 (1977)
483. Laitalainen T: Ann Acad Sci Fenn Ser A **195,** 1 (1982)
484. Viswanatha V, Rao GSK: Tetrahedron Lett 247 (1974)
485. Nagaoka H, Schmid G, Iio H, Kishi Y: Tetrahedron Lett **22,** 899 (1981)
486. Iio H, Nagaoka H, Kishi Y: J Am Chem Soc **102,** 7965 (1980)
487. Branca SJ, Smith AB: J Am Chem Soc **100,** 7767 (1978)
488. Smith AB, Pilla NN: Tetrahedron Lett **21,** 4961 (1980)
489. Marx JN, Cox JH, Norman LR: J Org Chem **37,** 4489 (1972)
490. Branca SJ, Lock RL, Smith AB: J Org Chem **42,** 3165 (1977)
491. Bell KH: Tetrahedron Lett 3979 (1968)
492. Bell KH: Tetrahedron Lett 397 (1967)
493. Ringold MJ, Rosenkranz G, Sondheimer F: J Org Chem **21,** 239 (1956)
494. Baran JS: J Am Chem Soc **80,** 1687 (1958)
495. Edwards JA, Ringold HJ, Djerassi C: J Am Chem Soc **82,** 2318 (1960)
496. Heller M, Bernstein S: J Org Chem **26,** 3876 (1961)
497. Meystre C, Frey H, Voser W, Wettstein A: Helv Chim Acta **39,** 734 (1956)
498. Bernstein S, Littell R: J Am Chem Soc **82,** 1235 (1960)
499. Koft ER, Smith AB: J Am Chem Soc **104,** 2659 (1982)

References

500. Barton DHR, Lindsey AS: J Chem Soc 2988 (1951)
501. Banerji JC, Barton DHR, Cookson RC: J Chem Soc 5041 (1957)
502. Fieser LF, Fieser M: Reagents for Organic Synthesis, John Wiley and Sons, New York, **1**, 996 (1967)
503. Barnes CS, Barton DHR, Fawcett JS, Thomas BR: J Chem Soc 2339 (1952)
504. Oliveto EP, Gerold C, Hershberg EB: J Am Chem Soc **76**, 6113 (1954)
505. Magerlein BJ: J Org Chem **24**, 1564 (1959)
506. Nussbaum AL, Popper TL, Oliveto EP, Friedman S, Wender I: J Am Chem Soc **81**, 1228 (1959)
507. Payne GB, Smith CW: J Org Chem **22**, 1680 (1957)
508. Hellman HM, Jerussi RA, Rosegay A: Ann NY Acad Sci **193**, 44 (1972)
509. Dittmann W, Kirchhof W, Stumpf W: Ann **681**, 30 (1965)
510. Granger R, Boussinesq J, Girard JP, Rossi JC, Vidal JP: Bull Soc Chim Fr 2806 (1969)
511. Granger R, Girard JP, Boussinesq J: CR Acad Sci Paris, Ser C, **263**, 1317 (1966)
512. Granger R, Boussinesq J, Girard JP, Rossi JC: Bull Soc Chim Fr, 2801 (1969)
513. Granger R, Boussinesq J, Girard JP, Rossi JC: Bull Soc Chim Fr, 1445 (1968)
514. Caspi E, Balasubrahmanyam SN: Tetrahedron Lett 745 (1963)
515. Caspi E, Shimizu Y, Balasubrahmanyam SN: Tetrahedron **20**, 1271 (1964)
516. Caspi E, Malhotra SM, Shimizu Y, Maheshwari K, Gasic MJ: Tetrahedron **22**, 595 (1966)
517. Hellman HM, Jerussi RA: Tetrahedron **20**, 741 (1964)
518. Christol H, Plenat F, Reliaud C: Bull Soc Chim Fr 1566 (1968)
519. Faulkner D, McKervey MA: J Chem Soc Chem Commun 3906 (1971)
520. Hellman HM, Rosegay A: Tetrahedron Lett 1 (1959)
521. Sonoda N, Tsutsumi S: Bull Chem Soc Japan **34**, 1006 (1961)
522. Sonoda N, Tsutsumi S: Bull Chem Soc Japan **36**, 1311 (1963)
523. Sonoda N, Tsutsumi S: Bull Chem Soc Japan **33**, 1440 (1960)
524. Sonoda N, Sasazima Y, Tsutsumi S: Technol Repts Osaka Univ **14**, 673 (1964)
525. Hellman HM, Jerussi RA, Lancaster J: J Org Chem **32**, 1967 (2148)
526. Biellmann JF, Rajic M: Bull Soc Chim Fr 441 (1962)
527. Giroud AM, Rassat A, Witz P, Ourisson G: Bull Soc Chim Fr 3240 (1964)
528. Caspi E, Balasubrahmanyam SN: J Org Chem **28**, 3383 (1963)
529. Ishii Y, Murai S, Sonoda N: Technol Repts Osaka Univ, 36, 623 (1976) Chem Abstr 86: 54738 (1977)
530. Smith CW, Holm RT: J Org Chem **22**, 746 (1957)
531. Kazlauskas R, Pinhey JT, Simes JJH, Watson TG: J Chem Soc Chem Commun 945 (1979)
532. Badger GM: J Chem Soc 764 (1947)
533. Jerchel D, Bauer E, Hippchen H: Chem Ber **88**, 156 (1955)
534. Seyhan M: Chem Ber **85**, 425 (1952)
535. Sakamoto T, Sakasai T, Yamanaka H: Chem Pharm Bull (Tokyo) **29**, 2485 (1981)
536. Liao TK, Wittek PJ, Cheng CC: J Heterocycl Chem **13**, 1283 (1976)
537. Forbis RM, Rinehart KC: J Am Chem Soc **95**, 5003 (1973)
538. Sakasai T, Sakamoto T, Yamanaka H: Heterocycles **13**, 235 (1979)
539. Maitte P: Ann Chim **9**, 431 (1954)
540. Colonge J, Boisde P: Bull Soc Chim Fr 1337 (1956)
541. Cain M, Campos O, DiPierro M, Mantei R, Gawishi A, Cook JM: Heterocycles **14**, 975 (1980)
542. Jongsma H, Kooreman HJ, Van Os JL: Ger Offen 2, 658, 977, Chem Abstr 87: 135078y (1977)
543. Wittek PJ, Liao TK, Cheng CC: J Org Chem **44**, 870 (1979)

544. Kende AS, Lorah DP, Boatman RJ: J Am Chem Soc **103**, 1271 (1981)
545. Cook DJ, Stamper M: J Am Chem Soc **69**, 1467 (1947)
546. Kaplan H: J Am Chem Soc **63**, 2654 (1941)
547. Boyd G, Doughty M, Kenyon J: J Chem Soc 2196 (1949)
548. Crowell JH, Bradt WE: J Am Chem Soc **56**, 1500 (1933)
549. Fu PP, Harvey RG: Chem Rev **78**, 317 (1978)
550. Weygand F, Kinkel KG, Tietjen D: Chem Ber **83**, 394 (1950)
551. Dygos JH, Chinn LJ: J Org Chem **38**, 4319 (1973)
552. Colonge J, Reymermier M: Bull Soc Chim Fr 188 (1956)
553. Sternson LA, Coviello DA: J Org Chem **37**, 139 (1972)
554. Berge DD, Kale AV, Sharma TC: Chem Ind (London) 787 (1980)
555. Berge DD, Kale AV: Chem Ind (London) 662 (1979)
556. Sosnovsky G, Krogh SA: Synthesis 703 (1978)
557. Sosnovsky G, Krogh JA, Umhoefer SG: Synthesis 722 (1979)
558. Lalezari I, Shafiee A, Yalpani M: Tetrahedron Lett 5105 (1969)
559. Lalezari I, Shafiee A, Yalpani M: Angew Chem Int Ed **9**, 464 (1970)
560. Lalezari I, Shafiee A, Yalpani M: J Heterocycl Chem **9**, 1411 (1972)
561. Meier H, Menzel I: J Chem Soc Chem Commun 1059 (1971)
562. Meier H, Voigt E: Tetrahedron **28**, 187 (1972)
563. Keay BA, Rodrigo R: J Am Chem Soc **104**, 4725 (1982)
564. Gugel VH, Meier H: Chem Ztg **103**, 155 (1979)
565. Sander WW, Chapman OL: J Org Chem **50**, 543 (1985)
566. Lalezari I, Shaffie A, Yalpani M: J Org Chem **36**, 2836 (1971)
567. Lalezari I, Shafiee A, Golgolab H: J Heterocycl Chem **10**, 655 (1973)
568. Lalezari I, Shargi N, Shafiee A, Yalpani M: J Heterocycl Chem **6**, 403 (1969)
569. Petersen H, Kolshorn H, Meier H: Angew Chem Int Ed Engl **17**, 461 (1978)
570. Meier H, Echter T, Petersen H: Angew Chem Int Ed Engl **17**, 942 (1978)
571. Lankey AS, Oliaruso MA: J Org Chem **36**, 3339 (1971)
572. Meier H, Layer M, Combrink W, Schniepp S: Chem Ber **109**, 1950 (1976)
573. Hanold N, Meier H: Chem Ber **118**, 198 (1985)
574. Gleiter R, Karcher M, Schäfer W: Tetrahedron Lett **26**, 1635 (1985)
575. Meier H: Synthesis 235 (1972)
576. Buhl H, Gugel H, Kolshorm H, Meier H: Synthesis 536 (1978)
577. Spencer HK, Lashmikantham MV, Cava MP, Garito AF: J Chem Soc Chem Commun 867 (1975)
578. Berg C, Bechgaard K, Andersen JR, Jacobsen OS: Tetrahedron Lett 1719 (1976)
579. Johannsen I, Bechgaard K, Mortensen K, Jacobsen C: J Chem Soc Chem Commun 295 (1983)
580. Engler EM, Patel VV, Andersen JR, Schumaker RR, Fukushima AA: J Am Chem Soc **100**, 3769 (1978)
581. Sosnovsky G, Krogh JA: Synthesis 654 (1980)
582. Shevchuk MI, Tolochko AF, Domkrovskii AV: Zh Org Khim **7**, 1692 (1971)
583. Kloosterziel HJ, Backer HJ: Rec Trav Chim Pays-Bas **71**, 1235 (1952)
584. El Sheikh SIA, Patel SM, Smith BC, Waller CB: J Chem Soc, Dalton Trans 641 (1977)
585. Ando F, Koketsu J, Ishii Y: Bull Chem Soc Japan **52**, 807 (1979)
586. Werbel LM, Dawson TD, Hooton JR, Dalbey TE: J Org Chem **22**, 452 (1957)
587. Suseela B Chem Ber **88**, 23 (1955)
588. Postowsky BY, Lugovkin BP, Mandryk GTh: Chem Ber **69**, 1913 (1936)
589. Attanasi O, Cagliotti L, Gasparrini F: J Chem Soc Chem Commun 138 (1974)
590. Arase A, Masuda Y: Chem Lett 1331 (1975)
591. Arase A, Masuda Y: Chem Lett 419 (1975)

References

592. Sharpless KB, Hori T, Truesdale LK, Dietrich CO: J Am Chem Soc **98,** 269 (1976)
593. Sharpless KB, Hori T: J Org Chem **41,** 176 (1976)
594. Kresze G, Wucherpfenning W: Angew Chem Int Ed Engl **6,** 149 (1964)
595. Sharpless KB, Singer SP: J Org Chem **41,** 2504 (1976)
596. Barton DHR, Britten-Kelly MR, Ferreira D: J Chem Soc Perkin Trans I, 1090 (1978)
597. Schaefer JP, Sonnenberg F: J Org Chem **28,** 1128 (1963)
598. Olah GA, Nojima M, Kerekes I: J Am Chem Soc **96,** 925 (1974)
599. Sheppard WA, Sharts CM: Organic Fluorine Chemistry, Benjamin WA, New York (1969)
600. Kent PW, Wood KR: British Pat 1,136,075 (1968), Chem Abstr 70: 88124x (1969)
601. Mori S, Yokoi T: Jpn Kokai Tokkyo Koho 79 52,016 (1976), Chem Abstr 91: 174805k (1979)

Subject Index